Insect Science: Evolution, Behavior and Management of Insects

Insect Science: Evolution, Behavior and Management of Insects

Edited by **Christopher Fleming**

SYRAWOOD
PUBLISHING HOUSE

New York

Published by Syrawood Publishing House,
750 Third Avenue, 9th Floor,
New York, NY 10017, USA
www.syrawoodpublishinghouse.com

Insect Science: Evolution, Behavior and Management of Insects
Edited by Christopher Fleming

International Standard Book Number: 978-1-68286-094-6(Hardback)

Contents

Preface

I am honored to present to you this unique book which encompasses the most up-to-date data in the field. I was extremely pleased to get this opportunity of editing the work of experts from across the globe. I have also written papers in this field and researched the various aspects revolving around the progress of the discipline. I have tried to unify my knowledge along with that of stalwarts from every corner of the world, to produce a text which not only benefits the readers but also facilitates the growth of the field.

Insects have been studied across the globe for decades. This book includes some of the vital pieces of work being conducted across the world, on various topics related to entomology. It strives to provide a fair idea about this discipline and to help develop a better understanding of the latest advances within this field. Included in this book are lucid explanations about important topics such as evolution of insects, systematics of social and pre-social insects such as ants, termites, wasps, etc. It aims to equip students and experts with the advanced topics in this area. This book is highly recommended for graduate and post graduate students, and academicians pursuing entomological studies and associated disciplines.

Finally, I would like to thank all the contributing authors for their valuable time and contributions. This book would not have been possible without their efforts. I would also like to thank my friends and family for their constant support.

Editor

The Spatial Distribution of mtDNA and Phylogeographic Analysis of the Ant *Cardiocondyla kagutsuchi* (Hymenoptera: Formicidae) in Japan

I Okita[1], K Murase[2,3], T Sato[3], K Kato[4], A Hosoda[5], M Terayama[2], K Masuko[6]

1 - Gifu University, Gifu, Gifu, Japan

2 - The University of Tokyo, Bunkyo-ku, Tokyo, Japan

3 - Tokyo University of Agriculture and Technology, Fuchu, Tokyo, Japan

4 - Shizuoka Prefectural Research and Coordination Office, Shizuoka, Shizuoka, Japan

5 - Hamamatsu Gakuin University Junior College, Hamamatsu, Shizuoka, Japan

6 - Senshu University, Kawasaki, Kanagawa, Japan

Keywords

ants, *Cardiocondyla*, male polymorphism, mitochondrial DNA, spatial distribution

Corresponding author

Kaori Murase
Graduate School of Natural Sciences
Nagoya City Univesity
Yamanohata 1, Mizuho-Cho, Mizuho-Ku
Nagoya, Aichi 467-8501, Japan
E-Mail: kmurase@nsc.nagoya-cu.ac.jp

Abstract

In this study, we investigated the geographical distribution of haplotypes of *Cardiocondyla kagutsuchi* Terayama in Japan using COI/II mitochondrial DNA. We also examined their genealogy with *C. kagutsuchi* in other areas and their close relative species. Four haplotypes were found. While two of them were found in a limited area (Ishigaki and Okinawa Islands) separately, the others were distributed widely across Honsyu, Shikoku, and Kyusyu areas in Japan. The newly invaded area by *C. kagutsuchi* in Japan was Shizuoka prefecture. Their haplotype of Shizuoka were the same as the two haplotypes of the Honsyu, Shikoku, and Kyusyu areas. The haplotype network showed that the two haplotypes were distant from each other. The distance between them was 33, even though the two haplotypes are distributed in the same area. From the phylogenetic tree that we constructed, we found that *C. strigifrons* was in the same clade as *C. kagutsuchi*.

Introduction

Many invasive ant species, such as the Argentine ant or the red imported fire ant, are not desirable because they outcompete and eliminate native ants (Suarez et al., 2008), as well as they cause serious agricultural damage and are harmful to humans (Heinze et al., 2006, Suarez et al., 2008). For such species, early detection and monitoring are important to the management of new invasions. However, the situation is not always simple. There are many invasive species that have been transferred without being noticed and have spread their distribution in new land. In such cases, to reveal their current distribution is important for the conservational actions regarding native species.

Cardiocondyla kagutsuchi Terayama (*C. kagutsuchi*) is one of such species. The genus *Cardiocondyla* is a common ant genus that belongs to the subfamily Myrmicinae. It is an invasive ant group, and is commonly known as "stealthy invaders" (Heinze et al., 2006). Approximately 50 species are currently recognized as belonging to this genus, most of which are distributed in the Old World tropics and subtropics, but a few of which occur in the temperate zone. Some species are also found widely separated in North America and the Pacific Islands, as a result of human introduction. Several species of this genus have a striking male polymorphism, with both winged and wingless forms (Terayama, 1999; Seifert, 2003; Heinze et al., 2005; Yamauchi et al., 2005). These males differ not only in morphology, but also in reproductive tactics (Yamauchi & Kawase, 1992; Yamauchi & Kinomura, 1993; Heinze et al., 1998; Anderson et al., 2003).

Recently, Seifert (2003, 2008) revised the taxonomy of this genus as a result of a morphological study that used the morphometrical method. However, questions remain around the status of *Cardiocondyla kagutsuchi* Terayama, 1999. This ant is known to have three morphologically different colony types: type 1 - produces both winged and wingless (ergatoid) males in each colony; type 2 - produces both short-winged (brachypterous) and wingless males; and type 3 - produces only wingless males (Yamauchi et al., 2005). Consequently, Yamauchi et al. (2005) renamed *C. kagutsuchi* sensu Seifert (2003) as "*C. kagutsuchi* - complex."

Types 1 and 3 have been found in Japan. Type 1 occurs only on Ishigaki Island, the Ryukyu Islands, whereas type 3 is distributed from Kanto District to Okinawa Island (Terayama et al., 1992; Terayama, 1999). Terayama (1999) considered both types to be different species due to the difference in morphology of the male caste and karyotype of the worker caste (Terayama, 1999). However, in the recent taxonomic revision of the genus, both types were considered conspecific due to the high morphological similarity between the worker castes (Seifert, 2003; Yoshimura et al., 2008).

As mentioned above, there have been many studies about its classification or male polymorphism. Such kind of studies is not good enough to understand all the aspects of the problems with invasive species. In the study of invasive species, it is important to estimate the invasion routes. When estimating the invasion routes, we need to know the haplotype distribution of the target species. Thus, we need to know the haplotype distribution of *C. kagutsuchi* in order to promote the conservation biology of the native ant diversity in Japan. However, there was no study about its haplotype distribution. The first purpose of this study is to collect this ant from across Japan and investigate their mtDNA sequences to reveal the distribution of their haplotypes in Japan. The second purpose is to construct genealogical trees and a haplotype network to clarify their position in the genus *Cardiocondyla*.

According to "Japanese Ant Image Database" (http://ant.edb.miyakyo-u.ac.jp/J/index.html), *C. kagutsuchi* is not distributed in the eastern Japan without Kanagawa prefecture. We, however, have collected a lot of colonies of *C. kagutsuchi* in Shizuoka prefecture, located in the eastern Japan, in the last few years. Where do they come from? There has been no report of their existence in the area. They may have come from foreign countries. The third purpose is to infer the invasion routes of *C. kagutsuchi* in Shizuoka prefecture, using haplotype distribution of mtDNA across Japan and *C. kagutsuchi* in foreign countries.

Materials and Methods

Sampling

We collected all samples during the species' reproductive season (June and July) in 2010. This ant is known to inhabit open areas and nest in the soil (Terayama, 1999); therefore, we searched for these ants in green areas, vegetable fields, parks, near the seashore, etc. In each locality,

we spent at least one hour searching for the ant; if we did not find any ants within this time, we moved to a different place. Colonies or foraging workers of *C. kagutsuchi* were collected from 27 localities between Kanto and Ishigaki Islands in Japan. Nests were located by following foragers back to the nest entrance, and complete colonies and colony fragments were then collected by carefully digging up soil using a scoop. Some nests were found in gaps between tiles and under stones. All samples were collected near the seashore, with the exception of those collected from Shizuoka Prefecture and Ishigaki Island. Following collection, samples were stored in ethanol (99.5%). Only workers were used for DNA analysis. When we could sample male individuals in each locality, we examined their wing polymorphism.

DNA extraction, amplification, and purification

A 890-bp fragment of COI/II (including 58 base pairs of leucine tRNA), corresponding to positions 2284 to 3191 in the *Drosophila yakuba burla* mitochondrial genome (Clary & Wolstenholme, 1985), was used for phylogenetic analysis. DNA was extracted from individual workers using Landry's method (Cheung et al., 1993). The dried pellet was eluted in purified water and preserved at −80°C until further analysis.

We amplified COI/II using the primers C1-J-2195 and C2-N-3661 (Simon et al., 1994). Polymerase chain reactions (PCRs) were performed using SYBR Premix Ex Taq (TaKaRa) and conducted in a MyCycler (Bio-Rad) and MJ Mini Gradient Thermal Cycler (Bio-Rad). PCR consisted of 34 cycles of denaturing at 95°C for 1 min, annealing at 59°C for 1 min, and extension at 72°C for 2 min, with the exception of the initial denaturing step at 95°C for 4 min. Amplified DNA was purified using a PCR Purification Kit (Qiagen).

All products were sequenced with the 3730xl DNA Analyzer (Applied Biosystems) using the primers C1-J-2195 and C2-N-3661 (Simon et al., 1994). To obtain the complete sequences, some additional primers were also designed by Nukui and Okita (unpubl.):

- 3F [5′-CCTTTAATTAGAGGATACAC-3′] and 4F [5′-GGCAGATAAGTGCAAAGGAC-3′] (which were used with C1-J-2195).

- 1R [5′-TTCTATAGAGTGATTTTGGAGGAG-3′], 2R [5′-GGTATACCTCTGAGACC-3′] and 3R [5′-CAGCTCCTATAGAGAGAACATAG-3′] (which were used with C2-N-3661).

Phylogenetic analyses

We obtained 28 sequences and aligned these with a further 11 sequences obtained from GenBank using the CLUSTAL W algorithm (Thompson et al., 1994). The accession numbers of the 11 additional sequences were

DQ023083, DQ023084, DQ023085, DQ023086, DQ023087, DQ023088, DQ023091, DQ023094, DQ023102, DQ023108, and DQ023118 (Heinze et al., 2005). Six of the 11 sequences (DQ023083, DQ023084, DQ023085, DQ023086, DQ023087, and DQ023088) were sequences of *C. kagutsuchi* originating from Hawaii, Malaysia, and Indonesia; the other 5 sequences (DQ023091, DQ023094, DQ023102, DQ023108, and DQ023118) were sequences of *C. mauritanica, C. minutior, C. obscurior, C. strigifrons*, and *C. wroughtonii*, which were used as outgroups (Heinze et al., 2005). Phylogenetic relationships were inferred from the aligned COI/II using a distance and Bayesian analysis.

A neighbor-joining tree (Saitou & Nei, 1987) was constructed using the Kimura 2-parameter distance method (Kimura, 1980) in MEGA5, with bootstrap values estimated from 5000 replicates. In the Bayesian analysis, a single run consisted of 300,000 generations. We also constructed a haplotype network using TCS 1.21 (Clement et al., 2000) with the haplotypes identified by the sequences described above.

Results

We found four different haplotypes of *C. kagutsuchi* in Japan. The distribution of each of the four haplotypes is shown in Fig. 1. The number of sequences we obtained was 28, while the number of sampling locality was 27. This is because the two types of haplotype were found in sampling locality 5. The first was distributed in Kanagawa (1 locality), Shizuoka (4), Aichi (1), Wakayama (1), and Hyogo (1) (Honshu area); Kagawa (1) (Shikoku area); and Miyazaki (1) and Kagoshima

(1) (Kyushu area). The second was distributed in Shizuoka (8 localities) (Honshu area); Tokushima (2), Kagawa (1), and Kochi (1) (Shikoku area); and Kumamoto (1) (Kyushu area). The third was found only on Okinawa Island (2 localities), and the fourth was found only on Ishigaki Island (2 localities). The first and second haplotypes were distributed widely across the temperate zone in Japan (Honsyu, Shikoku, and Kyusyu areas) and, in some cases, were found very close together. In contrast, the third and fourth haplotypes were found in a limited area (a single island).

We constructed a neighbor-joining tree and a Bayesian tree to analyze the relationship between haplotypes and male polymorphism (Figs. 2, 3). Both trees showed that *C. wroughtonii, C. obscurior*, and *C. minutior* were outgroups, but *C. strigifrons* was in the same group as *C. kagutsuchi. Cardiocondyla mauritanica* was also an outgroup according to the neighbor-joining tree, but was in the same group as *C. kagutsuchi* according to the Bayesian tree. The accession numbers of the 4 new sequences were showed in each Figure.

We also constructed a haplotype network (Fig. 4). It contained four haplotypes found in Japan in this study. The distance between the haplotype 1 and 2 was 33. The distance between the haplotype 1 and 3 was two, the number is the smallest among all haplotype pairs. The distance between the haplotype 1 and 4 was 10. The distance between the haplotype 2 and 3 was 31. The distance between the haplotype 2 and 4 was 41, the number is the highest. The distance between the haplotype 3 and 4 was 10.

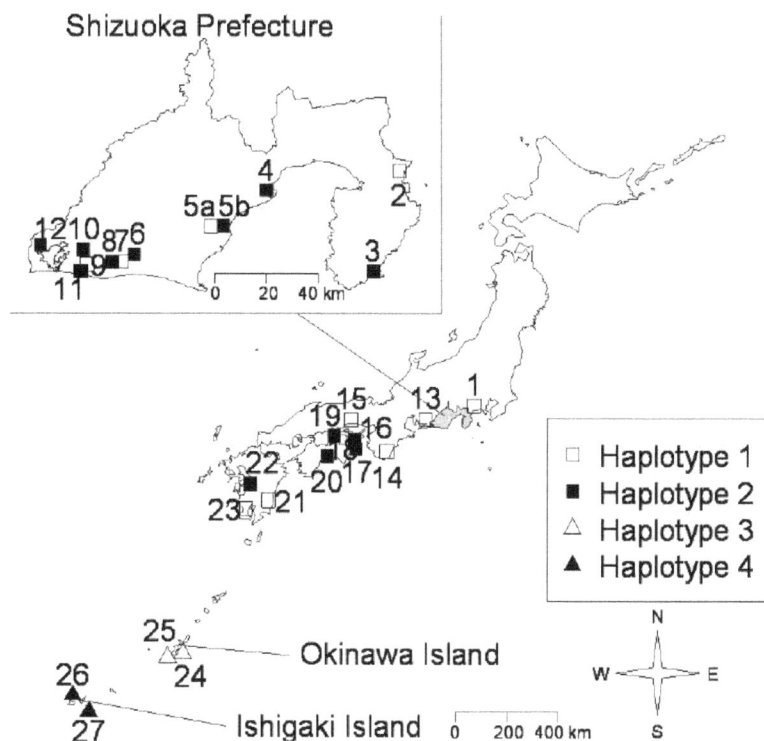

Fig. 1. Distribution of *C. kagutsuchi* in Japan. Sampling localities are numbered from 1 to 27. The haplotype of *C. kagutsuchi* in each locality is indicated by the corresponding symbol. Shizuoka Prefecture is enlarged because many sampling localities are included in it. Sampling locality 5 (in the enlarged part of the figure) is divided into 5a and 5b because the two types of haplotype were found.

K 1   Yugawara Kanagawa JPN
K 2   Atami Shizuoka JPN
K 5a Yaizu Shizuoka JPN
K 7   Iwata Shizuoka JPN
78 K 9   Hamamatsu Shizuoka JPN
K 13 Gamagori Aichi JPN   Haplotype 1
K 14 Nachikatsuura Wakayama JPN   (AB723724)
59 K 15 Aioi Hyogo JPN
K 18 Takamatsu Kagawa JPN
94 K 21 Miyazaki Miyazaki JPN
K 23 Kagoshima Kagoshima JPN
78 K 24 Yomitanson Okinawa JPN   Haplotype 3
100   78 K 25 Nanjo Okinawa JPN   (AB723726)
K Gombak Selangor MYS DQ023084
K Kuala_Lumpur KWP MYS DQ023083
K 26 Ishigaki Okinawa JPN   Haplotype 4
96   K 27 Ishigaki Okinawa JPN   (AB723727)
99 K Gombak Selangor MYS DQ023085
88 K Kuala_Lumpur KWP MYS DQ023086
K Bogor Java IDN DQ023087
S Bedugul Bali IDN DQ023108
K Mauna_Kea Hawaii USA DQ023088
K 3   Shimoda Shizuoka JPN
K 4   Shimizu Shizuoka JPN
97   K 5b Yaizu Shizuoka JPN
K 6   Fukuroi Shizuoka JPN
100 K 8   Iwata Shizuoka JPN
K 10 Hamamatsu Shizuoka JPN
K 11 Hamamatsu Shizuoka JPN   Haplotype 2
K 12 Hamamatsu Shizuoka JPN   (AB723725)
65 K 16 Naruto Tokushima JPN
K 17 Anan Tokushima JPN
K 19 Utazu Kagawa JPN
K 20 Kochi Kochi JPN
K 22 Yatsushiro Kumamoto JPN
100 MA Ta_Cenc Gozo MLT DQ023091
MI Itabuna Bahia BRA DQ023094
O Ishigaki Okinawa JPN DQ023102
100 W Nago Okinawa JPN DQ023118

0.01

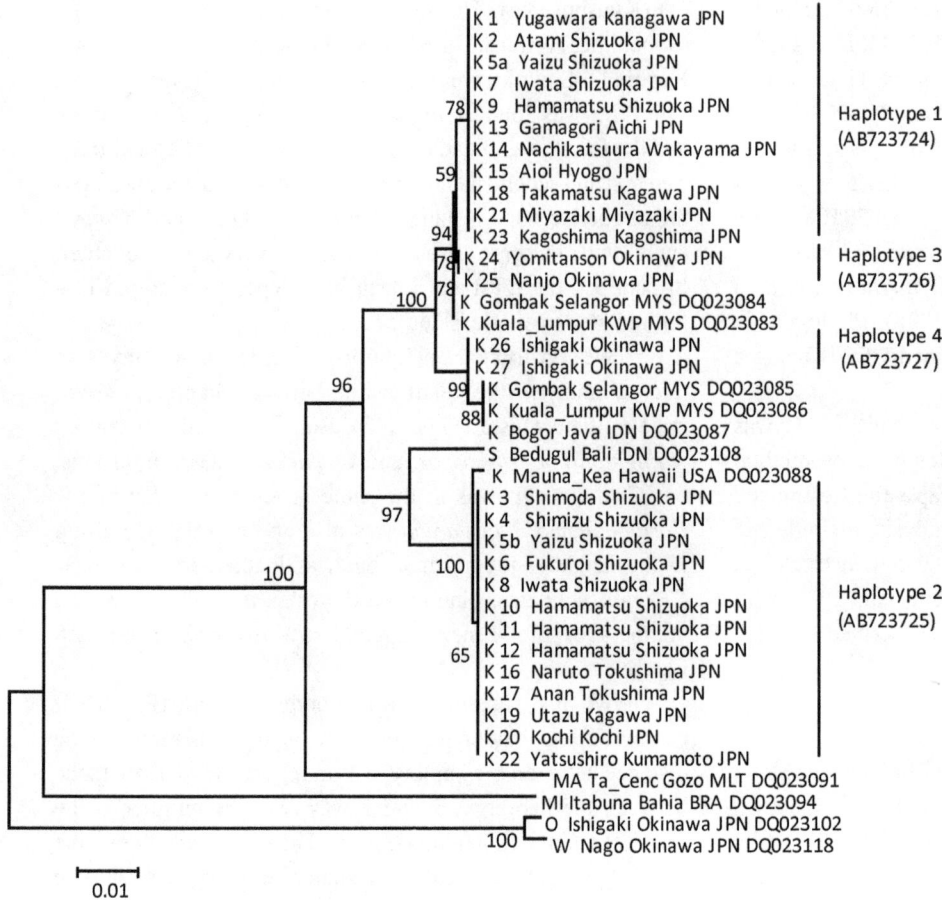

Fig. 2. Neighbor-joining tree based on the mitochondrial COI/II genes (890 bp) of the ant genus *Cardiocondyla*. Numbers indicate bootstrap values over 50% in 5000 pseudoreplications. Each sample name is given in the following order: (1) species name; (2); sampling locality number in Fig 1 if the sample is ours; (3) local area name; (4) prefecture or state name; (5) country name; and (6) accession number in GenBank if the sample is in the data of Heinze *et al.* (2005). The species names are as follows: K—*C. kagutsuchi*, S—*C. strigifrons*, MA—*C. mauritanica*, MI—*C. minutior*, O—*C. obscurior*, and W—*C. wroughtonii*. Sampling locality 5 is divided into 5a and 5b because the two types of haplotype were found. The haplotype numbers in Fig 1 are added in the right of the figure, with their accession numbers.

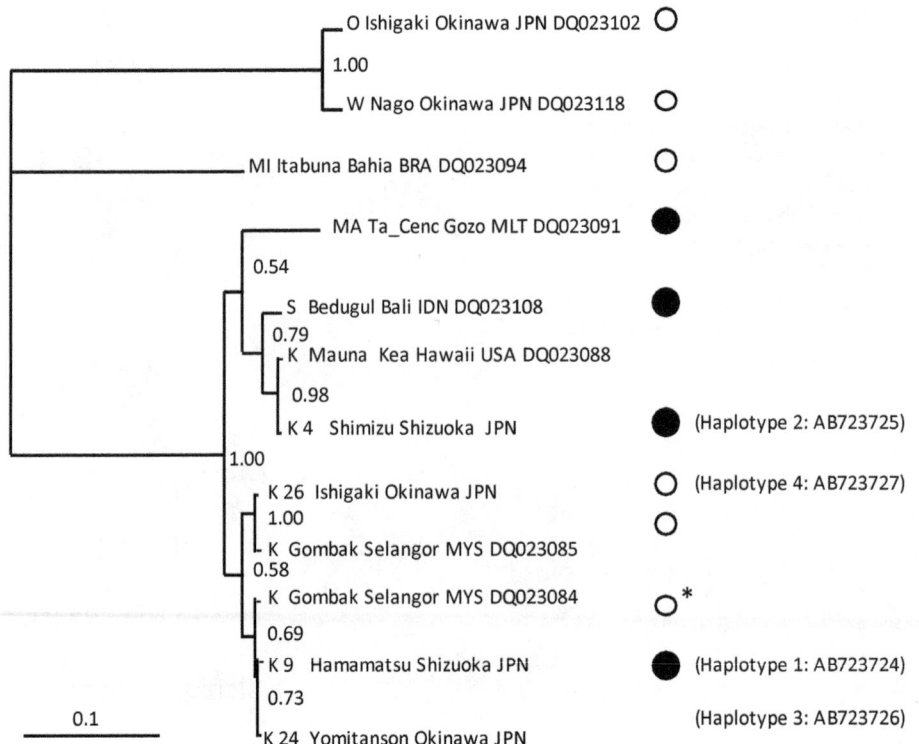

Fig. 3. Consensus tree of *Cardiocondyla* based on the mitochondrial COI/II using MrBayes (300,000 generations). Sample names are given in the same way as in Fig 2. White circles indicate that both winged and wingless males are present (type 1, dimorphic), a white circle with asterisk is for both short-winged and wingless males (type 2, dimorphic), and filled circles are for only wingless males (type 3, monomorphic), respectively. No males are known from *C. kagutsuchi* from Mauna Kea, Hawaii and Yomitanson, Okinawa. The haplotype numbers in Fig 1 are added in the right of the figure, with their accession numbers.

O Ishigaki Okinawa JPN DQ023102   ○
1.00
W Nago Okinawa JPN DQ023118   ○
MI Itabuna Bahia BRA DQ023094   ○
MA Ta_Cenc Gozo MLT DQ023091   ●
0.54
S Bedugul Bali IDN DQ023108   ●
0.79
K Mauna_Kea Hawaii USA DQ023088
0.98
K 4 Shimizu Shizuoka JPN   ● (Haplotype 2: AB723725)
1.00
K 26 Ishigaki Okinawa JPN   ○ (Haplotype 4: AB723727)
1.00
K Gombak Selangor MYS DQ023085   ○
0.58
K Gombak Selangor MYS DQ023084   ○*
0.69
K 9 Hamamatsu Shizuoka JPN   ● (Haplotype 1: AB723724)
0.73
K 24 Yomitanson Okinawa JPN   (Haplotype 3: AB723726)

0.1

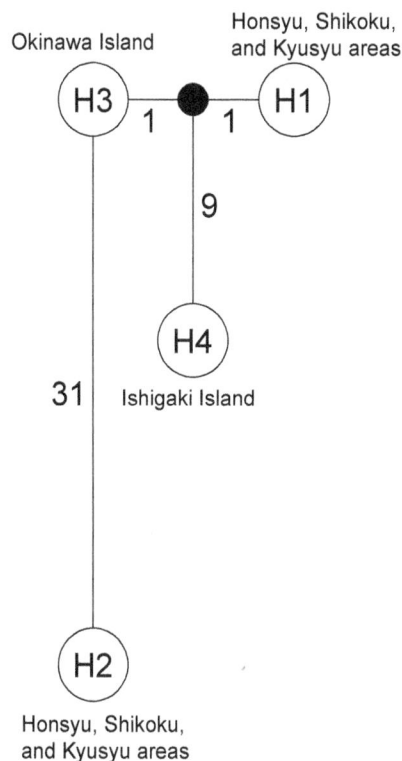

Fig. 4. Parsimonious network among four haplotypes. Open circles indicate identified haplotypes, and the filled circle indicates a hypothetical haplotype. "H1" means haplotype 1 in Fig 2, "H2" does haplotype 2, and so forth. The numbers between nodes indicate the number of nucleotide substitutions.

Discussion

At first, we want to discuss the first purpose. The distributions of the four different haplotypes are shown in Fig. 1. Haplotypes 1 and 2 were distributed widely across Japan, rather than being restricted to a single area, such as an island. This overlapped distribution pattern and this relatively low level of genetic variability across a wide area indicate that *C. kagutsuchi* is not native ant species in Japan and it may have extended its distribution quickly. What caused this distribution pattern and this low variability? Is it a result of this species being invasive, mating in its natal nest, or being a cryptic species, or is there some other reason? This would be a very interesting area for future study.

We want to discuss the second purpose. The haplotype 1 and 2 were placed in distant clades in phylogenetic trees (Figs. 2, 3) and they were considerably distant from each other in the haplotype network (Fig. 4). This suggests that the haplotype 1 and 2 are originated from other localities. The haplotype 1 is a near relation of DQ023083 and DQ023084, which were sampled in Malaysia, whereas the haplotype 2 is closely related to DQ023088, which was sampled in Hawaii. Further, the haplotype 2 is more closely related to DQ023108, *C. strigifrons*, than the other haplotypes of *C. kagutsuchi*. This might indicate that the two haplotypes are different species

rather than the same species. For the topic whether the two haplotypes are different species or not, their nuclear DNA would be needed to be examined because of the possibility of their crossing.

We want to discuss the third purpose. *C. kagutsuchi* that had expanded its distribution to the eastern Japan was neither the foreign new genotype nor the one in isolated islands like Okinawa: it was the haplotype 1 and 2. This result indicates that there is a strong possibility of its distribution expansion in Honsyu area in Japan. We should keep an eye on the routes it will use for its future expansion.

Most of the colonies in this study were sampled in its breeding season. In consequence, as for male dimorphism, Type 1, which is regarded as *C. kagutsuchi* sensu Terayama, 1999 and in which there is male dimorphism, was only found on Ishigaki Island and appeared as an independent clade in the phylogenetic analysis. For future study, whether male wing polymorphism and other ecological differences are related to genetic variations is an interesting issue.

Acknowledgments

We thank H. Nukui, H. Iyozumi, K. Kinomura, K. Yamauchi, T. Matsumoto, Y. Kawashima, F. Ikeda, R. Kawakami, H. Nakada, and K. Tsuchida for their helpful support. For letting us collect in their gardens and fields, we are also thankful to H. Matsuya, M. Matsuya, and the staff of Arakawa Garden Shop, Yuto Cultural Center, Lake-Hamana Garden Park and Takenaka Garden Shop. Lastly, we are grateful to the night class teachers of Iwata Minami High School for deep understanding and support for our study.

References

Anderson, C., Cremer, S. & Heinze, J. (2003). Live and let die: why fighter males of the ant *Cardiocondyla* kill each other but tolerate their winged rivals. Behav. Ecol., 14: 54–62. doi:10.1093/beheco/14.1.54

Cheung, W.Y., Hubert, N. & Landry, B.S. (1993). A simple and rapid DNA microextraction method for plant, animal, and insect suitable for RAPD and other PCR analyses. Genome Res., 3: 69–70.

Clary, D.O. & Wolstenholme, D.R. (1985). The mitochondrial DNA molecule of *Drosophila yakuba*: Nucleotide sequence, gene organization, and genetic code. J. Mol. Evol., 22: 252–271. doi:10.1007/BF02099755

Clement, M., Posada, D. & Crandall, K. A. (2000). TCS: a computer program to estimate gene genealogies. Mol. Ecol., 9: 1657–1660. doi:10.1046/j.1365-294x.2000.01020.x

Heinze, J., Cremer, S., Eckl, N. & Schrempf, A. (2006). Stealthy invaders: the biology of *Cardiocondyla* tramp ants. Insectes Soc., 53: 1–7. doi:10.1007/s00040-005-0847-4

Heinze, J., Hölldobler, B. & Yamauchi, K. (1998). Male competition in *Cardiocondyla* ants. Behav. Ecol. Sociobiol., 42: 239–246. doi:10.1007/s002650050435

Heinze, J., Trindl, A., Seifert, B. & Yamauchi, K. (2005). Evolution of male morphology in the ant genus *Cardiocondyla*. Mol. Phylogenet. Evol., 37: 278–288. doi:10.1016/j.ympev.2005.04.005

Kimura, M. (1980). A simple method for estimating evolutionary rates of base substitutions through comparative studies of nucleotide sequences. J. Mol. Evol., 16:111–120. doi: 10.1007/BF01731581

Saitou, N. & Nei, M. (1987). The neighbor-joining method: a new method for reconstructing phylogenetic trees. Mol. Biol. Evol., 4: 406–425.

Seifert, B. (2003). The ant genus *Cardiocondyla* (Insecta: Hymenoptera: Formicidae) - a taxonomic revision of the *C. elegans*, *C. bulgarica*, *C. batesii*, *C. nuda*, *C. shuckardi*, *C. stambuloffii*, *C. wroughtonii*, *C. emeryi*, and *C. minutior* species groups. Ann. Nat. Hist. Mus. Wien, B, 104: 203–338.

Seifert, B. (2008). *Cardiocondyla atalanta* FOREL, 1915, a cryptic sister species of *Cardiocondyla nuda* (MAYR, 1866) (Hymenoptera: Formicidae). Myrmecol. News, 11: 43–48.

Simon, C., Frati, F., Beckenbach, A., Crespi, B., Liu, H. & Flook, P. (1994). Evalution, weighting, and phylogenetic utility of mitochondrial gene sequences and a compilation of conserved polymerase reaction primers. Ann. Entomol. Soc. Am., 87: 651-701.

Suarez, A.V., Holway, D.A. & Tsutsui, N.D. (2008). Genetics and behavior of a colonizing species: The invasive Argentine ant. Am. Nat., 172: S72–S84. doi:10.1086/588638

Terayama, M. (1999). Taxonomic studies of the Japanese Formicidae. Part 6. Genus *Cardiocondyla Emery*. Mem. Myrmecol. Soc. Jap., 1, 99–107.

Terayama, M., Yamauchi, K. & Morisita, M. (1992). *Cardiocondyla* sp. 4. A Guide for the Identification of Japanese Ants (III), 32.

Thompson, J.D., Higgins, D.G. & Gibson, T.J. (1994). CLUSTAL W: Improving the sensitivity of progressive multiple sequence alignment through sequence weighting, position-specific gap penalties and weight matrix choice. Nucl. Acids Res., 22: 4673–4680. doi:10.1093/nar/22.22.4673

Yamauchi, K., Asano, Y., Lautenschläger, B., Trindl, A. & Heinze, J. (2005). A new type of male dimorphism with ergatoid and short-winged males in *Cardiocondyla* cf. *kagutsuchi*. Insectes Soc., 52: 274–281. doi: 10.1007/s00040-005-0803-3

Yamauchi, K. & Kawase, N. (1992). Pheromonal manipulation of workers by a fighting male to kill his rival males in the ant *Cardiocondyla wroughtonii*. Naturwissenschaften, 79: 274–276. doi: 10.1007/BF01175395

Yamauchi, K. & Kinomura, K. (1993). Lethal fighting and reproductive strategies of dimorphic males in *Cardiocondyla* ants (Hymenoptera: Formicidae). In T. Inoue & S.K. Yamane (Eds.), Evolution of insect societies (pp. 373–402). Tokyo: Hakuhinsha.

Yoshimura, M., Kubota, M., Onoyama, K. & Ogata, K. (2008). Taxonomic changes since the publication of Japanese Ant Image Database 2003. Ari, 31:13–28.

2

Methodology for Internal Damage Percentage Assessment by Subterranean Termites in Urban Trees

FJ ZORZENON, AE DE C CAMPOS

Instituto Biológico, Unidade Laboratorial de Referência em Pragas Urbanas, São Paulo, SP, Brazil.

Keywords
Isoptera, urban forest, tree management, falling risk, inspection

Corresponding author
Ana Eugênia de Carvalho Campos
Instituto Biológico
Unidade Laboratorial de Referência em
Pragas Urbanas
Av. Conselheiro Rodrigues Alves, 1252
São Paulo, SP, Brazil 04014-002
E-mail: anaefari@biologico.sp.gov.br

Abstract
One of the most important problems in urban trees is termite infestation. Simple observations of damages on outside trunks or dead branches and leaves do not always confirm infestations. Several trees may present severe termite damage internally that can only be observed through drilling. This paper presents a methodology to evaluate estimated percentages of internal damage caused by termites in urban trees. Tests were made on 1,477 plants in a neighborhood in the city of São Paulo, Brazil and 27% of them were infested by subterranean termites. The results showed that the methodology is simple to use, fast and inexpensive, and it allows assessment of termite internal damage which may help in making decisions on tree management. The trees did not show any phytosanitary problems along the 9 year study after being submitted to the new technique.

Introduction

Problems associated to urban trees are common and are mostly caused by inadequate management (Jim, 2001; Rodrigues et al., 2002; Nowak & Dwyer, 2007). Lack of planning to choose suitable plant species to the site, lack of knowledge on plant biology and physiology, lack of care when transplanting the trees, lack of space for plant growth, drastic pruning, infestations by wood boring, sucking and defoliating insects, termites, ants, besides plant diseases, and negligence are some of the problems that make urban trees die early (Zorzenon & Potenza, 2006).

Tree survival in different environments depends on the mechanical reliability of its structures, and problems in the trunk may cause falling risks followed by serious or fatal accidents (Niklas, 1992, Mattheck & Breloer, 1997; Pereira et al., 2007). Infestations by subterranean termites are common in Brazil (Constantino, 2002) and promote damping off, especially when attacks occur in street trees.

Studies have evaluated methods for assessing termite infestations in trees. Mattheck and Breloer (1994) developed a method called visual tree assessment (VTA). If symptoms are detected the related defect has to be confirmed and measured by deeper inspection, which includes measuring the speed of a sound wave traveling through a cross-section on the trunk and by drilling methods. The strength of the remaining healthy wood is determined with a fractometer. Lax and Osbrink (2003) refers to other nondestructive methods using a resistograph. Nicolotti et al. (2003) reported application of electric, ultrasonic, and georadar tomography for detection of decay in trees and their comparison with the penetrometer. Mankin et al. (2002) used a portable, low-frequency acoustic system to detect termite infestations in urban trees. The likelihood of infestation was rated independently by a computer program and an experienced listener that distinguished insect sounds from background noises. Mankin and Benshemesh (2006) used a geophone system to monitor activity of subterranean termites and ants in a desert environment with low vibration noise.

Therefore, the methodology must be used in quiet environments due to its sensibility and it is not adequate to urban environments.

Pereira et al. (2007) reported that nondestructive techniques have been developed by tomography investigations, where the impulse tomography enables transversal section reconstructions of the whole tree, through the energy that flows through the wood. However, they advise that the technique is still under development and needs further study. Osbrink and Cornelius (2013a) as well as Osbrink and Cornelius (2013b) on the other hand used an acoustic emission detector (AED) to evaluate the presence of termite infestations in trees and could determine the presence of the insects.

Since termite damages are common, pest control operators should have a range of methods which can be chosen for each condition. This paper proposes a simpler methodology for tree assessment to estimate the percentage of internal damages caused by subterranean termites in urban trees that could be used worldwide.

Material and Methods

The study was conducted from January 2004 to January 2013, in a neighborhood with 1,609 plant specimens, including trees and palm trees along eight kilometers of streets and avenues, in the city of São Paulo, Brazil (S 23° 35' 34.72" / W 46° 41' 57.60"). We identified 1,339 trees and 138 palm trees, totaling 1,477 plant specimens to test the proposed methodology.

Steel drills, measuring 6 mm diameter by 200 mm in length, and 10mm diameter by 320 to 400mm in length, normally used to drill wooden stakes and fiber cement tiles, were used with single or triple ends. A professional hammer driller (Bosch GSB 19-2 Model 650-watt) and an electric/gasoline generator (Branco Model BT2 950) were also used.

Small drills were used on trees with the Circumference at Breast Height (CBH) smaller than or equal to 40cm and the larger ones for the trees with CBH greater than 40cm. The CBH was measured at 1.30m height from the base of the plant (Daniel, 2006). All plants were georeferenced.

After perforation, holes were painted with Bordeaux mixture (copper sulfate, hydrated lime and water), and sealed with silicone rubber to prevent penetration of moisture and the entry of plant pathogens.

A borescope, which is a micro camera attached to a 6mm flexible stem joined to a LCD monitor, was used to visualize internal damage caused by termites, and to check for the presence of live termites inside the trunk of the trees.

To determine the size and exact location of the termite damage in the inner region of the trunk, three holes (n) were drilled at a 45 degree angle into the base of the trees to establish triangulation points (Fig. 1 – wider arrows). The intention was to reach the deepest regions of the heartwood for inner exploration to confirm termite infestations that are not externally

visible, and to estimate the percentage of damage. The percentage of internal damage ranged from 0% (no damage) to 90% (extensive damage); this range was used to determine the correlation between CBH and percentage of internal damage.

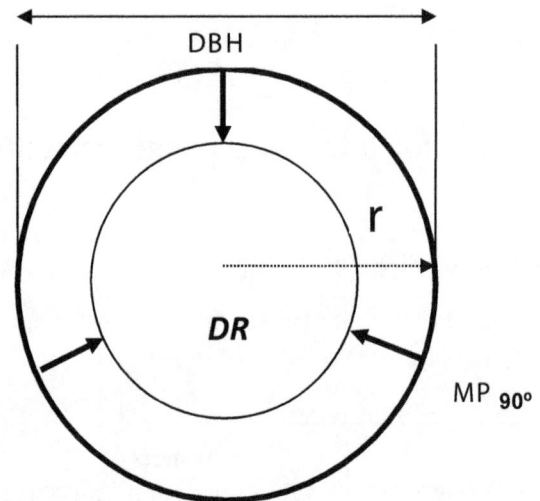

Figure 1 - Schematic cross section of tree trunk showing *DBH*, *r*, *DR*, and *MP*$_{90°}$. Holes drilled are indicated by wide arrows.

Termite assessments were conducted in 2004. Infested trees and palm trees were controlled, and plants were individually inspected each month during the nine year study to determine if the technique caused any other damage to the trees.

Estimated percentage of internal damage

To obtain the estimated percentage of internal termite damage a methodology was developed for assessing and adapting trigonometric formulas for converting values into percentage estimates of internal damage.

Percentage estimates were obtained by measuring the depths penetrated by the drills, to evaluate the difference in strength of a healthy tissue of the trunk (high resistance) to a termite-damaged tissue (low resistance) or hollow (no resistance). The estimated average was obtained from the sum of the three depth measurements.

Once the angle of the drill relative to the tree trunk is 45 degrees ($DP_{45°}$), the mean obtained values from the three holes were converted to the following formula to obtain $MP_{90°}$ (average measured depth of the drill at 90 degrees):

$$MP_{90°} = \frac{\left(\frac{\Sigma DP45°}{n}\right)}{\sqrt{2}}$$

$MP_{90°}$ = average measured depth of the drill at 90 degrees
$DP_{45°}$ = diagonal depth of the drill at 45 degrees
$\boldsymbol{n}$ = hole number

For those results of depth measurements in which the drill did not hit any cavity, the number to be considered in the sum of DP_{45}° should be a value obtained through r. $\sqrt{2}$.

For practical purposes it was assumed that the cross section of the trunk was a circular shape. It was also assumed that the transverse extent of internal damage was circular in shape. Therefore, to make the calculations of estimated percentages of internal damage, it is necessary to first obtain the trunk diameter:

CBH = Circumference at Breast Height (1.3 m)
DBH = Diameter at Breast Height (1.3 m)
$\pi = 3,1416$
r = radius of the circumference
PD = % of internal damages
DR = trunk damaged region

For obtaining the diameter:

$$DBH = \frac{CBH}{\pi}$$

Subsequently determine the radius of the circle:

$$r = \frac{DBH}{2}$$

Subtracting the $MP90^{\circ}$ (average measured depth of the drill at 90 degrees) of r (radius of the circumference), an estimated value of the injured area is obtained, where:

$$DR = r - MP90^{\circ}$$

Thus, to obtain the estimated percentage of the injured area (PD):

$$PD = \frac{DR \cdot 100}{r}$$

Results and Discussion

According to Juttner (1997) unique observations of external damage, such as tunnels, do not correspond to real termite infestations on urban trees. Numerous trees, apparently healthy, have serious internal damage that can only be determined after drilling.

The survey on the 1,477 plants revealed 27% (399) trees infested by subterranean termites. The practical methodology presented here proved to be feasible both in operational terms, regarding the use of low cost tools, and reliability when compared to the reported difficulties using other methods (Lax & Osbrink, 2003). Tomography, for example, is a noninvasive way to assess internal injuries caused by termites in trees (Pereira et al., 2007). Although technologically superior, the equipment has high cost (ca. US\$ 12,500.00) and must be calibrated according to the density of the wood, what makes operating it in the field difficult and time consuming. The methodology used in this work costs approximately US\$ 550.00.

The procedure used during this nine year study did not promote significant lesions on the assessed trees, and there were no interferences due to phytosanitary aspects. The injuries due to the holes drilled in the trees did not lead to diseases, and there was satisfactory healing after some months. It was not observed any tree death or impoverishment due to the adopted procedure. It is suggested that once the termite infestations are found trees must be treated, but trees with more than 60% of damages must be cut off, once they have risk of falling down.

The mathematical formulas for data conversion and obtention of internal damage estimated percentages were developed to be simple and easy to understand. They have the advantage of being easy to use by properly trained technicians.

Acknowlegments

The authors thank Dr. William H. Robinson for revising the English of the manuscript. We thank two other anonymous referees for their comments and criticisms on the manuscript.

References

Constantino, R. (2002). The pest termites of South America: taxonomy, distribution and status. Journal of Applied Entomology, 126: 355-365. doi: 10.1046/j.1439-0418.2002.00670.x

Daniel, O. (2006). Silvicultura. Apostila Cultivo de Árvores. Universidade Federal da Grande Dourados. http://pt.scribd.com/doc/70003302/Silvicultura-%E2%80%93-Omar-Daniel-Apostila-Cultivo-de-Arvores-Universidade-Federal-da-Grande-Dourados-Faculdade-de-CienciasAgrarias-de-Dourados-MS-Br. (Accessed date: 5 April, 2013).

Jim, C.Y. (2001). Managing urban trees and their soil envelopes in a contiguously developed city environment. Environmental Management, 28: 819-832. doi: 10.1007/s002670010264

Juttner, A.S. (1997). Termite Control - Drill n treat form control of Formosan termite in tree. Arbor Age. http://www.greenmediaonline.com/aa/1997/1097/1097dri.html(Accessed date: 15 June, 2004).

Lax, A.R., Osbrink, W.L.A. (2003). United States Department of Agriculture—Agriculture Research Service research on targeted management of the Formosan subterranean termite Coptotermes formosanus Shiraki (Isoptera: Rhinotermitidae). Pest Management Science, 59:788-800. doi: 10.1002/ps.721

Mankin, R.W., Osbrink, W.L., Oi, F.M., Anderson, J.B. (2002). Acoustic detection of termite infestations in urban trees. Journal of Economic Entomology, 95: 981-988. doi: 10.1603/0022-0493-95.5.981

Mankin, R.W., Benshemesh, J. (2006). Geophone detection of subterranean termite and ant activity. Journal of Economic Entomology, 99: 244-250.

Mattheck, C., Breloer, H. (1994). Field guide for visual tree assessment (VTA). Journal of Arboriculture, 18: 1-23.

Mattheck, C., Breloer, H. (1997). The body language of trees: a handbook for failure analysis. London: Her Majesty´s Stationery Office, 260p.

Nicolotti, G., Socco, I.V., Martinis, R., Godio, A., Sambuelli, L. (2003). Application and comparison of three tomographic techniques for detection of decay in trees. Journal of Arboriculture, 29: 66-78.

Niklas, K. (1992). Plant biomechanics: an engineering approach to plant form and function. University of Chicago Press, Chicago, IL. 622 p.

Nowak, D.J., Dwyer, J.F. (2007). Understanding the benefits and costs of urban forest ecosystems. In Kuser, J.E. (Ed.), Urban and Community Forestry in the Northeast (pp. 25-46). Springer Netherlands. doi: 10.1007/978-1-4020-4289-8_2.

Osbrink,W., Cornelius, M. (2013a). Utility of Acoustical Detection of *Coptotermes formosanus* (Isoptera: Rhinotermitidae). Sociobiology, 60: 69-76. doi: 10.13102/sociobiology.v60i1 .69-76

Osbrink,W., Cornelius, M. (2013b). Acoustic Evaluation of Trees for *Coptotermes formosanus* Shiraki (Isoptera: Rhinotermitidae) Treated with Imidacloprid and Noviflumuron in Historic Jackson Square, New Orleans, Louisiana. Sociobiology, 60: 77-95. doi: 10.13102/sociobiology.v60i1.77-95

Pereira, L.C., Silva Filho, D.F., Tomazello Filho, M., Couto, H.T.Z., Moreira, J.M.M.A.P., Polizel, J.L. (2007). Tomografia de impulso para avaliação do interior do lenho de árvores Revista da Sociedade Brasileira de Arborização Urbana, 2: 65-75.

Rodrigues, C.A.G., Bezerra, B.C., Ishii, I.H., Cardoso, E.L., Soriano, B.M.A., Oliveira, H. (2002). Arborização Urbana e Produção de Mudas de Essências Florestais Nativas em Corumbá, MS. Corumbá (MS): EMBRAPA.

Zorzenon, F.J., Potenza, M.R. (2006). Cupins: pragas em áreas urbanas. Boletim Técnico, Instituto Biológico, Brazil. 18.

Diversity of the Nests of Social Wasps (Hymenoptera: Vespidae: Polistinae) in the Northern Pantanal, Brazil

SM Almeida[1,3], SR Andena[2], EJ Anjos-Silva[3]

1 - Universidade do Estado de Mato Grosso, Nova Xavantina, Mato Grosso, Brazil.

2 - Universidade Estadual de Feira de Santana, Feira de Santana, Bahia, Brazil.

3 - Universidade do Estado de Mato Grosso, Cáceres, Mato Grosso, Brazil.

Keywords

IndVal, cambarazal, pombeiral, campo limpo, wetlands

Corresponding author

Evandson José dos Anjos-Silva
Lab. de Abelhas e Vespas Neotropicais
Universidade do Estado de Mato Grosso
Departamento de Biologia
Av. Tancredo Neves s/n, Cavalhada
Cáceres, MT, Brazil 78200-000
E-mail: beevandson@uol.com.br

Abstract

Some species of wasps demonstrate plasticity with diverse nesting habits according to the environmental conditions and substrates used for building the nests, while others are restricted to habitats with specific conditions and may exhibit some degree of fidelity. The aim of this study was to estimate species richness and abundance of nests of Polistini and Epiponini wasps in four landscape units in the Pantanal of Poconé, Retiro Novo Farm, southwestern Mato Grosso state, Brazil. The nests of social wasps were sampled in four plant physiognomies locally known as cambarazal, landizal, pombeiral and campo limpo from August 25, 2011 to April 11, 2012, being recorded 308 nests of eight genera and 14 species of social wasps. The highest number of nests belongs to *Polybia ruficeps xanthops* (32.69%), *Poly. sericea* (24.27%) and *Synoeca surinama* (15.21%). The highest species richness was recorded in cambarazal and the highest abundance of nests in pombeiral, while campo limpo showed the lowest richness and abundance of nests. The nests of *S. surinama* were associated with cambarazal and landizal (IndVal = 93.3, *P* = 0.001), while the nests of *Poly. ruficeps xanthops* and *Poly. chrysothorax* were associated with cambarazal, landizal and pombeiral (IndVal = 97, *P* = 0.001). There was lower abundance and lower species richness of wasps in campo limpo. These results demonstrate that the maintenance of forest environments in the Pantanal is essential for the establishment and maintenance of social wasp nests.

Introduction

The most known social insects are the bees and the ants (Grimaldi & Engel, 2005; Melo et al., 2012), which belong to Apidae and Formicidae, respectively. The term "wasp" is applied to all other groups within the taxon (Melo et al., 2012). Most people, usually, have the perception that species of wasps live in a nest, sharing it with the other members of the colony, however only a small portion of the taxon is eusocial, in the sense that overlapping of generations, cooperative care over offspring and reproductive division of labor occurs. Actually, in wasps the eusociality is restricted only to Polistinae and Vespinae, both within Vespidae. Polistinae, which comprises the tribes Polistini, Mischocyttarini and Epiponini, is widely distributed in Neotropics with 25 genera and around 900 species recorded (Richards, 1978; Carpenter & Marques, 2001).

The social wasps play a vital ecological role as pollinators (see Prezoto & Machado 1999; Vitali-Veiga & Machado 2001; Silva-Pereira & Santos, 2006) and predators and act as natural agents of biological control (Clapperton, 1999; Carpenter & Marques, 2001; Hunt, 2007). Also they have frequently been used to test evolutionary models for the origin of social behavior because of their different levels of sociality from solitary to eusocial (West-Eberhard, 1978, 1996; Itô, 1986; Spradbery, 1991 as cited in Noll et al., 2004).

The nests can be the size of a thimble or more than a meter long, as durable as hard felt or more fragile than egg shells, more regular and uniform than the much-celebrated honeybee comb or wildly chaotic with an intricate mazelike interior (Wenzel, 1998). Except for four Neotropical species that build nests of mud, the nests are all made of vegetable fiber without wax or plant resins (Wenzel, 1998; Hunt, 2007).

Soil or glandular secretion may be used to reinforce or repair nests, but they rarely constitute the primary building material for mature colonies (Wenzel, 1998).

In a general sense, according to the shape, nests can be classified as: a) Phragmocyttarous, where the initial comb is fixed on substratum and covered by an envelope. A second comb is built by adding new cells at the bottom of the first envelope, and also is covered by an envelope. Each envelope has an entrance to the respective combs, b) Astelocyttarous nest, where a single comb is built directly on substratum and covered by an envelope with one entrance; and c) Stelocyttarous nest, where a comb or combs are suspended by stalks. They can or cannot have envelope and are called of calyptodomous and gymmodomous, respectively (Saussure, 1853; Richards and Richards, 1951, Richards, 1978, Carpenter & Marques, 2001).

The preference for nesting substrates is different, depending on the physical and biological characteristics of the environment (Dejean et al., 1998; Cruz et al., 2006). Nests can be built on natural substrate, such as plants rocks, cavities, besides termites and human constructions (Carpenter & Marques, 2001). Some wasp species have great ecological plasticity and varied nesting habits according to environmental conditions and substrates for nest building (Santos & Gobbi 1998). However, other species have lower ecological plasticity and are restricted to habitats with specific conditions, and may have some allegiance to such environments (Heithaus, 1979; Dejean et al., 1998; Cruz et al., 2006; Silva Pereira & Santos, 2006; Santos et al., 2009a; Souza et al., 2010). Such characteristics may elect wasps as bioindicators of environmental quality, as pointed out by Souza et. al (2010).

The choice of sites for nesting is more characteristic and less diverse than those for foraging (Richards, 1978), and the diversity of these insects is associated more to the nesting habitat and not necessarily to foraging habitat (Simões et al., 2012). The vegetation structure can favor the nesting of social wasps either by increasing the availability of physical support for nests or the number and diversity of available food resources, by imposing lower variability in microclimate characteristics (Lawton, 1983).

Studies on social wasps have been conducted in Brazilian areas of Cerrado (Richards 1978; Henriques et al., 1992; Diniz & Kitayama, 1994; Diniz & Kitayama, 1998; Silva-Pereira & Santos, 2006; Santos et al., 2009a; Santos et al., 2009b; Silva et al., 2011), Amazon Forest (Silveira, 2002; Silveira et al., 2005, 2012; Silva & Silveira 2009; Somavilla, 2012), Atlantic Forest (Santos et al., 2007) and Caatinga (Santos et al., 2009b). In areas of Cerrado 130 species were recorded (Richards, 1978; Diniz & Kitayama, 1994), while in the Amazon 20 genera and over 200 species were recorded (Silveira, 2002; Silva & Silveira, 2009). In the Cerrado of Mato Grosso, there are records of 88 species in Nova Xavantina (Richards, 1978) and 36 species in Chapada dos Guimarães (Diniz & Kitayama 1998).

Despite the importance of social wasps for the mainte-

nance of ecosystems, this group still has little information in the literature (Prezoto et al., 2008) and for Pantanal of Mato Grosso there is shortage or no data on the social wasps. Given the importance of social wasps to ecological systems, the present study aimed at determining the species richness and abundance of Polistinae wasp nests in four different landscape units in the North Pantanal, Poconé, Mato Grosso.

Material and Methods

Study area

The Pantanal, considered the largest wetland in the world, is located in the middle of South America and covers parts of Brazil, Bolivia, Paraguay and Argentina. The pattern of flood inside the Pantanal is strongly influenced by rainfall ($\pm$ 1.250 mm per year). The climate in this floodplain is hot with average annual temperature around 25° C, a pronounced dry climate period from May to September, and a rainy season from October to April (Junk et al., 2006; Fantin-Cruz et al., 2010; Fernandes et al., 2010).

The Pantanal of Poconé is a subregion belonging to the Northern Pantanal, located in the state of Mato Grosso (Fernandes et al., 2010), and it was the region chosen to develop this study conducted in the Retiro Novo Farm (16°15'12" S; 56°22'12" W), municipality of Poconé (Fig 1).

In this area of the Pantanal, the Cerrado dominates the landscape in the form of natural grasslands, but has also fea-

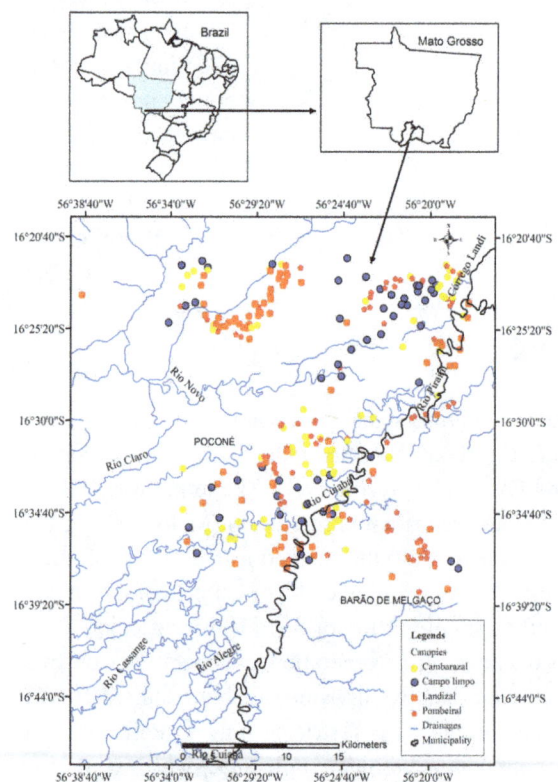

Fig 1. Study Area located at Pirizal District, Retiro Novo Farm, municipality of Poconé, Southwestern Mato Grosso state, Brazil.

tured forest formations, locally known as landizal, cambarazal and pombeiral (Nunes da Cunha et al., 2010) but can occur as forests mosaic areas as show in the Fig 1.

The cambarazal (Fig 2a) is a dense forest formation, semi-evergreen, in which *Vochysia divergens* Pohl (cambará), sometimes occurring in monospecific stands, dominates. *Vochysia divergens* is a characteristic species of the Pantanal. Dense stands develop in low-lying humid areas because this species is flood-tolerant and in multi-year wet periods spreads into surrounding savannas, creating serious problems for farmers. The expansion of cambarazal is counteracted by the wild fires of extremely dry years. Elongated and sinuous stretches of semi-evergreen forest in high-lying plains covered by cerrado vegetation point to the presence of landizal (Fig 2b). This can be explained by the fact the, in periodically drying drainage systems, floods last longer and water availability during low-water periods is better because of a high groundwater level (Nunes da Cunha et al., 2007). Landi refers to the regional name of *Calophyllum brasiliense* Cambess. (Calophyllaceae) trees, that forms the landizal plant life vegetation. The following species, which are also common in other seasonally flooded forests, are characteristic of this sub-type of forest: *Licania parvifolia* Huber (Chrysobalanaceae), *Erythroxylum anguifugum* Mart. (Erythroxylaceae), *Alchornea discolor* Poepp. (Euphorbiaceae), *C. brasiliense, Mouriri guianensis* Aubl. (Melastomataceae), *Ficus pertusa* L.F. (Moraceae), *Sorocea sprucei* (Baill.) Macbr. (Moraceae), *Eugenia florida* DC. (Myrtaceae), *Coccoloba ochereolata* Weed. (Polygonaceae), and *Triplaris gardneriana* Wedd. (Polygonaceae).

In pombeiral (Fig 2c), it is evident the abundance of *Combretum lanceolatum* Pohl ex Eichler. Scrubland with *C. lanceolatum* Pohl ex Eichler, which reaches a height of about four meters, is often monospecific. It is widespread in the Pantanal and occurs near permanent water bodies in areas subject to several months of inundation.

The campo limpo (Fig 2d), is characterized by low density of trees and shrubs, dominated by grasses and are open areas subject to periodic flooding, being the vegetation dominated by *Hyptis brevipes* (Lamiaceae), *Richardia grandiflora* (Rubiaceae) e *Axonopus purpusii* (Poaceae) (Nunes da Cunha et al., 2007; Nunes da Cunha et al., 2010).

Methodology

The search for nests of social wasps was performed from 25 August 2011 to 11 April 2012 in the landscape units known as landizal, pombeiral, cambarazal and campo limpo. In 68 days of field work (1024 hours of sampling), each of the four landscape units (17 days/locality) were visit using linear transects, about 3 km, from 7:00h to 17:00h, with a different landscape unit being sampled every day.

Sampling was carried out by two people who walked along the transect, extending about five meters on either sides of vegetation searching for nests. The nests and colony were

Fig 2. The four plant physiognomies found in the study area, Retiro Novo Farm, Pantanal de Poconé, Mato Grosso, Brazil. A – Cambarazal; B – Landizal; C – Pombeiral; D – Campo limpo, during flooded season.

taken off from the substrate. In nests that were located in substrates difficult to reach, a sample of 10 to 60 specimens were collected with telescopic entomological net.

The nests recorded in this inventory were georeferenced using GPS (Global Position System), being listed and individually marked with colored plastic tape. The height of nests above the ground was measured with the use of tape, recording the substrate used for nesting, such as termites, trunks, branches, thorns etc.

The specimens were identified following the keys of Richards (1978), Carpenter and Marques (2001), Andena and Carpenter (2012). Voucher specimens were deposited in the collection of the Laboratory for Neotropical Bees and Wasps (licence number # 18147), Department of Biology, Universidade do Estado Mato Grosso (UNEMAT), *Campus* Cáceres, in Mato Grosso; Museum of Zoology, Universidade Estadual de Feira de Santana (MZFS), Feira de Santana, Bahia; and Universidade Estadual Paulista Júlio de Mesquita Filho (UNESP), *Campus* São José do Rio Preto, São Paulo.

Data Analysis

The number of active colonies of wasps in the four plant physiognomies was used as a measure of abundance, each sampling day being considered a sample. We did a hierarchical ordering cluster analysis and classification from binary matrices (Cluster Analysis). Therefore, we used the Jaccard dissimilarity measure relating them to the plant physiognomies with wasps' species, through the connection method UPGMA (Legendre & Legendre, 2012).

To verify the occurrence of wasps' species that nest in specific physiognomies, the Indval method via indicspecies program package (R Development Core Team, 2011) was employed. Such a method combines the degree of specificity for a particular species to an ecological *status* such as, for example, the habitat type, and its fidelity within status, measured by the percentage of occurrence.

This analysis gives a value of 0 to 100%, where 0% is equivalent to no indication of an indicator species for a particular environment, and 100% indicates that the occurrence of particular species is characteristic to the environment (Dufrene & Legendre, 1997). The analysis significance was performed by Monte Carlo test with 10,000 randomizations by accepting $P < 0.05$ as significant.

Results

A total of 308 nests of social wasps belonging to 14 species, distributed in eight genera were recorded in the four landscape units studied, with *Polybia* accounting for 69.57% of the nests sampled and 42.85% of the species. Except *Polybia* and *Brachygastra*, the other six genera, *Agelaia*, *Apoica*, *Chartergus*, *Parachartergus*, *Polistes* and *Synoeca* were represented by only one species each (Table 1).

The species with the highest number of nests cataloged were *Poly. ruficeps xanthops* (32.69%), *Poly. sericea* (24.27%) and *S. surinama* (15.21%). The highest number of nests recorded in cambarazal and pombeiral was of *Poly. ruficeps xanthops* 40.48% and 42.7% of nests, respectively. In landizal was recorded the highest number of nests of *S. surinama* (N = 35, 39.78%) while in campo limpo was recorded the highest number of *Poly. sericea* (N = 21; 44.68%) (Table 1).

The highest wasp species richness was recorded in cambarazal (S = 13) and highest abundance of nests in pombeiral (N = 89), landizal (N = 88) and cambarazal (N = 84), while campo limpo had the lowest richness (S = 7) and abundance (N = 47) (Table 1; Fig 3).

Fourteen species recorded nesting in the four landscape units sampled; nests of three species were associated to three landscape units. The nests of *S. surinama* were associated with cambarazal and landizal (IndVal = 93.3, P = 0.001), whereas nests of *Poly. ruficeps xanthops* were associated with cambarazal, landizal and pombeiral (IndVal = 97, P = 0.001), as well as nests of *Poly. chrysothorax* (IndVal = 77, P = 0.014).

The cluster analysis shows that cambarazal, pombeiral and landizal are more related than campo limpo, which is a distinct branch. The Cophenetic Correlation Coefficient (CCC) was 0.80, and this value demonstrates that the distance matrix is well represented (Fig 3).

The nests of wasps cataloged in this study were located, in average, at 2.0m (SD = 5.70) above the ground, those of *Agelaia* sp. 1 (N = 2) were at lower height, constructed in soil cavities; and those of greater height, *C. globiventris* nests (N = 12; mean = 5.44; min = 1.96m, max = 7.71m, SD = 1.96), exposed in the vegetation. Among the plants used as support by the wasps, 10.23% of the nests were associated with *Bactris glauscescens* Drude (Arecaceae).

Regarding the substrate used for building the nests,

Table 1. Absolute frequency (N), relative frequency (%), average height of the wasps nests from de soil (HS) and standard deviation (SD) at Pirizal District. Retiro Novo Farm, Pantanal de Poconé, Mato Grosso, Brazil.

Tribe / Species	Cambarazal N (%)	Landizal N (%)	Pombeiral N (%)	Campo limpo N (%)	Total	HS (m) (mean ± SD)
Epiponini						
Agelaia sp. 1	1 (1.19)		1 (1.12)		2 (0.64)	0
Apoica pallens (Fabricius. 1804)	2 (2.38)	1 (1.13)			3 (0.97)	2.0 ± 0.69
Brachygastra augusti (de Saussure, 1854)	2 (2.38)				2 (0.64)	0.6 ± 1.09
Brachygastra lecheguana (Latreille, 1824)	5 (5.95)	2 (2.27)		6 (12.76)	13 (4.21)	1.4 ± 1.25
Chartergus globiventris de Saussure, 1854	3 (3.57)		1 (1.12)	7 (14.9)	11 (3.56)	5.9 ± 1.96
Parachartergus fraternus (Gribodo, 1892)	7 (8.33)	2 (2.27)	2 (2.25)	4 (8.51)	15 (4.85)	3.2 ± 1.57
Polybia chrysothorax (Lichtenstein, 1796)	4 (4.77)	8 (9.1)	10 (11.24)		22 (7.12)	1.4 ± 0.41
Polybia ignobilis (Haliday, 1836)	1 (1.19)			2 (4.25)	3 (0.98)	0.8 ± 0.25
Polybia jurinei de Saussure, 1854	1 (1.19)		2 (2.25)		3 (0.98)	1.4 ± 0.96
Polybia gr. *occidentalis*	1 (1.19)	1 (1.13)	5 (5.62)	4 (8.51)	11 (3.56)	0.7 ± 0.31
Polybia ruficeps xanthops Richards, 1978	34 (40.48)	26 (29.54)	38 (42.7)	3 (6.38)	101 (32.69)	1.5 ± 0.57
Polybia sericea (Olivier, 1791)	12 (14.28)	13 (14.78)	28 (31.46)	21 (44.68)	74 (24.27)	2.5 ± 5.86
Synoeca surinama (L., 1767)	11 (131)	35 (39.78)	1 (1.12)		47 (15.21)	1.7 ± 0.66
Polistini						
Polistes versicolor (Olivier, 1791)			1 (1.12)		1 (0.32)	1.1
Total of nests	84	88	89	47	308	-
Richness	13	8	10	7	14	-

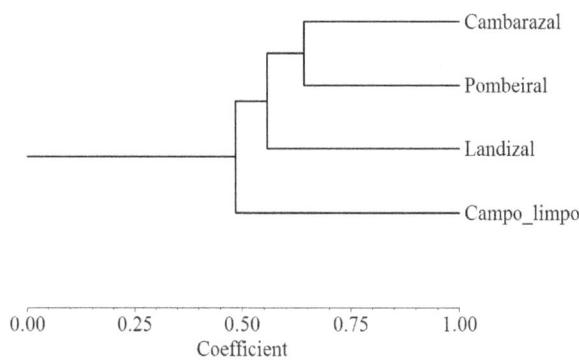

Fig 3. Dendrogram of Hierarchical Clustering Analysis classification according to vegetation types based on presence or absence of social wasp species (Hymenoptera: Vespidae) using UPGMA and Cophenetic Correlation Coefficient (CCC = 0.80), Retiro Novo Farm, Pantanal de Poconé, Mato Grosso state, from August, 2011 to April, 2012.

those of *Agelaia* sp. 1 and *Poly. ignobilis* were found in termite nests and fallen logs, while the nests of *S. surinama* were built directly on tree trunks. The nests of *C. globiventris, Par. fraternus* and *Poly. jurinei* were attached in branches of plants, while the nests of *Polistes versicolor* were among thorns of *B. glauscescens*, as recorded for *Poly. chrysothorax, Poly.* gr. *occidentalis,* and *B. lecheguana* and *B. augusti* also built their nests on the branches. The nests of *Poly. ruficeps xanthops* and *Poly. sericea* were found in pre-existing cavities, branches of plants and between spines of *B. glauscescens.*

Discussion

In this work 14 species of wasps, that represents 15.90% of the total of species recorded for the cerrado of Nova Xavantina, 88 species (Richards, 1978) and 38.88% when compared with the species collected in Chapada dos Guimarães, 35 species (Diniz & Kitayama,1998). Also in Chapada dos Guimarães Diniz & Kitayama (1994) found 100 nests, belonging to 30 species and 15 genera. Richards (1978) recorded 199 colonies comprising 51 species in 14 genera nesting in six habits in Nova Xavantina and Serra do Cachimbo. The species with the largest number of colonies was *Poly. ruficeps* (14%), followed by *Poly. occidentalis* (9%), *Poly. jurinei* (6%), *Poly. erythrotorax* (5%), *Poly. ignobilis* (5%), *Pa. fraternus* (4%) and *Epipona tatua* (4%). The genus *Polybia* represents 60% of the colonies reported from the two regions of Mato Grosso State, ranging from 67% in Nova Xavantina and Serra do Cachimbo to 45% in Rio Manso (Richards, 1978; Diniz & Kitayama, 1994).

The highest species richness of wasps in cambarazal may be related to vegetation structure and foliage density, once the cambarazal is a semi-deciduous forest and maintains much of the vegetation cover during the dry season (Nascimento & Nunes da Cunha, 1989). The campo limpo had the lowest species richness, while pombeiral and landizal showed intermediate values. Environments such as cambarazal, pom-

beiral and landizal are physiognomies with more structured vegetation and retain most of the vegetation cover in the dry season (Nunes da Cunha et al., 2007). Such environments may harbor more species of nesting social wasps in comparison to more open environments, such as grasslands, once the forest areas can provide greater protection, materials to build nests and substrates for attachment of wasp nests (Santos et al., 2009a; Santos et al., 2009b).

Studies on other animal groups, such as birds and beetles, in this region of the Pantanal, shows higher species richness in cambarazal (Pinho & Marini, 2012, Marques et al., 2010). In this study, conducted with birds in forestry environments in the area of the present survey, Pinho & Marini (2012) shows that the highest species richness and a greater number of bird nests were observed in cambarazal, suggesting some relationship with vegetation structure, food supply, protection from predators and suitable microclimate; being the cambarazal an important nesting habitat for several species that prefer forest habitats in this region of the Pantanal. The landizal in turn, despite maintaining much of the vegetation cover during the dry season is characterized by few species with low density in the understory (Pinho & Marini, 2012).

To edaphic beetles (Coleoptera), the highest abundance and species richness were also recorded in cambarazal, while the campo limpo had the lowest values of abundance and richness in this region (Marques et al., 2010). In this case, the type of vegetation determined the ecological condition such as microclimate, light, shelter, food supply, making it important to define the composition, richness and abundance of species of beetles (Marques et al., 2010).

Other studies that investigated the social wasps nesting in different landscape units in Brazil, also found higher species diversity in forest environments. For example, a study conducted in three landscape units of Cerrado (agricultural systems, campo sujo and Cerradão (arboreal plants), there were 19 nests of social wasps species, distributed in 13 genera, with the highest richness having been observed in Cerradão, and lower richness in agricultural systems (Santos et al., 2009a). A total of 319 nests belonging to 17 species of social wasps were recorded in three Caatinga vegetation types, the highest richness found in arboreal Caatinga, while agricultural systems and shrubby Caatinga showed similar species richness (Santos et al., 2009b).

In the present study, the highest number of wasp nests and the second highest number of wasp species have been recorded in pombeiral, invader vegetation of natural and artificial pastures, and that reduces the foraging ability of cattle, which led some farmers from Pantanal to try to eradicate it (Nunes da Cunha et al., 2007).

According to the Indval analysis, the nests of *S. surinama* were associated with landizal and cambarazal. Nests of *P. ruficeps xanthops* and *P. chrysothorax* in turn, were associated with landizal, cambarazal and pombeiral. However, no wasp species had nests associated with campo limpo in the study area, despite *Poly. sericea* has presented a greater number of nests (44.58%) in this plant physiognomy.

Some wasp species are found nesting in environments with specific conditions, with some fidelity to these sites (Heithaus, 1979; Santos et al., 2009a). In the riparian forest of Rio das Mortes, Minas Gerais, the Indval analysis showed that *Pseudopolybia vespiceps* and *Polybia fastidiosuscula* were associated with more conserved sites, while *Mischocyttarus drewseni* was associated with disturbed areas and can be used as environmental indicators (Souza et al., 2010). Nests of *Mischocyttarus* (Mischocyttarini) were not found in the present study. According to Silva et al. (2011) smaller colonies like *Mischocyttarus*, or those camouflaged or built inside cavities (e.g. *Agelaia* spp.) may easily go unnoticed.

In the present study, nests of *S. surinama* were not record in campo limpo, just in forestry environments. This result can be attributed to the fact of *Synoeca* presenting arboreal and sessile nests, which are built directly on the trunk of trees and occupy a large area of the substrate (Wenzel, 1998). Open areas with few trees and shrubs, such as campo limpo, may not provide the necessary conditions for establishing *Synoeca* nests. In Bahia, nests of *S. cyanea* were not observed in agricultural systems, although many individuals have been observed collecting nectar and water in such environments (Santos et al., 2009a).

In the Cerrado *lato sensu* in Mato Grosso, despite *S. surinama* and *P. sericea* having been sampled foraging in different habitats, such as gallery forest, Cerrado *sensu stricto*, campo sujo and campo úmido, both nested only in gallery forests and Cerrado *sensu stricto*, respectively (Diniz & Kitayama, 1994).

In this study, the wasp species with the highest number of nests recorded in campo limpo was *Poly. sericea*, species widely distributed throughout South America, especially in open areas like fields, and various types of Cerrado and dry forests (Richards, 1978). It is known that *B. lecheguana*, *Poly. sericea*, *Poly. ignobilis*, *Poly. occidentalis* have broad ecological tolerance and are generally dominant in open ecosystems and under adverse environmental conditions, being very important in simpler community structure and those subject to strict ecological conditions (Santos, 2000).

The nests of most social wasp species (85.71%) cataloged in this study were located at medium height compared to soils less than three meters, and only the nests of *C. globiventris and Pa. fraternus* were recorded at heights above. The behavior of nesting near the soil can benefit social wasps because this type of environment offers greater availability of substrates for nesting, low temperatures and high humidity (Raw, 1998).

Regarding the use of thorny plants as support for nests of social wasps, such as *B. glauscescens*, such behavior may be one of the strategies of social wasps to reduce nest predation, as noted in the study on the nest site selection in palm plants *B. simplifrons* and *Astrocaryum sciophilum* (Arecaceae) (Dejean et al., 1998).

The natural grasslands showed lower abundance and lower species richness in the study area, and none wasp species with nests associated with this vegetation type. These results show that maintaining forest environments in the Pantanal of Mato Grosso is essential for the establishment of social wasp nests and therefore, it is important for maintenance of wasp species in this region.

We hope that this work helps studies related to ecology, biology, distribution and abundance of social wasps in Northern Pantanal and stimulate additional surveys of social wasps in the Pantanal of Mato Grosso.

Acknowledgements

Financial support was provided by FAPEMAT (737955/2008; 285060/20010). We thank L.A. Castro for providing the map used in figure 1.

References

Andena, S.R. & Carpenter, J.M. (2012). A phylogenetic analysis of the social wasp genus *Brachygastra* Perty, 1833, and description of a new species (Hymenoptera: Vespidae: Epiponini). American Museum Novitates, 3753: 1-38.

Carpenter, J.M. & Marques, O.M. (2001). Contribuição ao Estudo dos Vespídeos do Brasil. Salvador, Universidade Federal da Bahia, Departamento de Fitotecnia. Série Publicações Digitais, v. 3, CD.

Clapperton, B.K. (1999). Abundance of wasps and prey consumption of paper wasps (Hymenoptera, Vespidae: Polistinae) in Northland, New Zealand. New Zealand Journal of Ecology, 23: 11-19.

Cruz, J.D., Giannotti, E., Santos, G.M.M., Bichara-Filho, C.C. & Rocha, A.A. (2006). Nest site selection and flying capacity of neotropical wasp *Angiopolybia pallens* (Hymenoptera: Vespidae) in the Atlantic Rain Forest, Bahia State, Brazil. Sociobiology, 47: 739-749.

Dejean, A., Corbara, B. & Carpenter, J.M. (1998). Nesting site selection by wasps in the Guianese rain forest. Insectes Sociaux, 45: 33-41.

Diniz, I.R. & Kitayama, K. (1994). Colony densities and preferences for nest habitats of same social wasps in Mato Grosso State, Brazil (Hymenoptera, Vespidae). Journal of Hymenoptera Research, 3: 133-143.

Diniz, I.R. & Kitayama, K. (1998). Seasonality of vespid species (Hymenoptera: Vespidae) in a central Brazilian cerrado. Revista de Biologia Tropical, 46: 109-114.

Dufrene, M. & Legendre, P. (1997). Species Assemblages and Indicator Species: the need for a flexible asymmetrical approach. Ecological Monographs, 67: 345-366.

Fantin-Cruz, I., Girard, P., Zeilhofer, P. & Collischonn, W.

(2010). Dinâmica de inundação. In I.M. Fernandes, C.A. Signor & J.M.F. Penha (Orgs.), Biodiversidade no Pantanal de Poconé. Cuiabá: Centro de Pesquisa do Pantanal. 196 p

Fernandes, I. M., Signor, C. A., Penha, J. (Orgs.) Biodiversidade no Pantanal de Poconé. Cuiabá: Centro de Pesquisa do Pantanal. 196 p

Grimaldi, D. & Engel, M.S. (2005). Evolution of the insects. Cambridge, New York, Melbourne: Cambridge University Press. 772 p

Heithaus, E.R. (1979). Community structure of neotropical flower visiting bees and wasps: diversity and phenology. Ecology, 60: 190-202.

Henriques, R.P.B., Diniz, I.R. & Kitayama, K. (1992). Nest density of some social wasp species in Cerrado Vegetation of Central Brazil (Hymenoptera: Vespidae). Entomologia Generalis, 17: 265-268.

Hunt, J.H. (2007). The evolution of social wasps. New York: Oxford University Press, 259 p

Itô, Y. (1986). On the pleometrotic route of social evolution in the Vespidae. Monitore Zoologico Italiano, 20: 241-262.

Junk, W.J., Nunes-da-Cunha, C., Wantzen, K.M., Petermann, P., Strüssmann, C., Marques, M.I. & Adis, J. (2006). Biodiversity and its conservation in the Pantanal of Mato Grosso, Brazil. Aquatic Science, 68: 278-309. doi: 10.1007/s00027-006-0851-4.

Lawton, J.H. (1983). Plant architecture and the diversity of phytophagous insects. Annals of the Entomological Society of America, 28: 23-39.

Legendre, P. & Legendre, L. (2012). Numerical Ecology. New York: Oxford, 853 p Marques, M.I., Souza, W.O., Santos, G.B., Battirola, L.D. & Anjos, K.C. (2010). Fauna de artrópodes de solo. In: I.M. Fernandes, C.A. Signor & J.M.F. Penha (Orgs.), Biodiversidade no Pantanal de Poconé (pp. 25-35). Cuiabá: Centro de Pesquisa do Pantanal.

Melo, G.A.R., Aguiar, A.P. & Garcete-Barrett, B.R. (2012). Hymenoptera. In J.A. Rafael, G.A.R. Melo, C.J.B. Carvalho, S.A. Casari & R. Constantino (Eds.), Insetos do Brasil: Diversidade e Taxonomia. (pp. 553-612). Ribeirão Preto: Holos Editora.

Nascimento, M.T. & Nunes-da-Cunha, C. (1989). Estrutura e composição florística de um Cambarazal no Pantanal de Poconé, MT. Acta Botanica Brasilica, 3: 3-23.

Noll F.B.,Wenzel, J.W. & Zucchi, R. (2004). Evolution of Caste in Neotropical Swarm-Founding Wasps (Hymenoptera: Vespidae; Epiponini). American Museum Novitates, 3467: 1-24.

Nunes-da-Cunha, C., Junk, W.J. & Leitão-Filho, H.F. (2007). Woody vegetation in the Pantanal of Mato Grosso, Brazil: a preliminary tipology. Amazoniana, 11: 159-184.

Nunes-da-Cunha, Rebello, C.L. & Costa, C.P. (2010). Vegetação e Flora: experiência pantaneira no sistema de grade. In I.M. Fernandes, C.A. Signor & J.M.F. Penha (Eds.), Biodiversidade no Pantanal de Poconé (pp. 37-57). Cuiabá: Centro de Pesquisa do Pantanal.

Pinho, J.B. & Marini, M.A. (2012). Using birds to set conservation priorities for Pantanal wetland forests, Brazil. Bird Conservation International, 22: 155-169. doi: 10.1017/S0959270911000207

Prezoto, F. & Machado, V.L.L. (1999). Ação de Polistes (Aphanilopterus) simillimus Zikán (Hymenoptera, Vespidae) no controle de Spodoptera frugiperda (Smith) (Lepidoptera, Noctuidae). Revista Brasileira de Zoologia, 16: 841-851.

Prezoto, F., Cortes, S.A.O. & Melo, A.C. (2008). Vespas: de vilãs a parceiras. Ciência Hoje, 48: 70-73.

R Development Core Team (2011) R: A language and environment for statistical computing. R Foundation for Statistical Computing, Vienna, Austria. Retrieved from: http://www.R-project.org/.

Raw, A. (1998). Social wasps (Hymenoptera, Vespidae) of the Ilha de Maracá. In: J.A. Ratter J.A. & W. Milliken (Eds.), Maracá: Biodiversity and environment of an Amazonian Rainforest (pp. 311-325). Chichester: John & Sons.

Richards O.W., & Richards, M.J. (1951). Observations on the social wasps of South America (Hymenoptera, Vespidae). Transactions of the Royal Entomological Society, 102: 1-170.

Richards, O.W. (1978). The social wasps of the Americas, excluding the Vespinae. London: British Museum (Natural History), 580 p

Santos, G.M.M. (2000). Comunidades de vespas sociais (Hymenoptera-Polistinae) em três ecossistemas do estado da Bahia, com ênfase na estrutura da guilda de vespas visitantes de flores de Caatinga. Tese de doutorado, Faculdade de Filosofia, Ciências e Letras de Ribeirão Preto/USP, 129 p

Santos, G.M.M. & Gobbi, N. (1998). Nesting habits and colonial productivity of Polistes canadensis canadensis (L.) (Hymenoptera-Vespidae) in a Caatinga area, Bahia State, Brazil. Journal of Advanced Zoology, 19: 63-69.

Santos, G.M.M., Bichara-Filho, C.C., Resende, J.J., Cruz, J.D. da & Marques, O.M. 2007. Diversity and community structures of social wasps (Hymenoptera: Vespidae) in three ecosystems in Itaparica Island, Bahia State, Brazil. Neotropical Entomology, 36: 180-185. doi: 10.1590/S1519-566X2007000200002.

Santos, G.M.M., Gobbi, J., Cruz, J.D. da, Marques, O.M. & Gobbi, N. (2009a). Diversidade de vespas sociais (Hymenoptera: Vespidae) em áreas de cerrado na Bahia. Neotropical Entomology, 38: 317-320. doi: 10.1590/S1519-566X2009000300003.

Santos, G..M.M., Bispo, P.C. & Aguiar, C.M.L. (2009b). Fluctuations in Richness and abundance of social wasps during the dry and wet seasons in three phyto-physiognomies at the tropical dry forest of Brazil. Environmental Entomology, 38: 1613-1617.

Saussure, H. de. 1853–58. Monographie des guêpes sociales ou de la tribu des vespiens. Paris: Masson. [1–96, 1853; 97–256, 1854]

Spradbery, J.P. (1991). Evolution of queen number and queen control. In: K.G. Ross and R.W. Matthews (Eds.), The Social Biology of Wasps (pp. 336-388). Ithaca, NY: Cornell Univ. Press.

Silva, S.S. & Silveira, O.T. (2009). Vespas sociais (Hymenoptera, Vespidae, Polistinae) de floresta pluvial Amazônica de terra firme em Caxiuanã, Melgaço, Pará. Iheringia, Série Zoologia, 99: 317-323.

Silva, S.S., Azevedo, G.G. & Silveira, O.T. (2011). Social wasps of two Cerrado localities in the northeast of Maranhão state, Brazil (Hymenoptera, Vespidae, Polistinae). Revista Brasileira Entomologia, 55: 597-602. doi: 10.1590/S0085-56262011000400017.

Silva-Pereira, V. & Santos, G.M.M. (2006). Diversity in bee (Hymenoptera: Apoidea) and social wasp (Hymenoptera: Vespidae, Polistinae) community in Campos Rupestres, Bahia, Brazil. Neotropical Entomology, 35: 165-174. doi: 10.1590/S1519-566X2006000200003.

Silveira, O.T. (2002). Surveying neotropical social wasps. An evaluation of methods in the "Ferreira Penna" Research Station (ECFPn), in Caxiuanã, PA, Brazil (Hymenoptera, Vespidae, Polistinae). Papéis Avulsos deZoologia, 42: 299-323. doi: 10.1590/S0031-10492002001200001

Silveira, O.T., Esposito, M.C, Santos, J.N. & Gemaque, F.E. (2005). Social Wasps and bees captured in carrion traps in a rainforest in Brazil. Entomological Science, 8: 33-39. doi: 10.1111/j.1479-8298.2005.00098.x

Silveira, O.T., Silva, S.S., Pereira, J.L.G. & Tavares, I.S. (2012). Local scale spatial variation in diversity of social wasps in an Amazonian rain forest, Caxiuanã, Pará, Brazil (Hymenoptera,

Vespidae, Polistinae). Revista Brasileira Entomologia, 56: 329-346. doi: 10.1590/S0085-56262012005000053.

Simões, M.H., Cuozzo, M.D. & Frieiro-Costa, F.A. (2012). Diversity of social wasps (Hymenoptera, Vespidae) in Cerrado biome of the southern of the state of Minas Gerais, Brazil. Iheringia, Série Zoologia, 102: 292-297.

Somavilla, A., Oliveira, M.L. & Silveira, O.T. (2012). Guia de identificação dos ninhos de vespas sociais (Hymenoptera: Vespidae: Polistinae) na Reserva Ducke, Manaus, Amazonas, Brasil. Revista Brasileira Entomologia, 56: 405-414. doi: 10.1590/S0085-56262012000400003.

Souza, M.M., Louzada, J., Serrão, J.E. & Zanuncio, J.C. (2010). Social wasps (Hymenoptera: Vespidae) as indicators of conservation degree of riparian forests in Southeast Brazil. Sociobiology, 56: 387-396.

Spradbery, J.P. (1991). Evolution of queen number and queen control. In: K.G. Ross and R.W. Matthews (Eds.), The Social Biology of Wasps (pp. 336-388). Ithaca, NY: Cornell Univ. Press.

Vitali-Veiga, M.J. & Machado, V.L.L. (2001). Entomofauna visitante de Gleiditsia triacanthos L. – Leguminosae durante o seu período de floração. Revista Bioikos., 15: 29-38.

Wenzel, J.W. (1998). A generic key to the nests of hornets, yellowjackets, and paper wasps worldwide (Vespidae, Vespinae, Polistinae). American Museum Novitates, 3224: 1-39.

West-Eberhard, M.J. (1978). Temporary queens in Metapolybia wasps: non-reproductive helpers without altruism? Science, 200: 441-443.

West-Eberhard, M.J. (1996). Wasp societies as microcosms for the study of development and evolution. In: S. Turillazzi & M.J. West-Eberhard (Eds.), Natural history and evolution of paper-wasps (pp. 290-317). Oxford: Oxford University Press.

Wilson, E.O. (1985). The sociogenesis of insect colonies. Science, 228 (4704): 1479-1485.

4

First-year nest growth in the leaf-cutting ants *Atta bisphaerica* and *Atta sexdens rubropilosa*

SRS Cardoso[1,2], LC Forti[1], NS Nagamoto[1], RS Camargo[1]

1 - Universidade Estadual Paulista (UNESP), Botucatu, São Paulo, Brazil
2 - Instituto Federal do Tocantins, Araguatins, Tocantins, Brazil

Keywords
initial chamber, leaf-cutting ant, nest architecture, nest foundation, excavation

Corresponding author
Nilson Satoru Nagamoto
Depto. Proteção Vegetal, FCA, UNESP
PO Box 237, Botucatu, SP, Brazil
18.610-307
E-Mail: nsnagamoto@yahoo.com

Abstract
Most ants, as leaf cutting ants, construct nests underground to maintain environmental conditions favorable to the development of immature and adult individuals. But there was little works about this, especially doing comparison of nest growth among leaf-cutting ant species. Thus, we studied the growth of nests of the leaf-cutting ants *Atta bisphaerica* and *Atta sexdens rubropilosa* from nest foundation until the appearance of a second chamber. To this end, we verified the measurements of the chamber recently constructed by the queen and monitored its growth in the initial phase of nest development. The nests were marked immediately after nuptial flight, with 40 nests of each species being dug at 45, 90, 135, 180, and 225 days afterwards. As a result, it was found a first time statistical demonstration that an ellipsoid chamber shape was verified in both species in the initial chamber at 45 days, and that, after, these chamber became spherical. In general, chamber size increased and format change was found in both species. The depth of the first chamber was found to increase significantly only in *A. bisphaerica*; this result means that this chamber growths downside rather than upside. The occurrence of a second chamber was verified from six months after nest foundation, in both species. Our study contributes to knowledge of the colony development for up to 1 year, by performing comparison of two leaf-cutting ants species.

Introduction

Most ant species construct their nests underground, which promotes protection for the colony and aids in the maintenance of environmental conditions, facilitating the development of immature and adult individuals (Hölldobler & Wilson, 1990). However, nest construction requires intensive labor by the ants to excavate the soil (Halley et al., 2005).

In the genus *Atta* F. 1805 (Hymenoptera: Formicidae) the adult nests are formed by numerous interconnected chambers, reaching several meters in depth, with deposition of soil on the surface, forming a pile of loose soil (Gonçalves, 1960; Araújo & Della Lucia, 1997). Such chambers may present variations in format, localization and dimensions depending on the species and function (fungus, waste and soil) (Moreira, 2001). Despite such knowledge, little is known about the initial growth of nests formed by this genus.

In most social insects the size of the nest is a function of its population; however the dynamic of construction and enlargement of underground nests is not fully known (Rasse & Deneubourg, 2001). In *Atta*, the work of nest foundation is performed by a single recently fecundated queen that constructs the first chamber (Autuori, 1942; Ribeiro, 1995; Camargo et al., 2011; Fröhle & Roces, 2012). After the first worker ants initiate foraging, they assume activities within and outside of the nest, including the excavation of new tunnels and chambers (Autuori, 1942; Amante, 1972); so, the nest enlargement is thus the responsibility of adult offspring. Furthermore, the initial chamber is also enlarged each day by the excavation effort of the worker ants (Jacoby, 1943). Some studies indicate that for several species of ants the nest size is adjusted to the population (Rasse & Deneubourg, 2001; Buhl et al., 2004; Mikheyev & Tschinkel, 2004; Camargo & Forti, 2014; Römer & Roces, 2014).

Although the genus *Atta* presents the ability to construct deep nests, the initial chamber is based in the topsoil layer at a depth of about 8.5 to 15 cm (Autuori, 1941), and is subject not only to variations in temperature and humidity, but also to the influence of strong and abundant rains that are typical of the period in which nest foundation occurs (Autuori, 1941; Bento et al., 1991).

Therefore, one of the most important aspects in the study of *Atta* is undoubtedly the knowledge of its initial development, which can be externally monitored through measurements of the nest openings and the soil mound (Mariconi, 1970) and internally by means of excavations, to ascertain, for example, the depth and size of the first chambers constructed. Although there have already been reports on the chamber format and initial chamber depth of *Atta* (Autuori, 1942; Jacoby, 1943, 1950; Mariconi, 1970), the literature on the foundation and initial development of *Atta* nests remains highly limited.

Due to the scarcity of biological information on nests of *Atta* species in their first year of life, the present paper studied the growth and morphology of nests in two species which exhibit contrasting adaptations: *Atta bisphaerica* Forel 1908, which nests in full sun and foraging predominantly grasses, and *Atta sexdens rubropilosa* Forel 1908, which nests in shaded places and forages dicots (Mariconi, 1970; Fowler et al., 1986; Nagamoto et al., 2009).

Material and Methods

The work was conducted in the city of Botucatu, Sao Paulo state, Brazil with the *A. bisphaerica* nests studied in a pasture area of Santana Farm (22°50.720' S and 48°26.155' W), and the *A. sexdens rubropilosa* nests being selected in an area of *Eucalyptus* plantations belonging to FCA/UNESP, Botucatu, SP (22°50.833' S and 48°26.476' W). Both areas present the Oxisol soil type (Dark Red Latosol). In each area 500 nests were randomly selected at the moment of nest foundation by the queen in October 2007. The nests were marked by labeled wire stakes.

The excavations were performed with using digging tools (such as shovels, picks and spatulas) by opening a trench at the side of the nest, carefully exposing the channel and chamber as described by Autuori (1942), Pereira-da-Silva (1979) and Pretto (1996), and then the biomass (fungus garden, ants and brood) was discarded. Forty nests of each species were excavated at 45, 90, 135, 180 and 225 days after nest foundation (a total of 400 nests). The measurements, including depth, length, width, and height were obtained for each excavated nest (Figure 1).

Format of the initial chamber

Based on the geometric aspect, some authors have defined the format of the chamber for the cultivation of fungi, as typically spherical or ellipsoid (Jacoby, 1950; Silva Junior et al., 2013). Here we purpose, for the first time (to our best knowledge), to classify the leaf cutting ant chamber statistically, taking into account that the ellipse and sphere posses well known coordinates: if the Cartesian coordinates x, y and z are equal, it is a sphere; if there are differences, is ellipsoid. For this, we compared length (L), width (W) and height (H), at each excavation period, using Analysis of Variance (ANOVA) (α=0.05).

Volume of the chamber

The volume of the chamber was monitored through comparison of each chamber measurements (L, W, H) among the periods studied: volume = $3/4\pi LWH$ (ellipsoid volume). For this, only the measurements of nests with alive queen were taken into account. The volume data among periods among a species, and in each period between these species, were submitted to Analysis of Variance (ANOVA) (α=0.05),

Figure 1. The measurements of depth (D), length (L), width (W) and height (H) at a nest of *Atta bisphaerica*, that was cement-moulded (as in Moreira et al., 2004) to enable a better view.

with the means compared by Bonferroni's test. To evaluate the difference in the depth of the chamber between the two species, their measurements in each period were compared and the data obtained were submitted to Student's t test ($\alpha = 0.05$).

Results

Format of the initial chamber

The initial chamber presented two distinct formats: first ellipsoid and spherical in the last periods of evaluation. *A. bisphaerica* showed an ellipsoid format until 90 days after foundation, with the dimensions of chamber differing significantly in 45 days ($p < 0.001$) and 90 days ($p = 0.006$) (Table 1). From 135 to 225 days after the foundation, the measurements did not differ from each other, indicating a spherical format.

In *A. sexdens rubropilosa*, it was verified a difference among the dimensions of chamber in 45 days ($p < 0.001$). The change in format occurred after 90 days because ceased to exist significant differences between length, width and height ($p > 0.050$) (Table 1).

Table 1. Summary statistic for values of ANOVA of the initial fungus chambers measurements (length, width and height), classification of the leaf-cutting ant initial chamber, statistically, into ellipsoid or spherical: if the Cartesian coordinates x, y and z are not significantly different, it is spherical; if there are significant differences, is ellipsoid. Comparisons in each excavation period (day) for each species.

Day	Atta bisphaerica			Atta sexdens rubropilosa		
	P value	F	DF	P value	F	DF
45	<0.001*	40.363	98	<0.001*	15.338	80
90	0.006*	5.523	65	0.060^{ns}	3.085	32
135	0.137^{ns}	2.040	83	0.226^{ns}	1.596	23
180	0.120^{ns}	2.263	35	0.830^{ns}	0.189	14
225	0.631^{ns}	0.470	26	0.366^{ns}	1.127	11

*Significant (= ellipsoid chambers); ^{ns}non significant (= spherical chambers)

In *A. bisphaerica*, at 45 days, the dimensions are (cm ± SD): the length (4.28 ± 0.65) differed from width (3.04 ± 0.37) and height (3.33 ± 0.68). For day 90, the length (5.14 ± 1.21) differed only from width (3.95 ± 0.93). Later, the highest growth in width and height in relation to height, resulted in no significant differences between these measurements ($p > 0.050$).

In *A. sexdens rubropilosa*, for day 45, the length (3.66 ± 0.57) differed from width (2.89 ± 0.80) and height (2.76 ± 0.52). Then, for both species, the width and height was the chamber measurements that grew most from 45 to 225 days.

Volume

The size of the initial chamber was growing significantly in both species (Fig. 2). In general, the mean measurements of the initial chamber were greater for *A. bisphaerica* than for *A. sexdens rubropilosa*, with volume differing significantly in many periods evaluated (Table 2). At 45, 135 and 225 days after foundation the volume was greater in *A. bisphaerica* than *A. sexdens rubropilosa*, significantly differing, $p < 0.001$, $p = 0.014$ and $p = 0.035$, respectively (Table 2).

Depth

The depth of the initial chamber was growing significantly for *A. bisphaerica* (Fig. 3), but not in *A. sexdens rubropilosa*. In general, chamber depth significantly differed between the ant species (Table 3), with *A. bisphaerica* showing deeper chambers (Fig. 3).

Second chamber

From 180 days, a new tunnel emanating down from the first chamber, it was always detected. The occurrence of a second chamber was verified from six months after nest foundation, in both species. For *A. bisphaerica* the mean measurements of the second chamber were: depth of 104.3 ± 35.0 cm, vol. of 29.4 ± 28.7 cm³. For *A. sexdens rubropilosa*, a second chamber occurred in only two nests, with the first

Table 2. Students's *t* test summary statistic for values of the initial fungus chamber volume between leaf-cutting ant species.

Day	Atta bisphaerica vs Atta sexdens rupropilosa		
	P value	t	DF
45	<0,001*	4,249	58
90	0,133^{ns}	1,541	31
135	0,014*	2,601	34
180	0,589^{ns}	-0,552	15
225	0,035*	2,411	11

*Significant difference in *t* test; ^{ns} non significant difference in *t* test.

Table 3. Students's *t* test summary statistic for values of the initial fungus chamber depth between leaf-cutting ant species.

Day	Atta bisphaerica vs Atta sexdens rupropilosa		
	P value	t	DF
45	<0,001*	5,372	58
90	0,002*	3,463	31
135	0,009*	2,751	34
180	<0,001*	4,831	15
225	0,062^{ns}	2,081	11

*Significant difference in *t* test; ^{ns} non significant difference in *t* test.

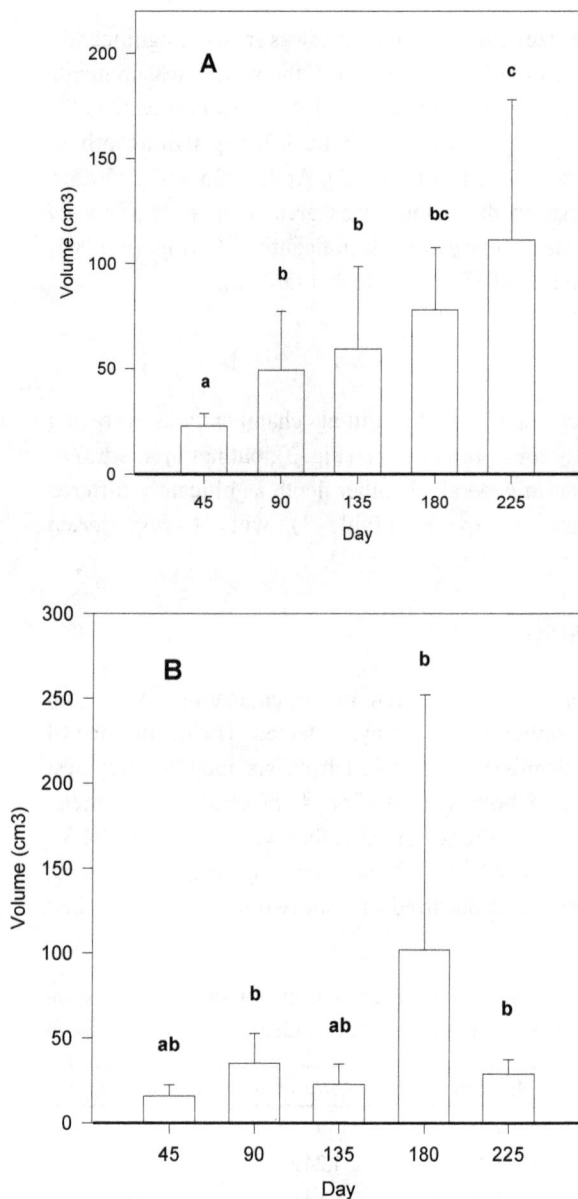

Fig 2. Mean volumes of the initial chamber of *Atta bisphaerica* (A) and *Atta sexdens rubropilosa* (B) as a function of time since foundation. Means followed by the same letter are not significantly different (Bonferroni test).

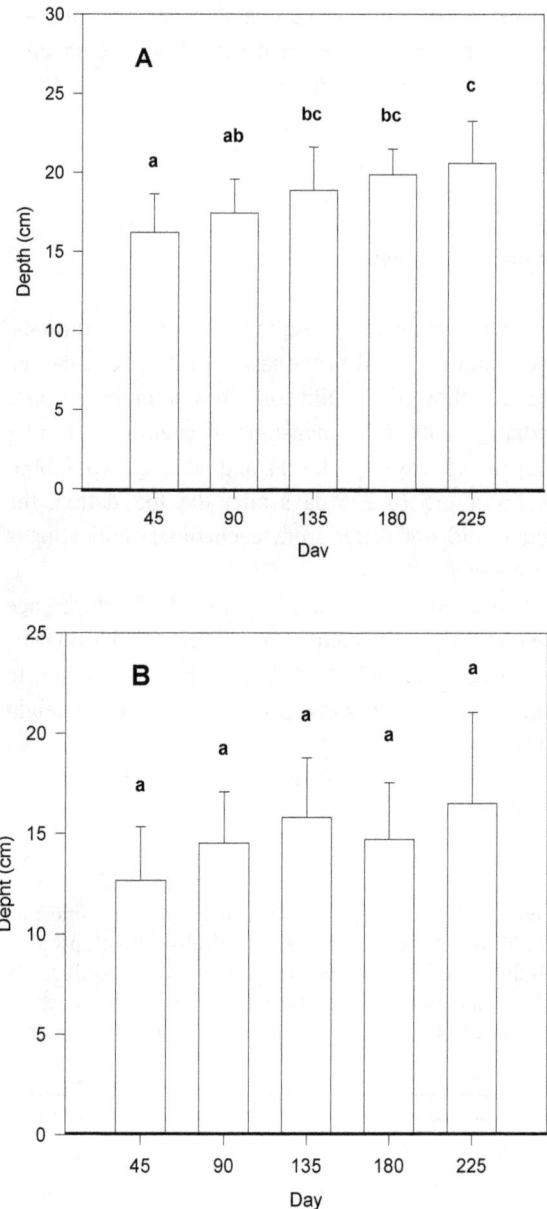

Fig 3. Mean chamber depths of the initial chamber of *Atta bisphaerica* (A) and *Atta sexdens rubropilosa* (B) as a function of time since foundation. Means followed by the same letter are not significantly different (Bonferroni test).

presenting the following measurements: depth of 39 cm, and vol. of 73.8 cm³. In the second nest, the depth was found to be less than 150 cm (an accurate measurement was not made due to some excavation difficulties).

Discussion

Format of the initial chamber

In the present study the occurrence of two shapes, ellipsoid and spherical, was observed in both studied species. The chamber founded by the queen initially presented an ellipsoid format, with changing to a spherical shape proceeded by the excavation by the worker ants. These workers excavated the chamber in all dimensions, but more in width and height than in depth.

The queen body is longer than wider or higher (Moser et al., 2004), so the chamber may also be optimized in this format to just accommodate it and to avoid unnecessary lost energy in digging process by the queen (Camargo et al., 2011). Later, it turns into spherical format, which would be more appropriate (optimized) for growing the fungus by the workers (Römer & Roces, 2014).

Growth

In both species, the size of the initial chamber increased as a function of time until digging of the second chamber started, approximately 180 days after nest foundation. The enlargement of the chamber probably occurred because of colony population growth, which can explain why *A. sexdens rubropilosa* chambers grew less: although not quantified, *A. sexdens rubropilosa* had fewer ants and smaller symbiotic fungus amounts in each period studied compared to *A. bisphaerica*. These results corroborate Buhl et al. (2004), who verified that the digging activity and volume increase in the initial chamber in *Messor sanctar* Emery were adjusted to colony population size. Rasse and Deneubourg (2001) also verified that both the volume and the maximum rate of excavation are related to the size of the group, and that the digging activity diminished when a *Lasius niger* L. nest reaches a particular volume. Similarly, Mikheyev and Tschinkel (2004) confirmed that total nest volume is highly correlated with the number of worker ants in *Formica pallidefulva* Wheeler. In leaf cutting ants, it is known that the fungus garden acts as a three-dimensional pattern for the final size of the chamber, that is, according to its growth, the ants increase the size of the chambers, as discussed by Fröhle and Roces (2009), Camargo et al. (2013) and Camargo and Forti (2014), for *Acromyrmex lundi* and *A. sexdens rubropilosa*, respectively.

Depth

The initial chamber depth significantly differed between the ant species, with *A. bisphaerica* showing deeper chambers than *A. sexdens rubropilosa*. It is known that *Atta bisphaerica* Forel 1908, presents nests in full sun and foraging predominantly grasses, and *Atta sexdens rubropilosa* Forel 1908 presents nests in shaded places and forages dicots (Mariconi, 1970; Fowler et al., 1986; Nagamoto et al., 2009), therefore differing in nesting habits and foraging strategies. Probably, the differences in nest depth between species are correlated to soil temperature, because shading alters soil temperature regimes by locally diminishing soil temperature (Rosenberg et al., 1983). In this perspective, we can hypothesize that nest exposed in grassland should have deeper fungus chamber than nest under shade of trees or inside the woods, given that soil temperature is negatively correlated with soil depth (Rosenberg et al., 1983). For leaf cutting ants, is indicated that soil moisture and temperature acts together: (*i*) Bollazzi et al. (2008) verified that workers' thermopreferences lead to the construction of superficial nests in cold soils, and subterranean ones in hot soils; and (*ii*) Pielström & Roces (2014) verified that soil moisture also varies according to soil depth, and demonstrably affects the digging behavior of leaf cutting ants.

Second chamber

The excavation of the second chamber by the worker ants occurred 6 months after the nest foundation in both species. This result differs from Mariconi (1970) who describes a 15-month period since foundation for digging the new channel and second chamber of *Atta capiguara*. However, the results approximate those of Jacoby (1943) who verified in *A. sexdens rubropilosa* the construction of a second chamber at 4 and 5 months after nest foundation. At 180 days of age, despite the presence of a second chamber, symbiotic fungus, workers and queens were still found in the first chamber. For this reason, it is suggested that in this period the new chamber had not been completely constructed, since according to Jacoby (1943), the fungus is established only in finalized chambers. In addition, Camargo and Forti (2013) argued that the structural growth of the nests occurred vertically, without lateral expansion of tunnels and chambers in the first year. The 3-month-old nests had a chamber with an average depth of 15 cm. In the course of 1 year, the nests expanded to have three to four chambers and were 3–4 m deep. The lateral expansion occurred after 1 year, when the nests grew laterally and reached large dimensions when adult ants were 3 years old. For example, adult nests of *A. laevigata* are 7 m deep, with a number of chambers ranging from 1149 to 7864 and a volume from 0.03 to 51 liters (Moreira et al., 2004).

Conclusion

It may be suggested from the present study that the worker ants actively dig the initial chamber, causing an increase in its size and influencing its depth to favor the chances of establishment of the colony. Although the adult nests of *A. sexdens rubropilosa* present different depth, chamber number and nest structure from *A. bisphaerica*, the initial nests are similar in shape as well as the time for construction of the second chamber. However, the initial chambers of *A. bisphaerica* present greater size and depth than those of *A. sexdens rubropilosa*. Besides, the chamber depth of *A. bisphaerica* is deeper than *A. sexdens rubropilosa*.

Acknowledgments

We wish to thank FAPEMA for the fellowship granted to the first author and FAPESP for financial support (2008/07032-7). L.C. Forti gratefully acknowledges the support of CNPq (472671/2008-1).

References

Amante, E. (1972). Influência de alguns fatores microclimáticos sobre a formiga saúva *Atta laevigata* (F. Smith, 1958), *Atta sexdens rubropilosa* Forel, 1908, *Atta bisphaerica* Forel, 1908 e *Atta capiguara* Gonçalves, 1944 (Hymenoptera, Formicidade), em formigueiros localizados no estado de São Paulo. Ph.D. Thesis, Escola Superior de Agricultura Luiz de Queiroz, USP. Piracicaba, SP, Brazil.

Araújo, M.S. & Della Lucia, T.M.C. (1997). Caracterização de ninhos de *Acromyrmex laticeps nigrosetosus* Forel, em povoamento de eucalipto em Paraopeba (MG). Anais da Sociedade Entomologica do Brasil, 26: 205-207. doi: 10.1590/S0301-80591997000100029

Autuori, M. (1941). Contribuição para o conhecimento da saúva (*Atta* spp. – Hymenoptera: Formicidae). I – Evolução do sauveiro (*Atta sexdens rubropilosa* Forel, 1908). Arquivos do Instituto Biológico, 12: 197-228.

Autuori, M. (1942). Contribuição para o conhecimento da saúva (*Atta* spp. –Hymenoptera - Formicidae). II – O sauveiro inicial (*Atta sexdens rubropilosa*, Forel, 1908). Arquivos do Instituto Biológico, 13: 67-86.

Bento, J.M.S., Della Lucia, T.M.C., Muchovej, R.M.C. & Vilela, E.F. (1991). Influência da composição química e da população microbiana de diferentes horizontes do solo no estabelecimento de sauveiros iniciais de *Atta laevigata* (Hymenoptera: Formicidae) em laboratório. Anais da Sociedade Entomológica do Brasil, 20: 307-316.

Bollazzi, M., Kronenbitter, J. & Roces, F. (2008). Soil temperature, digging behaviour, and the adaptive value of nest depth in South American species of *Acromyrmex* leaf-cutting ants. Oecologia, 158: 165-175.

Buhl, G.J., Deneubourg, J.L. & Theraulaz, G. (2004). Nest excavation in ants: group size effects on the size and structure of tunneling networks. Naturwissenschaften, 91: 602-606. doi: 10.1007/s00114-004-0577-x

Camargo, R.S. et al. (2011). Digging effort in leaf-cutting ant queens (*Atta sexdens rubropilosa*) and its effects on survival and colony growth during the claustral phase. Insectes Sociaux, 58: 17-22. doi: 10.1007/s00040-010-0110-5.

Camargo, R.S. & Forti, L.C. (2013). Queen lipid content and nest growth in the leaf cutting ant (*Atta sexdens rubropilosa*) (Hymenoptera: Formicidae). Journal of Natural History, 47: 65-73, doi: 10.1080/00222933.2012.738836

Camargo, R.S., Lopes, J.F. & Forti, L.C. (2013). O jardim de fungo atua como um molde para a construção das câmaras em formigas cortadeiras? Ciência Rural, 43: 565-570.

Camargo, R.S. & Forti, L.C. (2014). What is the stimulus for the excavation of fungus chamber in leaf-cutting ants? Acta Ethologica, 17: 1-5.

Fowler, H.G., Forti, L.C.; Pereira-da-Silva, V. & Saes, N.B. (1986). Economics of grass-cutting ants. In C.S. Lofgren & R.K. Vander Meer (Eds.), Fire ants and leaf-cutting ants: biology and management (pp.18-35). Boulder: Westview Press.

Fröhle, K. & Roces, F. (2009). Underground agriculture: the control of nest size in fungus-growing ants. In: G. Theraulaz et al. (Eds.), From insect nest to human architecture (pp. 95-104). Venice: European Centre for Living Technology.

Fröhle, K. & Roces, F. (2012). The determination of nest depth in founding queens of leaf-cutting ants (*Atta vollenweideri*): idiothetic and temporal control. Journal of Experimental Biology, 215: 1642-1650. doi: 10.1242/jeb.066217.

Gonçalves, C.R. (1960). Distribuição, biologia e ecologia das saúvas. Divulgação Agronômica, 1: 2-10.

Halley, J.D., Burd, M. & Wells, P. (2005). Excavation and architecture of argentine ant nests. Insectes Sociaux, 52: 350-356. doi: 10.1007/s00040-005-0818-9

Hölldobler, B. & Wilson, E.O. (1990). The ants. Cambridge: Harvard University Press.

Jacoby, M. (1943). Observações e experiências sobre *Atta sexdens rubropilosa* Forel visando facilitar seu combate. Boletim do Ministério da Agricultura, 32 (5): 1-54.

Jacoby, M. (1950). A arquitetura do ninho. In M. Jacoby (Ed.). A saúva: uma inteligência nociva (pp. 21-31). Rio de Janeiro: Serviço de Informação Agrícola.

Mariconi, F.A.M. (1970). As saúvas. São Paulo: Agronômica Ceres.

Mikheyev, A.S. & Tschinkel, W.R. (2004). Nest architecture of the ant *Formica pallidefulva*: structure, costs and rules of excavation. Insectes Sociaux, 41: 30-36. doi: 10.1007/s00040-003-0703-3

Moreira, A.A. (2001). *Atta bisphaerica* Forel, 1908 (Hym: Formicidae): arquitetura de ninhos e distribuição de isca nas câmaras. Ph.D. Thesis, Faculdade de Ciências Agronômicas, UNESP. Botucatu, SP, Brazil.

Moreira, A.A., Forti, L.C., Andrade, A.P.P., Boaretto, M.A.C., Lopes, J.F.S. (2004). Nest architecture of *Atta laevigata* (F. Smith, 1858) (Hymenoptera: Formicidae). Studies on Neotropical Fauna and Environment, 39: 109-116.

Moser, J.C., Reeve, J.D., Bento, J.M.S., Della Lucia, T., Cameron, R.S. & Heck, N.M. (2004). Eye size and behaviour of day-and night-flying leafcutting ant alates. Journal of Zoology, 264: 69-75.

Nagamoto, N.S., Carlos, A.A., Moreira, S.M., Verza, S.S., Hirose, G.L. & Forti, L.C. (2009). Differentiation in selection of dicots and grasses by the leaf-cutter ants *Atta capiguara*, *Atta laevigata* and *Atta sexdens rubropilosa*. Sociobiology, 54: 127-138.

Pereira-da-Silva, V. (1979). Dinâmica populacional, biomassa e estrutura dos ninhos iniciais de *Atta capiguara* Gonçalves, 1944 (Hymenoptera: Formicidae) na região de Botucatu, SP. Livre Docência Thesis, Instituto de Biociências, UNESP. Botucatu, SP, Brazil.

Pielström, S. & Roces, F. (2014). Soil moisture and excavation behaviour in the Chaco leaf-cutting ant (*Atta vollenweideri*): digging performance and prevention of water inflow into the nest. PLoS ONE 9(4): e95658. doi:10.1371/journal.pone.0095658

Pretto, D.R. (1996). Arquitetura dos túneis de forrageamento e do ninho de *Atta sexdens rubropilosa* Forel, 1908 (Hymenoptera - Formicidae), dispersão de substrato e dinâmica do inseticida na colônia. M.Sc. Dissertation, Faculdade de Ciências Agronômicas, UNESP. Botucatu, SP, Brazil.

Rasse, P.H. & Deneubourg, J.L. (2001). Dynamics of nest excavation and nest size regulation of *Lasius niger* (Hymenoptera: Formicidae). Journal of Insect Behavior, 14 (4): 433-449. doi: 10.1023/A:1011163804217

Ribeiro, F.J.L. (1995). A escavação do solo pela fêmea da saúva (*Atta sexdens rubropilosa*). Psicologia-USP, 6: 75-93.

Römer, D. & Roces, F. (2014). Nest enlargement in leaf-cutting ants: relocated brood and fungus trigger the excavation of new chambers. PLoS ONE, 9 (5): e97872. doi:10.1371/journal.pone.0097872

Rosenberg, N.J., Blad, B.L., & Verma, S.B. (1983). Microclimate-the biological environment. New York: Wiley.

Silva Junior, M.R., Castellani, M.A., Moreira, A.A., D'Esquivel, M., Forti, L.C. & Lacau, S. (2013). Spatial distribution and architecture of *Acromyrmex landolti* Forel (Hymenoptera, Formicidae) nests in pastures of Southwestern Bahia, Brazil. Sociobiology, 60: 20-29.

5

No Morphometric Distinction between the Host *Constrictotermes cyphergaster* (Silvestri) (Isoptera: Termitidae, Nasutitermitinae) and its Obligatory Termitophile *Corotoca melantho* Schiødte (Coleoptera: Staphylinidae)

HF CUNHA; JS LIMA; LF SOUZA; LGA SANTOS; JC NABOUT

Universidade Estadual de Goiás, UnUCET, Anápolis, GO, Brazil

Keywords
Coexistence, morphological mimicry, termite, termitophile

Corresponding author
Hélida Ferreira da Cunha
UEG/ UnUCET/ BR 153
Fazenda Barreiro do Meio n° 3105
CEP 75132-903, Anápolis-GO, Brazil
E-Mail: cunhahf@ueg.br

Abstract
Different species may live in termite nests, cohabiting in close association with the host colony or occupying nest cavities without direct contact with the host. The strategy of termitophile organisms to become integrated into termite societies include appeasement through chemical, morphological and/or behavioral mimicry. We investigated the hypothesis that there is a morphological mimicry between the obligate termitophile *Corotoca melantho* (Coleoptera: Staphylinidae) and workers of its termite host *Constrictotermes cyphergaster* (Isoptera: Termitidae). Pictures of thirty-one *C. cyphergaster* workers and *C. melantho* individuals were taken in top and side views and converted into thin-plate splines. Four homologous landmarks and five semilandmarks (reference points) were marked on the head and abdomen of both species and digitized. The body shape of both species are morphometrically similar, so there is no discrimination between specimens of termitophile beetles and worker of termite hosts. Body size of termite hosts is responsible for 20% to 30% of body shape variation, while the body size of termitophiles beetle affects near 50% to 60% body shape. However, termitophiles body shape had a greater variation than worker termites. This is the first study to compare morphological similarity among termites and termitophiles using morphometric geometry. Our results indicated the existence of a morphological mimicry between *C. cyphergaster* and *C. melantho*.

Introduction

The controlled temperature and humidity within termite nests create an internal atmosphere that attracts other animals seeking shelter, protection and food (Noirot & Darlington, 2000). The nest may specially represent a food source for some inquiline termites and termitophile invertebrates (Kistner, 1969). Different species can live in association with termite nests, cohabiting with the host species or occupying nest cavities without direct contact with the host colony. Kistner (1969) termed termitophiles the animals that live at least one phase of their life inside termite nests. In the Brazilian savannah (Cunha & Brandão, 2000; Cunha & Morais, 2010; Lopes & Oliveira, 2005; Costa et al., 2009; Cristaldo et al., 2012) and Brazilian Amazonia (Carrijo et al., 2012) different termitophile groups have been reported in termite nests: Acarina, Anura, Araneae, Blattaria, Chilopoda, Coleoptera, Diplopoda, Haplotaxida, Heteroptera, Hymenoptera, Lepidoptera, Orthoptera, Opiliones and Scorpiones. Strategies of termitophiles integration into the social life of termites can include appeasement by chemical (Wilson, 1971), morphological (Kistner, 1969) or behavioral mimicry (Kistner, 1979). Hydrocarbons are used among social insects in species recognition, colony or castes (Howard & Blomquist, 2005) and are detected by antennal contact. The hydrocarbons of some social insect nest inquilines are very similar to that of their hosts (Howard et al., 1982). Some Staphylinidae (Coleoptera) species maintain mutualistic interactions with their hosts through chemical mimicry (Rosa, 2012) and exocrine glands whose exudate can be licked by the workers, which, in turn, regurgitate stomach content for beetles (Pasteels & Kistner, 1971).

Some Staphylinidae beetles have the ability of overlapping their abdomen on the thorax to reduce body length and develop physogastry (Kistner, 1979; 1982). Physogastry is the development of the reproductive and/or glandular system of termite hosts, in which the glands may produce chemical messages of termitophile acceptance (Krikken, 2008). Most termitophile physogastry beetles belong to the family Staphylinidae (Kistner, 1979; 1982). In this family, some species have cuticular hydrocarbons similar to those of their hosts (Rosa, 2012). Sometimes, physogastry is followed by a subsequent secondary sclerotization of some or all of the expanded membrane, as is the case for *Corotoca melantho*, *Spirachtha eurymedusa* and *Termitoiceus* sp. nov. (Seevers, 1957; Kistner, 1982; Jacobson & Pasteels, 1985; Kistner, 1990; Cristaldo et al., 2012). The termitophiles have reduced body size and lateral abdominal projections, and seem to mimic the morphology of worker termites. Kistner (1968, 1969) explained that the mimicry between termitophiles and termites is based on palpation and not on sight. Krikken (2008) described several pronotum trichomes which presumably function in the communication among termitophiles and the termite hosts. In addition, the mentum morphology of both insect groups suggests a trophallaxis relationship.

Corotoca melantho is a species with ovoviviparous development (Liebherr & Kavanaugh, 1985), which may facilitate its survival within the termite nests, once females deposit larvae ready for pupation. The *C. melantho* have a short life cycle, and are thus quickly in touch with worker termites, from which they acquire food by trophallaxis (Costa & Vanin, 2010). Staphylinidae beetles with appendages on the curved and physogastric abdomen are common among species that coexist with ants and termites (Kistner, 1979). The termitophiles may request regurgitated food from the workers (Costa & Vanin, 2010), but may also feed on fecal matter and remains of dead animals (Costa-Lima, 1952).

Considering the interaction between the termite host and termitophile beetle, we investigate the hypothesis of a body shape similarity between the termitophile beetle *Corotoca melantho* Schiødte (Coleoptera: Staphylinidae) and workers of the termite *Constrictotermes cyphergaster* (Silvestri) (Isoptera: Termitidae).

Material and Methods

Examined Material

Twenty-four *C. cyphergaster* nests cohabited by *C. melantho* were collected in an area of Cerrado *sensu stricto*, a savannah biome, at the Parque Estadual da Serra de Caldas (PESCAN). PESCAN is located between Caldas Novas and Rio Quente (17°46'56" S and 48°42'35" W) in the state of Goiás, in the Central region of Brazil. The termite nests were removed from the trees, fragmented into small pieces and a sample of 10% was made to estimate the number of inhabiting individuals. The beetles were identified by comparison with the biological collection of the Zoology Museum of the University of São Paulo (MZUSP). The samples were fixed in 80% ethanol and stored in the laboratory at UnUCET/ UEG.

Morphometric Data

A total of 31 *C. cyphergaster* workers and 31 cohabiting *C. melantho* individuals were collected from 24 nests. Pictures of all specimens were taken in top and side views, with a 3MP

digital camera integrated into a stereomicroscope with 10X fixed eyepieces and 30X magnification. The pictures were saved as JPEG images with a 2048x1536 resolution and transformed into TPS (*thin-plate spline*) files using TPSutil version 1.44 (Rohlf, 2009a). Four homologous landmarks represented as Cartesian coordinates (X, Y) were marked on the head of termites and beetles in top and side views (Fig 1), and five semilandmarks (type 3 Bookstein coordinates) were marked on the abdomen of termites and beetles in top and side views (Fig 1). These semilandmarks are not truly homologous because in *C. melantho* the abdomen overrides the thorax. The landmarks and semilandmarks were digitized with the software TPSDig version 2.12 (Rohlf, 2008b).

Fig 1. Distribution of landmarks and semilandmarks in side and top view for *Corotoca melantho* and *Constrictotermes cyphegaster* workers. Four landmarks were marked on the basis of the antennas and the occipital area of the head in top view (a and b). Four landmarks were marked on the basis of the antennas, postoccipital area, the mandible tip and vertex of the head in side view (c and d). Five semilandmarks were marked representing the first and penultimate tergites and the apical tip of last tergite of the abdomen in top view (a and b). Five semilandmarks were marked representing the first, middle and ultimate tergites and the middle sternite of the abdomen in side view (c and d).

The shape variables were obtained overlapping landmarks and semilandmarks through the general Procrustes alignment using the software TPSRelw version 1.46 (Rohlf, 2008c). This method calculates an average configuration (consensus) that minimizes the sum of squares of the distances between the points of each configuration and the reference configuration (landmarks).

The deformation analysis obtained by TPSRelw generates a new set of variables – *partial warps* – calculated as the orthogonal coordinates of each eigenvector. The relative warp method is based on the Principal Component Analysis of partial warp scores, run with α = 0 to give the same weight to partial warps at different scales, and results in uniform components. The Thin-plate spline graphs were obtained using the software TPSSplin version 1.20 (Rohlf, 2004d). TPS softwares (Thin-Plate Splines) were obtained free of cost at the Stony Brook University Morphometric Website (http://life.bio.sunysb.edu/morph).

Data Analysis

We used centroid size, which is the square root of the summed square distances between each anatomical landmark/semilandmark and the shape's centroid to compare body shapes. The average sizes of the centroid of the species (*C. melantho* and

C. cyphergaster) were compared through a t-test (*P*<0.05). A regression analysis of the shape coordinates (dependent variable) and the centroid size (independent variable) (*P*<0.05) was carried out to evaluated if there is any effect of the size on the shape.

To distinguish shape variables (in top and side views) from termitophile beetles and worker termite hosts a Simple Discriminant Analysis was used, once a single function is needed to distinguish the two clusters. The parity of the means of the two groups was tested using Hotelling's T^2 (*P*<0.05) (Doornik & Hansen, 2008) with 10000 permutations, to meet the multinormality assumption. The percentage of correctly classified individuals was estimated using the Jackknife method to avoid bias in the initially classified group. The Discriminant Analysis carried out uses a linear model to visually confirm or reject the hypothesis that the two species are morphologically distinct (Legendre & Legendre, 2004). All statistical analyses were carried out using R (R Development Core Team 2014).

Results

There were in average 0.002 *C. melantho* individuals for each *C. cyphergaster* worker. A total of two, four or six specimens of termitophile beetles were found for 67% of the *C. melantho* recorded in the sampled nests. However, 18 termitophiles were found in one of the largest termite nests.

The body shape, initially represented by the centroid size of *C. cyphergaster* workers and the termitophiles is similar in side view ($t_{df=60}$= 2.0707; P= 0.0431), but differed in top view ($t_{df=60}$= -1.1847; P= 0.2412). The thin plate splines show deformations in the variations of head and body shape (Fig 2).

The termitophile beetles are slightly smaller than worker termites. Still, the difference in body size did not affect body shape for either species. Body size of *C. cyphergaster* has no effect on body shape in top view (r^2= 0.3378; $F_{(gl=14, 16)}$= 0.5830; P= 0.8418) or in side view (r^2= 0.2102; $F_{(gl=14, 16)}$= 0.3041; P= 0.9847). Body size of *C. melantho* has no effect on body shape in top view (r^2= 0.5469; $F_{(gl=14, 16)}$= 1.3790; P= 0.2664) or in side view (r^2= 0.6299; $F_{(gl=14, 16)}$= 1.9450; P= 0.1014).

The body shape of both species is similar in side view (T^2= 1.1516; df= 14, 47; P= 0.3317) and top view (T^2= 1.2941; df= 14, 47; P= 0.2712). A total of 70.97% of the specimens were correctly classified in side view and 69.35% of specimens were correctly classified in top view, according to the discriminant analysis (Fig 3). However, after applying the Jackknife method, the accuracy in the classification of individuals into each group dropped to 50% and 56.45%, in top and side views, respectively.

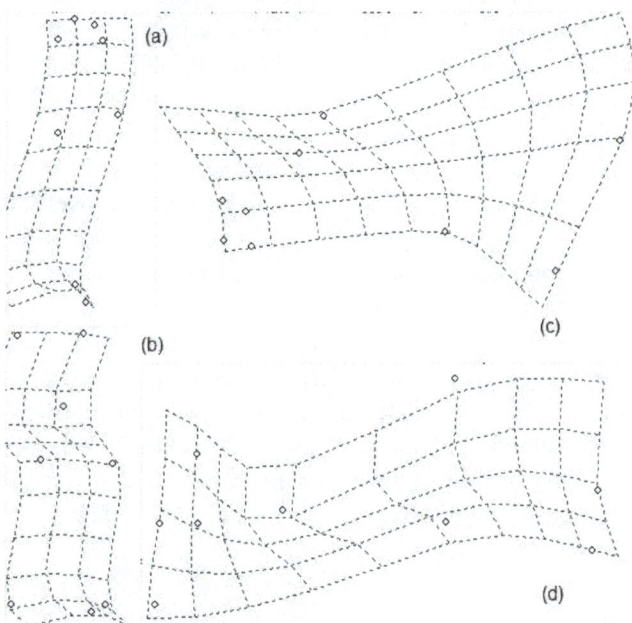

Fig 2. Thin plate spline representation of principal trends (relative warps) in the variation of the body shape. Grid show the deformation of the average shape in opposite directions of the first relative warps for *Corotoca melantho* and *Constrictotermes cyphergaster* in top view (a and b) and in lateral view (c and d).

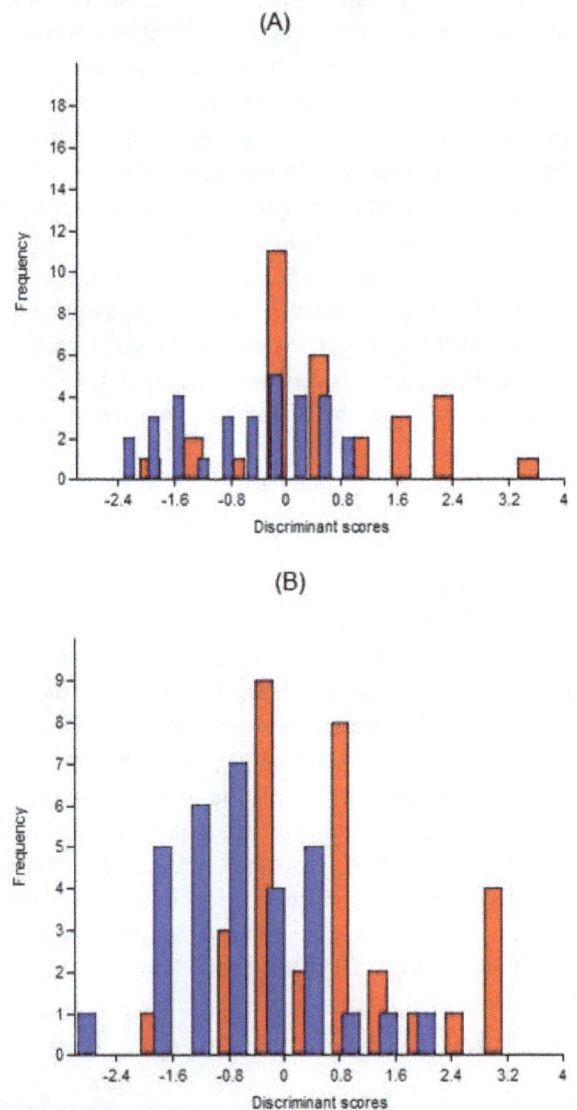

Fig 3. Frequency of discriminant scores for *Corotoca melantho* (red) and *Constrictotermes cyphergaster* (blue) in side view (A) and in top view (B).

Discussion

We found that the *C. melantho* population is higher in larger *C. cyphergaster* termite nests, as already described by Cristaldo et al. 2012. Rosa (2008) reported Staphylinidae beetles as termitophiles in nests of 104 termite species. However, studies regarding termitophile Staphylinidae in Brazil are restricted to the Borgmeier (1923, 1935, 1954, 1959 *apud* Rosa, 2008), Costa-Lima (1952), Cunha & Brandão (2000), Costa et al., (2009) and Cristaldo et al. (2012). In the field, beetles were observed walking inside the nest, near termites, with no antagonistic reaction from its worker termite hosts (personal communication). The diversity of staphylinids observed to be cohabiting termite nests suggests the evolution of a specific interaction between some termite pairs and staphylinid species, which include adaptations such as morphological mimicry (our results) and chemical mimicry (Rosa, 2012).

No difference in *C. cyphergaster* workers and *C. melantho* body shape was observed in the morphometric analysis (see fig 2 and 3). The body shape can be represented by centroid size and shape variables (the partial warps and uniform components). The new shape variables, obtained by overlapping the nine landmarks and semilandmarks by Procrustes alignment, showed that the termitophiles and workers of termite hosts are morphometrically similar. From 20% to 30% of the body shape variation is due to variation on termite host body size. However, body size for termitophiles affects from 50% to 60% of the body shape (see r^2 of the regression between body size and body shape). The Discriminant Analysis confirmed that the specimens of termitophile beetles and termite host worker are indistinguishable (see fig 3). Figure 3 shows higher variation among termitophiles than among the termite hosts. Despite the visual difference on head shape between workers of termite hosts and termitophiles, the average shape is similar in both species (as corroborated by discriminant function).

To our knowledge, morphometric geometric analysis has never been used to compare morphological similarity between termites and termitophiles. Traditional morphometric estimates the variation and covariation of distance measures between pairs of points – length and width of morphological structures (Monteiro & Reis, 1999; Zelditch et al., 2012). According to these authors, geometric morphometry enables analyzing the overall shape of the individual regardless of its size, and locates and describes the regions of shape changes, represented by anatomical points in homologous structures. Although the results indicate that *C. melantho* can mimic the morphology of workers of *C. cyphergaster*, we cannot say that both species co-evolved, because such a statement would require a phylogeny analysis of both species in search of reciprocal adaptations. The interaction between *C. melantho* and *C. cyphergaster* should be investigated by additional analyses, such as behavioral experiments inside the nests and phylogenetic comparisons.

Acknowledgments

Our study has been continuously supported by different grants of the Conselho Nacional de Desenvolvimento Científico (CNPp n° 475484/2011-8), Fundação de Amparo à Pesquisa do Estado de Goiás (FAPEG) and Coordenação de Aperfeiçoamento de Pessoal de Nível Superior (CAPES) (Auxpe 2036/2013). JSL thanks CNPq by productivity grant (306719/2013-4). HFC thanks the University Research and Scientific Production Support Program (PROBIP/UEG).

References

Carrijo, T.F., Gonçalves, R.B. & Santos, R.G. (2012). Review of bees as guests in termite nests, with a new record of the communal bee, *Gaesochira obscura* (Smith, 1879) (Hymenoptera, Apidae), in nests of *Anoplotermes banksi* Emerson, 1925 (Isoptera, Termitidae, Apicotermitinae). Insectes Sociaux, 59: 141-149, doi: 10.1007/s00040-012-0218-x

Costa. C. & Vanin, S.A. (2010). Coleoptera Larval Fauna Associated with Termite Nests (Isoptera) with Emphasis on the Bioluminescent Termite Nests from Central Brazil. Psyche, 2010, Article ID 723947, 12 pages, doi:10.1155/2010/723947

Costa, D.A., Carvalho, R.A., Lima-Filho, G.F. & Brandão D. (2009). Inquilines and Invertebrate Fauna Associated with Termite Nests of *Cornitermes cumulans* (Isoptera, Termitidae) in the Emas National Park, Mineiros, Goiás, Brazil. Sociobiology, 53 (2B): 443-453.

Costa-Lima, A. (1952). Insetos do Brasil, Coleoptera, Rio de Janeiro: Escola Nacional de Agronomia. (pp. 313-323).

Cristaldo, P.F., Rosa, C.S., Florencio, D.F., Marins, A. & DeSouza, O. (2012). Termitarium volume as a determinant of invasion by obligatory termitophiles and inquilines in the nests of *Constrictotermes cyphergaster* (Termitidae, Nasutitermitinae). Insectes Sociaux, 59: 541-548. doi 10.1007/s00040-012-0249-3

Cunha, H.F. & Brandão, D. (2001). Invertebrates associated with the Neotropical termite *Constrictotermes cyphergaster* (Isoptera: Termitidae, Nasutitermitinae). Sociobiology, 37: 593-599.

Cunha, H.F. & Morais, P.P.A.M. (2010). Relação Espécie-Área em Cupinzeiros de Pastagem, Goiânia-GO, Brasil. EntomoBrasilis, 3: 60-63.

Doornik, J. A. & Hansen, H. (2008). An omnibus test for univariate and multivariate normality. Oxford Bulletin of Economics and Statistics, 70: 927-939. doi: 10.1111/j.1468-0084.2008.00537.x

Howard, R.W., McDaniel, D.C.A. & Blomquist, L.J. (1982). Chemical mimicry as an integrating mechanism for three termitophiles associated with *Reticulitermes virginicus* (Banks). Psyche, 89: 157-67.

Howard, R.W. & Blomquist, G. (2005). Ecological, behavioral, and biochemical aspects of insect hydrocarbons. Annual Review of Entomology 50: 371-393. doi: 10.1146/annurev.ento.50.071803.130359

Jacobson, H.R. & Pasteels, J.M. (1985). A new termitophilous species of *Termitoptocinus* Silvestri from New Guinea and a redescription of the genus (Coleoptera, Staphylinidae, Aleocharinae, Corotocini). Indo-Malayan Zoology, 2: 319-323.

Kistner, D.H. (1968). Revision of the African species of the termitophilous tribe Corotocini (Coleoptera: Staphylinidae). I. A new genus and species from Ovamboland and its zoogeographic significance. Journal of the New York Entomological Society, 76: 213-221.

Kistner, D.H. (1969). The biology of termitophiles. In K. Krishna & F.M. Weesner (Eds.), Biology of termites. Vol. I. (pp. 525-557). New York, Academic Press.

Kistner, D.H. (1970). Australian termitophilous associated with *Microcerotermes* (Isoptera: Amitermitinae). Pacific Insects, 12 : 9-15.

Kistner, D.H. (1979). Social and evolutionary significance of social insect symbionts. In H.R. Hermann (Eds.), Social Insects, Vol. 1. (pp. 339-413). Academic Press, New York, San Francisco, and London.

Kistner, D.H. (1982). The social insects' bestiary. In H.R. Herman (Ed.) Social Insects, vol. 3. (pp. 1-244). Academic, New York.

Kistner, D.H. (1990). The integration of foreign insects into termite societies or why do termites tolerate foreign insects in their societies. Sociobiology, 17: 191-215.

Krikken, J. (2008). Two new species from Kenya in the physogastric termitophilous genus *Termitoderus* Mateu 1966 (Coleoptera Scarabaeidae Aphodiinae). Tropical Zoology, 21: 153-162.

Liebherr, J.K. & Kavanaugh, D.H. (1985). Ovoviviparity in carabid beetles of the genus *Pseudomorpha* (Insecta: Coleoptera). Journal of Natural History, 19: 1079-1086. doi: 10.1080/00222938500770681.

Legendre, P. & Legendre. L. (2004). Numerical Ecology. 3rd ed. Amsterdan: Elsevier Science.

Lopes, S.M. & Oliveira, E.H. (2005). Espécie nova de *Ischnoptera* Burmeister, 1838 (Blattaria: Blattellidae: Blattellinae) do estado de Goiás, Brasil, coletada em ninho de cupim. Biota Neotropica, 5: 71-74. doi.org/10.1590/S1676-06032005000100008

Monteiro, L.R. & Reis, S.F. (1999). Princípios de morfometria

geométrica. São Paulo: Holos 188p.

Noirot, C. & Darlington, J.P.E.C. (2000). Termite nests: architecture, regulation and defense, In T. Abe, D.E. Bignell & M. Higashi (Eds.), Termites: Evolution, Sociality, Symbioses, Ecology. (pp. 121-139). Netherlands: Kluwer Academic Publishers.

Pasteels, J.M. & Kistner, D.H. (1971). Revision of the termitophilous subfamily Trichopseniinae (Coleoptera: Staphylinidae). II. The remainder of the genera with a representational study of the gland systems and a discussion of their relationships. Miscellaneous Publications of the Entomological Society of America, 7: 351-399.

R Development Core Team (2014). R: A language and environment for statistical computing. Available at: http://www.R-project.org/. Accessed date: 18.VI.2014.

Rohlf, F.J. (2009a). TPSUtil version 1.44. Stony Brook, Departament of Ecology and Evolution, State University of New York. Available online at: http://life.bio.sunysb.edu/morph Accessed 30.V.2009.

Rohlf, F.J. (2008b). TPSDig version 2.12. Stony Brook, Departament of Ecology and Evolution, State University of New York. Available online at: http://life.bio.sunysb.edu/morph Accessed 30.V.2009.

Rohlf, F.J. (2008c). TPSRelw version 1.46. Stony Brook, Departament of Ecology and Evolution, State University of New York. Available online at: http://life.bio.sunysb.edu/morph Accessed 30.V.2009.

Rohlf, F.J. (2004d). TPSSplin version 1.20. Stony Brook, Departament of Ecology and Evolution, State University of New York. Available online at: http://life.bio.sunysb.edu/morph Accessed 30.V.2009.

Rosa, C.S. (2008). Interactions between termites (Insecta: Isoptera) and termitophiles. MSc Thesis (Animal Biology), Universidade Federal de Viçosa, Brazil.

Rosa, C.S. (2012). Interspecific interactions in termite mounds. PhD Dissertation (Entomology), Universidade Federal de Viçosa, Brazil.

Seevers, C.H. (1957). A monograph on the termitophilous Staphylinidae (Coleoptera). Fieldiana: Zoology, 40: 1-334.

Wilson, E.O. (1971). The Insect Societies. The Belknap Press of Harvard University Press, Cambridge, Mass. 548 p.

Zelditch, M.L., Swiderski, D.L., & Sheets, H.D. (2012). Geometrics Morphometrics for Biologists: a primer. Academic Press of Elsevier, 2nd. Edition.

6

An Updated Guide to the Study of Polyandry in Social Insects

R Jaffé

Instituto de Biociências, Universidade de São Paulo (USP), São Paulo-SP, Brazil

Keywords
effective paternity, multiple mating, multiple paternity, polyandry

Corresponding author
Rodolfo Jaffé
Laboratório de Abelhas
Departamento de Ecologia
Instituto de Biociências
Universidade de São Paulo – USP
Rua do Matão 321, 05508-090
São Paulo-SP, Brazil
Email: r.jaffe@ib.usp.br

Abstract
In spite of the importance of understanding the adaptive significance of polyandry in the social Hymenoptera (ants, bees and wasps), little consensus exists regarding the terminology employed, the use of different paternity estimates, the calculation of such estimates and their associated error measures, and the way paternity should be treated in comparative studies. Here I summarize previous methodological contributions to the study of polyandry in social insects, hoping that such a compendium will serve as an updated guide to future researchers. I first revise the estimates describing queen mating behavior and paternity outcomes in polyandrous social insects, outlining appropriate methods for calculating them. I then address the errors associated to paternity estimates and explain how to account for them. Finally I discuss in which cases paternity should be treated as a continuous or a categorical variable, and provide an insight into the distribution of paternity across the social Hymenoptera. This technical review highlights the importance of standardizing research methods to prevent common errors, raise confidence in the reported data, and facilitate comparisons between studies, to help shed light into many unanswered questions.

1. Why study polyandry in social insects?

Understanding the adaptive significance of multiple mating by social insect queens (polyandry) has been a central goal of many social insect researchers for the past three decades (Page & Metcalf 1982; Crozier & Fjerdingstad 2001; Boomsma et al. 2009; Kraus & Moritz 2010; Palmer & Oldroyd 2000). Single paternity resulting from monandry (single mating) is currently regarded as a crucial precondition for the evolution of eusociality in the Hymenoptera (ants, bees and wasps), since it maximizes genetic relatedness between colony members (Boomsma 2009; Hughes et al. 2008a). Polyandry, on the other hand, dilutes the relatedness between group members because it generates half-sib families within a colony. In consequence, the benefits gained through inclusive fitness can be reduced (Hamilton 1964), and may not outweigh the cost associated with the maintenance of sterile worker behaviors (Page & Metcalf 1982). Polyandry has nevertheless evolved independently in ants, in bees and in wasps (Hughes et al. 2008a).

Among the many hypotheses that have been proposed to explain the evolution of polyandry in the social Hymenoptera, the genetic diversity or genetic variance hypothesis enjoys most current support (Palmer & Oldroyd 2000; Crozier & Fjerdingstad 2001). The increase in genetic diversity within colonies, resulting from the co-occurrence of worker offspring from different fathers, has been suggested as the most plausible explanation for the evolution and maintenance of polyandry in this group of insects. High genetic diversity among colony members has been shown to increase productivity and broaden tolerance to environmental changes (Mattila & Seeley 2007; Oldroyd & Fewell 2007), increase resistance to pathogens (Hughes & Boomsma 2006; Seeley & Tarpy 2007; Schmid-Hempel 1998; Baer & Schmid-Hempel 1999), and enhance an efficient division of labor (Hughes et al. 2003; Jaffé et al. 2007; Smith et al. 2008). Paternity frequency has been found to be negatively correlated with the number of queens per colony (Hughes et al. 2008b; Keller & Reeve 1994), which suggests polyandry and polygyny (multiple queens) are alternative mechanisms to increase genetic diversity in social insect colonies. Colony size, has also been found positively correlated to paternity frequency, indicating that larger colonies might profit more from genetic diversity (Schmid-Hempel 1998; Bourke 1999), or alternatively, that queens heading large colonies need to mate with several males to obtain enough sperm (sperm limitation hypothesis) (Cole 1983; Kraus et al. 2004; but see also Jaffé et al. 2014). Finally, a recent study found a negative association between paternity frequency and paternity biases, showing that queens of highly polyandrous species maximize genetic diversity by equalizing paternity (Jaffé et al. 2012).

In addition to increasing within-colony genetic diversity, polyandry causes the co-occurrence of different ejaculates in the female's reproductive tract. This allows sexual selection to operate after copulation, either through the competition of ejaculates from different males to fertilize an egg (sperm competition) (Simmons 2001) or through the ability of females to influence which sperm fertilize their eggs (cryptic female choice) (Eberhard 1996). Post-copulatory sexual selection is known to be a significant evolutionary force, shaping the evolution of male and female traits across taxa (Andersson & Simmons 2006). Understanding the consequences of polyandry for the evolution of male and female traits is thus crucial to gain a complete understanding of the reproductive biology of social insects (Kvarnemo and Simmons 2013). Moreover, this knowledge is essential to design effective breeding programs for commercial species.

2. An updated guide to the study of polyandry in social insects

In spite of the importance of understanding the adaptive significance of polyandry in social insects, little consensus exists regarding the terminology employed, the use of different paternity estimates, the calculation of such estimates and their associated error measures, and the way paternity should be treated in comparative studies. Methodological consensus and standardization is important, because it could prevent common errors, raise confidence in the reported data, and facilitate comparisons between studies. Thus, my aim here is to gather and summarize previous methodological contributions to the study of polyandry in social insects, hoping that such a compendium will serve as an updated guide to future researchers. I first revise the estimates describing queen mating behavior and paternity outcomes in polyandrous social insects, providing definitions and estimation methods. I then address the errors associated to paternity estimates and explain how to account for them. Finally, I discuss in which cases paternity should be treated as a continuous or a categorical variable, and provide an insight into the distribution of paternity across the social Hymenoptera.

2.1. Mating frequency, paternity frequency, effective paternity or paternity skew?

Paternity in the social Hymenoptera is usually reported as observed and effective paternity. While observed paternity (K_{obs}), or paternity frequency, is the number of males siring offspring of a single queen, effective paternity (m_e) is paternity weighted by the proportion of offspring sired by each male (Nielsen et al. 2003). Mating frequency, often confused with paternity frequency, measures the actual number of males that copulated with a single queen, even if some of them failed to sire any offspring (Table 1). The distinction between these terms is important, because they reflect the outcome of different evolutionary processes.

Social insect research has focused on the study of effective paternity (m_e), because this estimate reflects the average genetic relatedness among the workers of colonies headed by a single queen (Pamilo 1993; Boomsma & Ratnieks 1996). Effective paternity is indeed the most informative estimate for studies addressing the impact of polyandry on colony relatedness and reproductive conflicts (Wenseleers & Ratnieks 2006; Sanetra & Crozier 2001). However, effective paternity is not a good indicator of the actual number of males the queens mated with. Observed paternity provides a more accurate estimate of the queen's real mating frequency, even

Table 1: Estimates describing queen mating behavior and paternity outcomes in polyandrous social insects.

Estimate	Definition	Estimation method
Copulation frequency	Number of times a single queen copulates *	Observation
Mating frequency	Number of males copulating with a single queen †	Observation
Insemination frequency	Number of males inseminating a single queen	Genotyping of sperm from spermatheca
Observed paternity (K_{obs}) or paternity frequency	Number of males fertilizing eggs and siring offspring of a single queen	Genotyping of eggs, pupae or worker offspring
Effective paternity (m_e)	Observed paternity weighted by the proportion of offspring sired by each male.	See $\hat{k}_{e3}$ estimate (Nielsen et al. 2003)
Paternity skew	Degree of paternity bias among the offspring of polyandrous queens.	See B-index (Nonacs 2000)

* Note that a queen may copulate several times with the same male.

† Note that insemination may or may not be involved.

though processes of post-copulatory sexual selection (sperm competition and cryptic female choice) might prevent some sperm from fertilizing eggs and siring offspring (Simmons 2001), and thus the actual mating frequency of a queen might be higher than the observed paternity. Because of this, observed paternity should always be provided along with effective paternity estimates. Mating frequency and insemination frequency should also be provided if available, although they are usually more difficult to assess (Baer 2011).

The degree of paternity biases among the offspring of polyandrous females (paternity skew), is another key quantitative measure needed to address mechanisms of post-copulatory sexual selection and sexual conflict (Den Boer et al. 2010; Jaffé et al. 2012; Baer et al. 2006). Levels of paternity skew may also be the outcome of kin-selection processes, as high paternity skew can bias paternity toward one or a few males, thus increasing genetic relatedness among the offspring of polyandrous queens (Jaffé et al. 2012; Cole 1983). Providing paternity skew along with paternity estimates, is thus essential.

2.2. General methodological considerations and useful software

Paternity is most commonly deduced from molecular markers, by grouping offspring sharing the same father and assigning them to patrilines. A large body of data from studies employing genetic markers have accumulated during the past decade (Boomsma et al. 2009), and this trend is likely to remain or increase with new technological developments that allow massively parallel and multiplexed sample sequencing (Ellegren 2013; Allendorf et al. 2010). Studies estimating paternity from worker genotypes should be careful not to sample offspring from different colonies occurring together in one colony. For instance, there is growing awareness of "worker drifting" between colonies of social Hymenoptera (Lopez-Vaamonde et al. 2004). A sample of honeybee workers collected at a nest entrance, for example, is usually composed of workers from the study colony as well as drifter workers from other colonies (Neumann et al. 2000). Depending on the genotypes of these samples, drifter workers could be misinterpreted as workers from a different patriline, thus inflating paternity estimates. It is therefore important to avoid sampling drifter workers, as this would simplify analyses and yield accurate paternity estimates. An easy way to avoid sampling drifter workers is to collect freshly emerged workers inside the colonies, or to sample pupae or even eggs (Paxton et al. 2003).

In cases where it is not possible to avoid sampling drifter offspring, or where colonies have more than one queen, sibship reconstruction analyses can be performed to assign workers into queens and patrilines within queens. A particularly useful software to perform this kind of analyses is COLONY, a free program implementing maximum likelihood method to assign sibship and parentage jointly, using individual multilocus genotypes at a number of co-dominant or dominant marker loci (Jones & Wang, 2010). COLONY can be found here: http://www.zsl.org/science/software/colony

MateSoft (Moilanen et al. 2004) is a free software developed to estimate paternity statistics in haplodiploid organisms (like the social Hymenoptera), based on co-dominant genetic marker data. The genetic data can be queen genotypes, genotypes of worker offspring from a single queen, or genotypes of sperm stored in the queen's spermatheca. A particularly appealing feature of this program is that it can also deduce parental genotypes and provide a likelihood probability for alternative genotypes. MateSoft can be found here: http://www.bi.ku.dk/staff/jspedersen/matesoft/

Another approach to deduce paternity from worker genotypes is to estimate effective paternity based on the genetic relatedness between workers (rww). Relatedness between worker offspring of a queen that mated with a single male is $rww = 0.75$, while relatedness between worker offspring of a highly polyandrous queen approaches $rww = 0.25$. Hence, effective paternity (m_e) can be obtained from the relatedness between the workers of a single queen, following Pamilo (1993): $m_e = 0.5 / (gww - 0.25)$, where gww is the pedigree relatedness. This approach assumes that the regression worker-worker relatedness (rww) is identical to the pedigree relatedness (gww), and hence should only be applied under no inbreeding. KINGROUP is a free open source program implementing a maximum likelihood approach to pedigree relationships reconstruction and kin group assignment (Konovalov et al. 2004). It allows estimating relatedness between offspring and includes a number of features originally found in the program KINSHIP (Goodnight & Queller 1999), which is no longer updated and only runs in the Classic Macintosh OS platform. KINGROUP can be found here: https://code.google.com/p/kingroup/

Among the different skew indexes, the B-index (Nonacs 2000) has been proposed as the standard estimate to be used in future studies, since it can be easily calculated from paternity data and shows very robust statistical properties (Nonacs 2003). The B-index can be calculated using the skew calculator available here: https://www.eeb.ucla.edu/Faculty/Nonacs/PI.html

2.3. Non-detection and non-sampling errors

Paternity estimates from genetic data are affected by two main types of error: Non-detection and non-sampling errors. The Non-detection Error (NDE) is the probability of two fathering males having identical haplotypes by chance. NDE is determined by the number of markers employed and their level of polymorphism and is an indicator of the resolution of these markers. It should always be reported along with paternity estimates to provide a quantitative measure of accuracy (high $NDEs$ imply a low detection power, and thus paternity estimates might be underestimated). NDE can also be calculated when estimating the number of matrilines in a colony (or the number of reproductive females). Formulae for calculating NDEs are summarized in Table 2.

The Non-sampling Error (*NSE*) estimates the number of males siring offspring remaining undetected because of an insufficient sampling. *NSE* will be affected by the level of paternity skew, and hence needs to be estimated based on the paternity shares of each male and the number of workers analyzed. In species with an observed paternity $K_{obs} = 1$, *NSE* should be calculated to indicate the probability of failing to detect a second male, which fertilized some of the queen`s eggs but remained undetected because none of its offspring were sampled. A low *NSE* would indicate that queens are indeed monandrous, as no further males remained undetected because of sampling effects. In species with an observed paternity $K_{obs} > 1$, *NSE* should be calculated to estimate the number of males remaining undetected due to insufficient sampling. To this end, a given frequency distribution (Binomial, Poisson, etc.) can be fitted to the real distribution of workers among patrilines. The expected frequencies for each category (number of sired workers) can then be computed, and the number of males remaining undetected because of an insufficient sampling estimated as the expected frequency for the zero or less than one category (Cornuet & Aries 1980; Human et al. 2013). In this case, *NSE* can be accounted for by adding the number of undetected males to observed paternity estimates. Nielsen´s effective paternity estimate (Nielsen et al. 2003) is already corrected for sample size, so there is no need for additional corrections for that estimator. Approaches for calculating NSEs are summarized in Table 2.

2.4. Is paternity a continuous or a categorical variable?

Traditionally, paternity has been treated as a categorical variable in the social Hymenoptera, with species being grouped into different paternity or polyandry categories. Boomsma and Ratnieks (1996) first proposed four paternity categories, based on the frequency of multiple paternity among study queens and the mean value of effective paternity

in the study population. Later studies proposed variations of these original categories, and employed either observed paternity (referred to as mating frequency), effective paternity or the frequency of multiple paternity among study queens, as grouping characteristics (see Table 3). To date, however, there is no consensus on how many categories should be used, how to establish the limits between them or which characteristics to use for assigning species into paternity categories.

A recent study retrieved paternity data for 87 polyandrous species of social Hymenoptera (Jaffé et al. 2012). This data set shows that paternity is not normally distributed, with nearly half of all polyandrous species (N = 46) showing mean observed paternities below 2 (Fig. 1). Two polyandry categories could be created based on this distribution: low polyandry (mean observed paternity below 2) and high polyandry (mean observed paternity above 2). By so doing, the high polyandry category would merge about half of all polyandrous species (N = 41), with mean observed paternities ranging from 2 to 55. Clearly, informative variance would be lost by this grouping, as selective forces differ considerably between species with small colonies and queens that mate with a few males (such as bumble bees), and species with huge colonies and highly polyandrous queens (such as honeybees). Nevertheless, such categorization based on the real frequency distribution of paternity across species is more parsimonious than the creation of categories based on arbitrary assumptions and lacking consensus across studies.

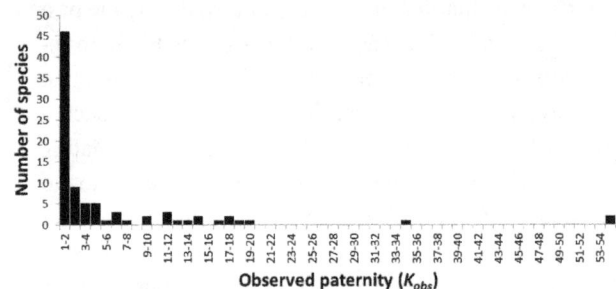

Fig. 1: Frequency distribution of mean observed paternity (K_{obs}) in 87 polyandrous species of social Hymenoptera (data taken from Jaffé et al. 2012).

Table 2: Non-detection and non-sampling errors.

Type of error	Level	Formula	Reference
Patriline non-detection	Population	$(\sum q_i^2)(\sum r_i^2)...(\sum z_i^2)$	Boomsma & Ratnieks 1996
Patriline non-detection	Colony	$(q_i)(r_i)...(z_i)$ [†]	Foster et al. 1999
Matriline detection probability	Locus	N/A [€]	Richards et al. 2005
Patriline non-sampling for $K_{obs} = 1$	Colony	$(1 - p)^n$ [‡]	Foster et al. 1999
Patriline non-sampling for $K_{obs} > 1$	Colony	N/A [£]	Human et al. 2013

* q_i are the allele frequencies at the first locus, r_i the allele frequencies at the second locus, and z_i are the allele frequencies at the last locus. This calculation assumes all loci are unlinked and under Hardy-Weinberg equilibrium.

† See corrections for ambiguity in identification of paternal alleles (Foster et al. 1999).

€ Not applicable. See different cases depending on the inheritance of distinct grandparental alleles (Richards et al. 2005).

‡ p is the proportion of offspring sired by the second male (usually set to three values: p = 0.50, p = 0.25 and p = 0.10) and n is the number of worker offspring analyzed.

£ Not applicable. Frequency distribution fitting. For an example see section 4.3.3.5 in (Human et al. 2013).

Table 3: Paternity categories as reported in the literature.

Category	Original description *	Reference
Single paternity	Double mating absent or very rare; population-wide effective mating frequency < 1.05	Boomsma & Ratnieks 1996
Single-Double paternity	Double mating occurs in ca 20%-50% of queens; effective mating frequency 1.05-1.25	
Single-Multiple paternity	Mating frequency above two occurs regularly; effective mating frequency 1.4-2	
Multiple paternity	Mating frequency usually greater than two; effective mating frequency > 2	
Polyandry	Multiple mating	Oldroyd & Fewell 2007
Extreme polyandry	Mating number > 6	
Monandry	N/A[†]	Hughes et al. 2008b
Facultative low polyandry	Effective mating frequencies of < 2	
Moderate polyandry	Effective mating frequencies of 2–10	
Extreme polyandry	Effective mating frequencies of > 10	
Monandry	N/A[†]	Hughes et al. 2008a
Facultative low polyandry	< 2 effective mates	
High polyandry	> 2 effective mates	
Singly mated	N/A[†]	Boomsma et al. 2009
Facultatively multiply mated	Usually ≥ 50% Singly mated with a variable minority of queens mated to 2-5 males	
Obligately multiply mated	Almost always ≥ 2 and often ≥ 5 matings per queen	

* Note that the word "mating" in the original descriptions actually refer to paternity.

[†] Not applicable

Phylogenetic studies assessing the transition from monandry to polyandry could benefit from categorizing species by their paternity frequency to perform ancestral state reconstruction analyses (Hughes et al. 2008a). Similarly, studies aiming to detect adaptations to polyandry, or the consequences of polyandry for the evolution of other relevant traits, could profit from comparing the traits of interest between highly polyandrous, lowly-polyandrous and monandrous species (Den Boer et al. 2010). However, efforts aiming to detect biologically meaningful associations between paternity frequency and other continuous traits or ecological factors would maximize detection power and avoid losing meaningful variance by employing the actual paternity estimates, as most sexual selection studies do (Simmons 2001). Even though Boomsma (2013) argued that facultative and obligate polyandry appear to be mutually exclusive lineage-specific syndromes, ecological factors could still have shaped the evolution of paternity frequency within as well as across clades (Arnqvist & Nilsson 2000). Hence, treating paternity as a continuous variable could prove more informative to unravel such factors.

3. Future perspectives

The study of polyandry in social insects offers exciting opportunities for future research. Efforts are still needed to understand, for example, how paternity skew has been shaped by the interplay between kin selection and sexual selection (Jaffé et al. 2012). Likewise, the mechanisms and adaptations by which queens and males influence paternity outcomes are still largely unknown (Baer et al. 2001; den Boer et al. 2009; Den Boer et al. 2010), because sexual selection has been considerably understudied in the social insects (Boomsma et al. 2005). Also, very little is known about the conflicts mediating paternity of sexual offspring (Moritz et al. 2005; Hughes and Boomsma 2008), as most studies have analyzed paternity in worker offspring. Finally, understanding the consequences of polyandry for the evolution of male and female traits could substantially improve current breeding programs of commercial species, such as honeybees. For instance, incorporating male selection into current honeybee breeding programs, which thus far focus exclusively on queen or colony traits (Bienefeld et al. 2008), could substantially increase breeding efficiency by improving drone and sperm quality, assuring high queen mating success, and speeding up the whole process of selecting desirable traits.

Standardizing research methods could aid such future research efforts by preventing common errors, raising confidence in the reported data, and facilitating comparisons between studies. A first step towards this standardization could be to employ a similar terminology and to report comparable paternity estimates, along with their associated error measures. The unification of available software into a common open source platform such as R (R Core Team 2013), could also facilitate analyses as well as enhance collaborative work. Finally, it is very important to make paternity data available through open access data bases or data repositories, so that they can be used in comparative studies and re-analyzed when new analytical tools become available.

Acknowledgements

I thank Vera L. Imperatriz-Fonseca for general support, Jacobus J. Boomsma, Sheina Koffler, Robin F. A. Moritz and two anonimous referees for constructive criticism, and the Fundação de Amparo à Pesquisa do Estado de São Paulo for funding (FAPESP 2012/13200-5).

References

Allendorf, F.W., Hohenlohe, P.A. & Luikart, G. (2010) Genomics and the future of conservation genetics. Nature Reviews; Genetics, 11:697-709. doi:10.1038/nrg2844

Andersson, M. & Simmons, L.W. (2006) Sexual selection and mate choice. Trends in Ecology and Evolution, 21 :296-302. doi:10.1016/j.tree.2006.03.015

Arnqvist, G. & Nilsson, T. (2000) The evolution of polyandry: multiple mating and female fitness in insects. Animal Behavior, 60:145-164. doi:10.1006/anbe.2000.1446

Baer, B. (2011) The copulation biology of ants (Hymenoptera: Formicidae). Myrmecological News, 14:55-68

Baer, B., Armitage, S.A.O. & Boomsma, J.J. (2006) Sperm storage induces an immunity cost in ants. Nature, 441 (7095):872-875. doi:10.1038/nature04698

Baer, B., Morgan, E.D. & Schmid-Hempel, P. (2001) A non-specific fatty acid within the bumblebee mating plug prevents females from remating. Proceedings of the National Academy of Sciences, USA, 98 (7):3926-3928. doi:10.1073/pnas.061027998

Baer, B. & Schmid-Hempel, P. (1999) Experimental variation in polyandry affects parasite loads and fitness in a bumblebee. Nature, 397 (6715):151-154. doi:10.1038/16451

Bienefeld, K., Ehrhardt, K. & Reinhardt, F. (2008) Bee breeding around the world - Noticeable success in honey bee selection after the introduction of genetic evaluation using BLUP. American Bee Journal, 148 (8):739-742

Boomsma, J.J. (2009) Lifetime monogamy and the evolution of eusociality. Philosofical Transactions of the Royal Society B, 364(1533): 3191-3207. doi:10.1098/rstb.2009.0101

Boomsma, J.J. (2013) Beyond promiscuity: mate-choice commitments in social breeding. Philosofical Transactions of the Royal Society B, 368(1613): 20120050 doi:10.1098/rstb.2012.0050

Boomsma, J.J., Baer, B. & Heinze, J. (2005) The evolution of male traits in social insects. Annual Review of Entomology, 50:395-420. doi:10.1146/annurev.ento.50.071803.130416

Boomsma, J.J., Kronauer, D.J.C. & Pedersen, J.S. (2009) The evolution of social insect mating systems. In: Gadau J, Fewell J, Wilson EO (eds) Organization of insect societies: from genome to sociocomplexity. Harvard University Press, Cambridge, Mass., pp 3-25

Boomsma, J.J. & Ratnieks, F.L.W. (1996) Paternity in eusocial Hymenoptera. Philosofical Transactions of the Royal Society B, 351 (1342): 947-975. doi:10.1098/rstb.1996.0087

Bourke, A.F.G. (1999) Colony size, social complexity and reproductive conflict in social insects. Journal of Evolutionary Biology, 12(2) :245-257. doi:10.1046/j.1420-9101 .1999.00028.x

Cole, B.J. (1983) Multiple mating and the evolution of social behavior in the Hymenoptera. Behavioral Ecology and Sociobiology, 12(3):191-201. doi:10.1007/bf00290771

Cornuet, J.-M. & Aries, F. (1980) Number of sex alleles in a sample of honeybee colonies. Apidologie, 11(1):87-93. doi:10.1051/apido:19800110

Crozier, R.H. & Fjerdingstad, E.J. (2001) Polyandry in social Hymenoptera - disunity in diversity? Annales Zoologici Fennici, 38 (3-4):267-285

Den Boer, S.P.A., Baer, B. & Boomsma, J.J. (2010) Seminal fluid mediates ejaculate competition in social insects. Science, 327 (5972):1506-1509. doi:10.1126/science.1184709

den Boer, S.P.A., Boomsma, J.J. & Baer, B. (2009) Honey bee males and queens use glandular secretions to enhance sperm viability before and after storage. Journal of Insect Physiology, 55 (6):538-543. doi:10.1016/j.jinsphys.2009.01.012

Eberhard, W.G. (1996) Female control: sexual selection by cryptic female choice. Monographs in behavior and ecology. Princeton University Press, Princeton, NJ.

Ellegren, H. (2013) Genome sequencing and population genomics in non-model organisms. Trends in Ecology and Evolution, in press (0). doi:10.1016/j.tree.2013.09.008

Foster, K.R., Seppä, P., Ratnieks, F.L. & Thorén, P.A. (1999) Low paternity in the hornet Vespa crabro indicates that multiple mating by queens is derived in vespine wasps. Behavioral Ecology and Sociobiology, 46 (4):252-257

Goodnight, K.F. & Queller, D.C. (1999) Computer software for performing likelihood tests of pedigree relationship using genetic markers. Molecular Ecology, 8 (7):1231-1234. doi:10.1046/j.1365-294x.1999.00664.x

Hamilton, W.D. (1964) Genetical Evolution of Social Behaviour I & II. Journal of Theoretical Biology, 7 (1):1-16, 17-52

Hughes, W.O.H. & Boomsma, J. (2008) Genetic royal cheats in leaf-cutting ant societies. PNAS, 105 (13):5150-5153. doi:10.1073/pnas.0710262105

Hughes, W.O.H. & Boomsma, J.J. (2006) Does genetic diversity hinder parasite evolution in social insect colonies? Journal of Evolutionary Biology, 19 (1):132-143. doi:10.1111/j.1420-9101.2005.00979.x

Hughes, W.O.H., Oldroyd, B.P., Beekman, M. & Ratnieks, F.L.W. (2008a) Ancestral monogamy shows kin selection is key to the evolution of eusociality. Science, 320 (5880):1213-1216. doi:10.1126/science.1156108

Hughes, W.O.H., Ratnieks, F.L.W. & Oldroyd, B.P. (2008b) Multiple paternity or multiple queens: two routes to greater intracolonial genetic diversity in the eusocial Hymenoptera. Journal of Evolutionary Biology, 21 (4):1090-1095. doi:10.1111/j.1420-9101.2008.01532.x

Hughes, W.O.H., Sumner, S., Van Borm, S. & Boomsma, J.J. (2003) Worker caste polymorphism has a genetic basis in Acromyrmex leaf-cutting ants. PNAS, 100 (16):9394-9397. doi:10.1073/pnas.1633701100

Human, H., Brodschneider, R., Dietemann, V., Dively, G., Ellis, J.D., Forsgren, E., Fries, I., Hatjina, F., Hu, F.L., Jaffé, R., Bruun Jensen, A.B., Köhler, A., Magyar, J.P., Özkýrým, A., Pirk, C.W.W., Robyn Rose, R., Strauss, U., Tanner, G., Tarpy, D.R., van der Steen, J.J.M., Vaudo, A., Fleming Vejsnæs, F., Wilde, J., Williams, G.R. & Zheng, H.Q. (2013) Miscellaneous standard methods for Apis mellifera research. Journal of Apicultural Research, 52 (4):1-56. doi:10.3896/IBRA.1.52.4.10

Jaffé, R., Garcia-Gonzalez, F., den Boer, S.P.A., Simmons, L.W. & Baer, B. (2012) Patterns of paternity skew among polyandrous social insects: What can they tell us about the potential for sexual selection? Evolution, 66 (12):3778-3788. doi:10.1111/j.1558-5646.2012.01721.x

Jaffé, R., Kronauer, D.J., Kraus, F.B., Boomsma, J.J. & Moritz, R.F.A. (2007) Worker caste determination in the army ant Eciton burchellii. Biology Letters, 3 (5):513-516. doi:10.1098/rsbl.2007.0257

Jaffé, R., Pioker-Hara, F., Santos, C., Santiago, L., Alves, D., M. P. Kleinert, A., Francoy, T., Arias, M. & Imperatriz-Fonseca, V. (2014) Monogamy in large bee societies: a stingless paradox. Naturwissenschaften:1-4. doi:10.1007/s00114-014-1149-3

Jones O.R. & Wang J. (2010) COLONY: a program for parentage and sibship inference from multilocus genotype data. Molecular Ecology Resources, 10: 551-555. doi:10.1111/j.1755-0998.2009.02787.x

Keller, L. & Reeve, H.K. (1994) Genetic Variability, Queen Number, and Polyandry in Social Hymenoptera. Evolution, 48 (3):694-704

Konovalov, D.A., Manning, C. & Henshaw, M.T. (2004) KINGROUP: a program for pedigree relationship reconstruction and kin group assignments using genetic markers. Molecular Ecology Notes, 4 (4):779-782. doi:10.1111/j.1471-8286.2004.00796.x

Kraus, F.B. & Moritz, R.F. (2010) Extreme polyandry in social Hymenoptera: evolutionary causes and consequences for colony organisation. In: Animal Behaviour: Evolution and Mechanisms. Springer, pp 413-439

Kraus, F.B., Neumann, P., van Praagh, J. & Moritz, R.F.A. (2004) Sperm limitation and the evolution of extreme polyandry in honeybees (Apis mellifera L.). Behavioral Ecology and Sociobiology, 55 (5):494-501. doi:10.1007/s00265-003-0706-0

Kvarnemo, C. & Simmons, L.W. (2013) Polyandry as a mediator of sexual selection before and after mating. Philosophical Transaction of the Royal Society B, 368 (1613). doi:10.1098/rstb.2012.0042

Lopez-Vaamonde, C., Koning, J.W., Brown, R.M., Jordan, W.C. & Bourke, A.F. (2004) Social parasitism by male-producing reproductive workers in a eusocial insect. Nature, 430 (6999):557-560. doi:10.1038/nature02769

Mattila, H.R. & Seeley, T.D. (2007) Genetic diversity in honey bee colonies enhances productivity and fitness. Science, 317(5836): 362-364. doi:10.1126/science.1143046

Moilanen, A., Sundstroem, L. & Pedersen, J.S. (2004) MATESOFT: a program for deducing parental genotypes and estimating mating system statistics in haplodiploid species. Molecular Ecology Notes, 4: 795-797. doi:10.1111/j.1471-8286.2004.00779.x

Moritz, R.F.A., Michael, H., Lattorff, G., Neumann, P., Kraus, F.B., Radloff, S.E. & Hepburn, H.R. (2005) Rare royal families in honeybees, Apis mellifera. Naturwissenschaften, 92 (11):548-548. doi:10.1007/s00114-005-0054-1

Neumann, P., Moritz, R.F.A. & Mautz, D. (2000) Colony evaluation is not affected by drifting of drone and worker honeybees (Apis mellifera L.) at a performance testing apiary. Apidologie, 31:67-79. doi:10.1051/apido:2000107

Nielsen, R., Tarpy, D.R. & Reeve, H.K. (2003) Estimating effective paternity number in social insects and the effective number of alleles in a population. Molecular Ecology, 12: 3157-3164. doi:10.1046/j.1365-294X.2003.01994.x

Nonacs, P. (2000) Measuring and Using Skew in the Study of Social Behavior and Evolution. American Naturalist, 156 (6):577-589. doi:10.1086/316995

Nonacs, P. (2003) Measuring the reliability of skew indices: is there one best index? Animal Behavior, 65 (3):615-627. doi:10.1006/anbe.2003.2096

Oldroyd, B.P. & Fewell, J.H. (2007) Genetic diversity promotes homeostasis in insect colonies. Trends in Ecology and Evolution, 22:408-413. doi:10.1016/j.tree.2007.06.001

Page, R.E. & Metcalf, R.A. (1982) Multiple mating, sperm utilization, and social evolution. American Naturalist, 119: 263-281

Palmer, K.A. & Oldroyd, B.P. (2000) Evolution of multiple mating in the genus Apis. Apidologie, 31 (2):235-248.

doi:10.1051/apido:2000119

Pamilo, P. (1993) Polyandry and allele frequency differences between the sexes in the ant *Formica aquilonia*. Heredity, 70 (5):472-480. doi:10.1038/hdy.1993.69

Paxton, R.J., Bego, L.R., Shah, M.M. & Mateus, S. (2003) Low mating frequency of queens in the stingless bee *Scaptotrigona postica* and worker maternity of males. Behavioral Ecology and Sociobioloy, 53:174-181. doi:10.1007/s00265-002-0561-4

R Core Team (2013) R: A Language and Environment for Statistical Computing. 3.0.2 edn. R Foundation for Statistical Computing, Vienna, Austria

Richards, M.H., French, D. & Paxton, R.J. (2005) It's good to be queen: classically eusocial colony structure and low worker fitness in an obligately social sweat bee. Molecular Ecology, 14 (13):4123-4133. doi:10.1111/j.1365-294X.2005.02724.x

Sanetra, M. & Crozier, R.H. (2001) Polyandry and colony genetic structure in the primitive ant *Nothomyrmecia macrops*. Journal of Evolutionary Biology, 14 (3):368-378. doi:10.1046/j.1420-9101.2001.00294.x

Schmid-Hempel, P. (1998) Parasites in social insects. Princeton University Press, Princeton, USA.

Seeley, T.D. & Tarpy, D.R. (2007) Queen promiscuity lowers disease within honeybee colonies. Proceedings of the Royal Society of London B Biol Sci, 274(1606):67-72. doi:10.1098/rspb.2006.3702

Simmons, L.W. (2001) Sperm competition and its evolutionary consequences in the insects. Monographs in behavior and ecology. Princeton University Press, Princeton, USA.

Smith, C.R., Toth, A.L., Suarez, A.V. & Robinson, G.E. (2008) Genetic and genomic analyses of the division of labour in insect societies. Nature Reviews; Genetics, 9 (10):735-748. doi:10.1038/Nrg2429

Wenseleers, T. & Ratnieks, Francis L.W. (2006) Comparative Analysis of Worker Reproduction and Policing in Eusocial Hymenoptera Supports Relatedness Theory. American Naturalist, 168 (6):E163-E179. doi:10.1086/508619

7

Brood hiding test: a new bioassay for behavioral and neuroethological ant research

A Szczuka, B Symonowicz, J Korczyńska, A Wnuk, EJ Godzińska

Nencki Institute of Experimental Biology, Warsaw, Poland.

Keywords

bioassay, brood, light, group size, *Formica polyctena*

Corresponding author

Ewa Joanna Godzińska
Laboratory of Ethology
Department of Neurophysiology
Nencki Institute of Experimental Biology
PAS, Pasteur St. 3, PL 02-093
Warsaw, Poland
E-Mail: e.godzinska@nencki.gov.pl

Abstract

We describe a new bioassay for behavioral and neuroethological ant research, the brood hiding test. A group of adult ants is taken out of the nest, confined together with brood and exposed to strong light. Ants may interact with brood, and, in particular, transport it to the provided shadowed area. The brood hiding test may be accompanied by administration of neuroactive compounds and/or by measurements of their levels in the brain and/or in specific brain structures. During pilot tests with workers of *Formica polyctena* the values of the score quantifying ant behavior were positively correlated with the group size.

Introduction

Adult workers of social Hymenoptera usually engage first in intranidal tasks and then switch to extranidal ones (Wilson, 1971; Hölldobler & Wilson, 2009). This transition is often called the transition nurse-forager (e.g., Heylen et al., 2008). However, older ant workers and/or foragers may retain the ability to engage in intranidal brood care (Lenoir, 1979; Sorensen et al., 1984; Seid & Traniello, 2006; Muscedere et al., 2009). Moreover, several studies revealed that ant foragers are more attracted to brood found outside the nest than nurses and show higher readiness to retrieve it to the nest (Weir, 1958; Walsh & Tschinkel, 1974; Lenoir, 1977; 1981). On the basis of these findings Lenoir (1977; 1981) proposed a bioassay acting as a reliable technique of identification of ant foragers: the brood-retrieving test. At the start of the test a simple artificial ant nest (a test tube equipped with a water reservoir and partly covered by a black paper tube to assure darkness in its humid zone) is inclined so that the brood falls on the dry cotton plug closing the other end of the tube. Workers inhabiting the nest may then retrieve brood back to the dark zone close

to the water reservoir. This test was successfully applied to identify about 80% of foragers in small colonies of *Lasius niger* L. However, it is less suitable for experiments carried out to evaluate the effects of various experimental treatments, as the ants are tested in their home nests, and neither the number of workers, nor the quantity and quality of brood can be easily controlled. Moreover, it is difficult to disentangle ant responses to humidity and illumination. Therefore, we developeda new bioassay for behavioral and neuroethological ant research, the brood hiding test. During the test a group of adult ants is taken out of the nest and confined together with brood in a container exposed to strong light. The ants may interact with brood and, in particular, transport it to the provided shadowed area.

Experimental settings recommended for the application of that test to study the behavior of workers of the wood ant *Formica polyctena* Först (Table 1) were chosen on the basis of the results of 20 pilot tests during which workers taken from a laboratory colony fragment (1-20) were confined together with 5 homocolonial pupae in various types of containers exposed to strong white light (Fig. 1). Single workers did not show brood hiding behavior and their interactions with brood were limited to antennal contacts. More advanced interactions with brood appeared as a function of increasing group size. The values of the score quantifying the outcome of each test (1-9) were highly significantly positively correlated with the number of workers tested together (Spearman rank correlation test: $r = 0.8536$, $P < 0.000001$).

Fig 1. The results of 20 min pilot tests (n = 20) investigating the responses of workers of the red wood ant *Formica polyctena* to homocolonial brood (5 pupae) exposed to strong white light (5000 lx) in various types of containers containing a shadowed area. Ant behavior was quantified as a score (1-9) denoting the most advanced form of interactions with brood observed during the test: 1: antennal contacts; 2: seizing; 3: attempt at transport (a short episode of transport no longer than 2- 3 s); 4. transport; 5. hiding of 1 pupa in the hadowed area; 6. hiding of 2 pupae; 7. hiding of 3 pupae; 8. hiding of 4 pupae; 9. hiding of 5 pupae. Two classes of nestmate workers were tested: nurses (taken out of artificial nest chambers from among workers employed in intranidal brood care), and foragers (taken from the foraging area of the same artificial nest).

These findings provide a new example of the crucial role of group size in the mediation of the expression of specific behavior patterns in social Hymenoptera. The phenomenon of critical (threshold) group size necessary for the expression of a particular behavior trait was first described by Chauvin (1954) and then reported in many other studies [reviewed in Wilson (1971) and Szczuka & Godzińska (2004a)]. Particularly clear-cut effects of group size on the expression of a specific behavior pattern were documented in a series of experiments during which workers of *F. polyctena* were confronted with dead adult houseflies (*Musca domestica* L.) offered to them in the foraging areas of their nests (Szczuka & Godzińska 1997; 2000; 2004a; 2004b). The values of the score quantifying the responses of ants to prey increased as a function of increasing group size, and prey retrieval was observed only in groups counting at least 30-40 workers. Several studies also reported similar relationships between worker group size and the degree of escalation of ant aggressive behavior (Roulston et al., 2003; Tanner, 2006; 2008; Buczkowski & Bennett, 2008).

During our pilot tests brood hiding scores tended to be lower in the case of foragers than in the case of nurses

(Fig. 1). This preliminary result was confirmed by another experiment in which brood hiding test was applied to compare the behavior of various subclasses of workers of *F. polyctena* (Szczuka et al. unpublished results). In that experiment foragers collected from the trails performed less well than workers from other experimental groups. As reported by Lenoir (1977; 1981), during the brood-retrieving tests brood is retrieved mostly by foragers. Responses to brood displayed by ants during the brood hiding test and the brood-retrieving test are thus mediated by at least partly different proximate mechanisms.

The brood hiding test may be used to investigate such questions as behavioral polymorphism within ant colonies, ontogeny of ant behavior, nestmate and species recognition, and behavioral differences between ants from various

Table 1. Experimental settings recommended for the application of the brood hiding test to study worker-brood interactions in the red wood ant *Formica polyctena* Först.

Element of experimental procedure	Description
Experimental arena	an open cylindrical glass container (5 cm high, inner diameter 10 cm) with the walls coated with Fluon®
Shadowed area	a small (25 mm x 30 mm x 5 mm) rectangular shelter made of aluminum foil with one of its longer side walls (facing the center of the arena) left open to allow the ants to enter the shadowed zone
Source of illumination	strong white light (5000 lx) produced by the lamp "Fotovita FV-10" (ULTRA-VIOL sp. j.).
Chemical cues left by nestmates	present (20 homocolonial ants are allowed to walk inside the container during 1 h preceding the first test carried out on a particular day)
Brood	5 homocolonial worker pupae
Time during which the ants are allowed to settle after the introduction into the test container and before the introduction of brood	30 min
Duration of the test	15 or 20 min
Recording of ant behavior	digital video camcorder
Analysis of behavioral recordings	software for the analysis of video recordings of behavior [for instance, "The Observer Video-Pro" (Noldus Information Technology)]

systematic groups related both to phylogenetic distance and differences in ecology. Groups of individuals tested together may be homogenous, but may also consist of individuals belonging to different castes and/or worker subclasses. Not only homocolonial, but also allocolonial and/or allospecific brood may be used, and the tested ants may be subjected to the situation of choice between various categories of brood. Behavior of the tested ants may be quantified by assigning a score to the outcome of each test, or by video recording the tests and analyzing the recordings by means of an appropriate software.

The brood hiding test may also be accompanied by the administration of neuroactive compounds delivered by various techniques including acute and chronic oral administration, injections, and topical (transcuticular) application. The variables quantifying ant behavior may also be analyzed as a function of levels of specific neuroactive compounds in the brain or specific brain structures. Such more complex versions of the brood hiding test may be applied to study neurobiological processes underlying such phenomena as task-related differences in responses to brood, effects of experience on expression of worker behavior, worker cooperation and communication, and inter-individual variability of behavior.

Acknowledgements

This study was supported by the project N N303 3068 33 (2007-2011) of the Ministry of Science and Higher Education (Poland).

References

Buczkowski, G. & Bennett, G. W. (2008) Aggressive interactions between the introduced Argentine ant, *Linepithema humile* and the native odorous house ant, *Tapinoma sessile*. Biological Invasions, 10: 1001-1013. doi: 10.1007/s10530-007-9179-9.

Chauvin, R. 1954. Aspects sociaux des grandes fonctions chez l'abeille. La theorie du superorganisme. Insectes Sociaux, 1: 123- 131.

Heylen, K., Gobin, B., Billen, J., Hu, T. – T., Arckens, L. & Huybrechts, R. S. (2008) A*mfor* expression in the honeybee brain: A trigger mechanism for nurse–forager transition. Journal of Insect Physiology, 54: 1400-1403. doi: 10.1016/j. jinsphys.2008.07015.

Hölldobler, B. & Wilson, E. O. (2009) The superorganism. New York: W. W. Norton.

Lenoir, A. (1977) Sur un nouveau test éthologique permettant d'étudier la division du travail chez la fourmi *Lasius niger* L. Comptes Rendus de l'Académie de Sciences, Paris, , Série D, 284: 2557-2559.

Lenoir, A. (1979) Le comportement alimentaire et la division du travail chez la fourmi *Lasius niger*. Bulletin Biologique de la France et de la Belgique, 113: 79-314.

Lenoir, A. (1981) Brood retrieving in the ant, *Lasius niger* L. Sociobiology, 6: 153-178.

Muscedere, M. L., Willey, T. A. & Traniello, J. F. A. (2009) Age and task efficiency in the ant *Pheidole dentata*: young minor workers are not specialist nurses. Animal Behaviour, 77: 911-918. doi: 10.1016/j.anbehav.2008.12.018.

Roulston, T. H., Buczkowski, G. & Silverman, J. (2003) Nestmate discrimination in ants: effect of bioassay on aggressive behavior. Insectes Sociaux, 50: 151-159.

Seid, M. A. & Traniello, J. F. A. (2006) Age-related repertoire expansion and division of labor in *Pheidole dentata* (Hymenoptera: Formicidae): a new perspective on temporal polyethism and behavioral plasticity in ants. Behavioral Ecology and Sociobiology, 60: 631-644.

Sorensen, A. A., Busch, T. M. & Vinson, S. B. (1984). Behavioral flexibility of temporal subcastes in the fire ant, *Solenopsis invicta,* in response to food. Psyche, 91: 319-331.

Szczuka, A. & Godzińska, E. J. (1997). The effect of past and present group size on responses to prey in the ant *Formica polyctena* Först. Acta Neurobiologiae Experimentalis, 57: 135-150.

Szczuka, A. & Godzińska, E. J. (2000) Group size: an important factor controlling the expression of predatory behaviour in workers of the wood ant *Formica polyctena* Först. Biological Bulletin of Poznan, 37: 139-152.

Szczuka, A. & Godzińska, E. J. (2004a). The role of group size in the control of expression of predatory behavior in workers of the red wood ant *Formica polyctena* (Hymenoptera: Formicidae). Sociobiology, 43: 295-325.

Szczuka, A. & Godzińska, E. J. (2004b). The effect of gradual increase of group size on the expression of predatory behavior in workers of the red wood ant *Formica polyctena* (Hymenoptera: Formicidae). Sociobiology, 43: 327-349.

Tanner, C. J. (2006). Numerical assessment affects aggression and competitive ability: a team-fighting strategy for the ant *Formica xerophila*. Proceedings of the Royal Society of London B., 273: 2737-2742. doi: 10.1098/rspb.2006.3626.

Tanner, C. J. (2008). Aggressive group behavior in the ant *Formica xerophila* is coordinated by direct nestmate contact. Animal Behaviour, 76: 1335-1347.

Walsh, J. P. & Tschinkel, W. R. (1974) Brood recognition by contact pheromone in the imported fire ant, *Solenopsis invicta*. Animal Behaviour, 22: 695-704.

Weir, J. S. (1958) Polyethism in workers of the ant *Myrmica* (Part II). Insectes Sociaux, 5: 315-339.

Wilson, E. O. (1971) The insect societies. Cambridge, MA: Belknap/Harvard University Press.

8

Survey of Subterranean Termite (Isoptera: Rhinotermitidae) Utilization of Temperate Forests

NS Little[1], NA Blount[2], MA Caprio[2], JJ Riggins[2]

1 - USDA-ARS, Southern Insect Management Research Unit, Mississippi, USA.

2 - Mississippi State University, Mississippi, USA.

Key words

Formosan subterranean termites, *Coptotermes formosanus*, timber stumps, hardwood, softwood

Corresponding author

Nathan S. Little
USDA-ARS, Southern Insect Managem. Research Unit
141 Experiment Station Road
Stoneville, Mississippi, USA
38776
E-mail: nathan.little@ars.usda.gov

Abstract

Both native and invasive subterranean termites (Isoptera: Rhinotermitidae), including the Formosan subterranean termite, are well known pests of urban areas, but little is known about their distribution or impact in forest ecosystems of the southeastern United States. Recently harvested timber stumps were mechanically inspected for the presence of subterranean termites in multiple locations across southern Mississippi and eastern Louisiana. A systematic line plot cruise with 100 x 200m spacing and 1/20th ha plots was implemented, and all stumps with a diameter greater than 7.6cm were inspected. In total, 7,413 stumps were inspected for the presence of subterranean termites, and 406 of those contained native subterranean termites (*Reticulitermes* spp.). Light traps were also placed at 8 sites to detect the presence of subterranean termite alates. While no invasive Formosan subterranean termites were found during mechanical inspection of tree stumps, alates were captured in light traps at three sites. The proportion of stumps infested with subterranean termites was negatively correlated with the number of stumps in each plot. Although 6.27% of pine stumps and 1.86% of hardwood stumps contained subterranean termites, no correlation was found between subterranean termite presence and type of stump (pine or hardwood) inspected. Subterranean termite presence in stumps ranged from 0.94% to 14.97% depending on site.

Introduction

Termites are ecologically and economically important insects, which can be found in nearly all forest ecosystems. Subterranean termites (Isoptera: Rhinotermitidae) play important roles in cellulose decomposition, nutrient cycling, and soil mineralization across a multitude of environments (Harris, 1966; Wood & Sands, 1978; Black & Okwakol, 1997). Although they are known to cause significant economic damage to wooden structures in urban areas throughout the U.S., native subterranean termites (*Reticulitermes* spp.) provide a valuable service to forest ecosystems as the predominant invertebrate decomposers of woody materials (La Fage & Nutting, 1978). The role of subterranean termites in forest ecosystem nutrient cycles is extremely important, and their presence can contribute up to 22% of total nitrogen input in tropical forest ecosystems (Yamada et al., 2006).

The non-native Formosan subterranean termite (*Coptotermes formosanus* Shiraki), is reported to infest 17 species

of living trees (Chambers et al., 1988) and at least forty other species of plants (Lai et al., 1983). Formosan subterranean termites are known to cause significant economic damage to standing timber in many regions of the world (Harris, 1966). Greaves et al. (1967) reported that *Coptotermes* spp. were responsible for up to 92% of tree losses in virgin eucalyptus forests and approximately 64% of losses in younger forests, which were predisposed to fire stress. While studies have investigated subterranean termite ecology in tropical forest ecosystems, little is known about their ecology in temperate forests.

Loblolly pine (*Pinus taeda* L.) is a predominant species in the southern U.S., comprising 45% of commercial forest-land and contributing $30 billion annually to the economy (Schultz, 1999). Considering the large industry for loblolly pine in the southeastern U.S. and the lack of research on Formosan subterranean termites in forested areas, the impact of this invasive insect could be substantial. For instance, native subterranean termites, which are ubiquitously reported in the literature to rarely infest living trees, have been observed readily utilizing blue-stained portions of bark beetle-attacked trees before any foliage chlorosis could be detected (Little, 2013). While Formosan termites commonly infest living trees in urban settings, loblolly pines are more frequently infested than other tree species (Osbrink et al., 1999; Guillot et al., 2010). Additionally, Morales-Ramos & Rojas (2001) observed that Formosan subterranean termites consumed loblolly pine wood at a higher rate relative to other wood species in laboratory no-choice tests.

Biological differences between Formosan and native subterranean termites create the possibility that Formosan subterranean termites could impact forest ecosystems and individual trees differently than our native subterranean termites. For example, Formosan subterranean termite colonies are considerably larger and they are more aggressive feeders than native species (Su & Scheffrahn, 1988). Formosan subterranean termites also exhibited higher survival and wood consumption rates than native *Reticulitermes* spp. in laboratory assays (Smythe & Carther, 1970). They can also nest above-ground with no connection to the soil, which allows them to utilize resources previously inaccessible to native species. La Fage (1987) noted that in certain instances in New Orleans, LA, Formosan termites completely replaced native termites as the predominant termite species. More recently, Su (2003) confirmed that Formosan subterranean termites were outcompeting native *Reticulitermes* spp. in multiple urban locations.

Despite these concerns, little research has been done on the distribution or impact of Formosan subterranean termites in non-urban forested environments of the U.S. The first indication that Formosan subterranean termites were established in forested areas of the southeastern U.S. was when Sun et al., (2007) reported that Formosan subterranean termite alate catches were higher in forested settings than in urban areas. In addition, alate catches increased throughout the four year trapping study, which may have indicated that Formosan subterranean termite populations were well established and colony expansion was occurring in Mississippi (Sun et al., 2007).

Native and Formosan subterranean termite ecology in local forested settings of the southern U.S. has yet to be quantified. We hypothesized that elevated light trap captures of Formosan subterranean termite alates in rural forested areas by Sun et al. (2007) would be explained by infestations of living trees in surrounding forest stands. Therefore, the objectives of this study were to determine the extent of subterranean termite utilization of living trees within localized forested settings, quantify subterranean termite presence in forested areas, determine if the proportion of stumps infested with subterranean termites was correlated with stump density or size, and quantify the effect of tree type (hardwood or softwood) on subterranean termite utilization of woody resources.

Material and Methods

Stump Inspections

Eleven sites encompassing 476.8ha throughout southern Mississippi and Louisiana were inspected during 2011 and 2012 (Fig 1). There were nine sites surveyed in four Mississippi counties (Pearl River, Harrison, Jackson, and Lamar) and two additional sites were surveyed in St. Tammany Parish, Louisiana (Table 1). According to Sun et al. (2007), the four Mississippi counties inspected during this study were among the top five counties in the state for Formosan subterranean termite alate captures. Additionally, Formosan subterranean termites have been found in St. Tammany Parish, Louisiana since the late 1980s (La Fage, 1987; Messenger et al., 2002).

Clear cut pine plantations were inspected for subterranean termite presence in residual stumps within three weeks of harvest because they offered the best opportunity to discover the presence of subterranean termites in living trees without damaging standing timber. A variety of methods have been used over the years to detect subterranean termites in living trees (e.g. acoustic devices, dogs, CO_2 detectors, coring, etc.); however, most have limitations, and are often less effective than physical inspections with hatchets and shovels (Osbrink et al., 1999). Site selection criteria (recent timber harvest and proximity to established Formosan subterranean termite populations) led to sporadic site availability; therefore sites were inspected as they became available. To limit the possibility of post-harvest colonization of stumps, all inspections took place within three weeks of timber harvest; however, this only oc-

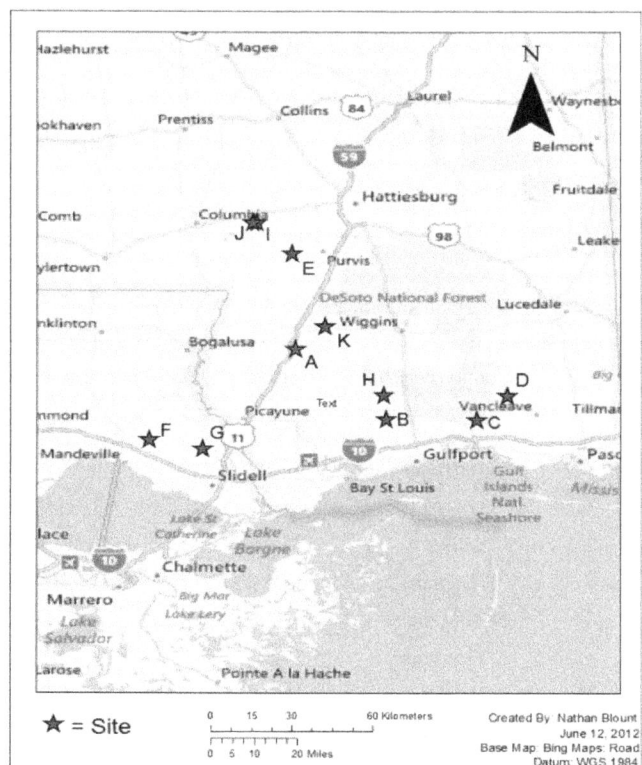

Fig 1. Map of inspecion sites.

curred for one site. The majority of sites were inspected while logging was still occurring, with the balance being inspected within a week of harvest. A 2.5% systematic line plot cruise (Avery & Burkhart, 2002) was implemented on post-harvest sites that were 20ha and larger to insure adequate sampling across each site and to limit sampling bias (with the exception of the first site inspected, A). Sites smaller than 20ha received a 5% cruise. Plots were 1/20th ha and circular, with 100m between plots and 200m between transects (Fig 2). A global positioning system (GPS) unit (GPSMAP 60CSx®, Garmin Ltd., Olathe, KS) was used to record plot centers.

All stumps greater than 7.6cm in diameter and located within plots were inspected with shovels and hatchets. Stumps less than 7.6cm in diameter were not sampled based

on findings from prior studies, which indicated that native subterranean termites primarily occur in coarse woody debris with diameters above 7.6cm (Wang & Powell, 2001; Wang et al., 2003). One-quarter of each stump was inspected for subterranean termite presence from the cut surface of the stump down to 15cm below the soil line. This method has been previously used to successfully detect infestations of Formosan subterranean termites in living trees throughout urban areas, even when visual symptoms were not present (Osbrink et al., 1999). If termites were present in stumps, they were immediately identified to genera, *Reticulitermes* (native) or *Coptotermes* (Formosan), using the distinct morphological characteristic of head shape in the soldier caste (Gleason & Koehler, 1980). Stump diameter was recorded for each occurrence of termites in stumps, which was subsequently labeled as hardwood or pine (softwood). Each stump was then marked and labeled on a GPS unit.

Alate Survey

Light traps were used during spring of 2012 to confirm the presence of Formosan subterranean termite alates near sites selected for physical inspections. Methods similar to Sun et al. (2007) were used to construct alate traps to ensure proper design. Traps were constructed using a 1.52m (5-foot) t-post, white polyvinyl chloride (PVC) pipe, 20ga 2.54cm (1-inch) poultry netting (Garden Plus), a solar powered light emitting diode (LED) light (model SPS2-P1-BK-T24, LG Sourcing Inc., N. Wilkesboro, NC), and a glue board (Trapper® LTD, Bell Laboratories, Inc., Madison, WI) (Fig 3).

To avoid non-target catches of vertebrates attempting to prey on captured insects, poultry netting was wrapped around the glue board and t-post to achieve cage dimensions of 16.5 (9.5-inches) x 35.6cm (14-inches). A wire top was fashioned to give easy access for glue board replacement. The t-post was driven into the ground to insure firm placement of the trap. The LED light came on approximately 20 min after sunset, and ran for six hours throughout the trapping season,

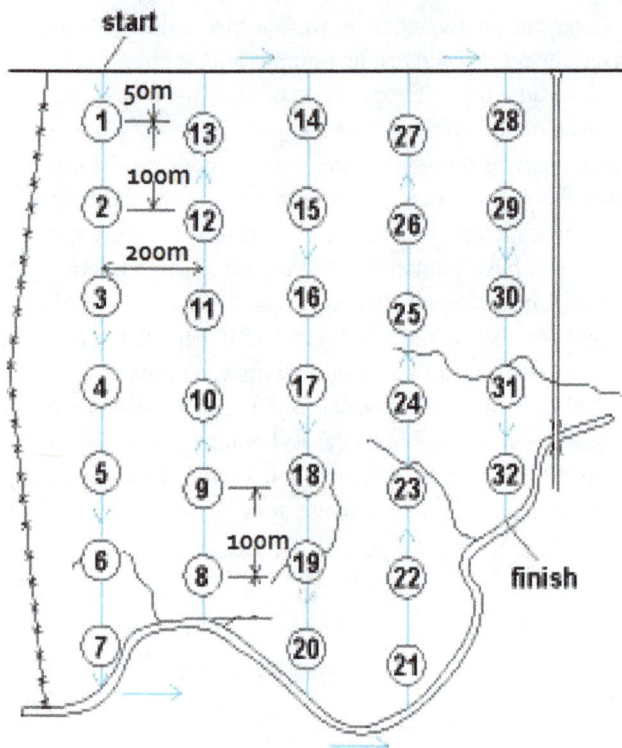

Fig 2. Example of systematic line plot cruise.

Table 1. Summary of Sites Inspected for This Study

Site	County/Parish	Hectares	% Cruise	GPS Coordinates	Soil Conditions	Site Designation
A	Pearl River	83.0	5	N30° 46' 58" W89° 29' 56"	Dry	Upland
B	Harrison	89.4	2.5	N30° 31' 25" W89° 11' 47"	Dry	Upland
C	Harrison	117.4	2.5	N30° 31' 43" W88° 53' 22"	Wet	Lowland
D	Jackson	25.9	2.5	N30° 36' 43" W88° 47' 17"	Dry	Upland
E	Lamar	13.4	5	N31° 8' 23" W89° 30' 52"	Moderate	Upland
F	St. Tammany	39.3	2.5	N30° 27' 10" W89° 59' 37"	Moderate	Lowland
G	St. Tammany	19.8	5	N30° 25' 1" W89° 48' 45"	Moderate	Lowland
H	Harrison	8.5	5	N30° 37' 3" W89° 12' 30"	Moderate	Lowland
I	Lamar	19.4	5	N31° 15' 13" W89° 37' 49"	Moderate	Upland
J	Lamar	39.3	2.5	N31° 15' 11" W89° 38' 20"	Moderate	Upland
K	Pearl River	21.4	5	N30° 52' 5" W89° 24' 16"	Moderate	Upland

Fig 3. Formosan subterranean termite alate light trap.

since peak Formosan subterranean termite alate flight activity occurs at dusk (Bess, 1970).

Formosan subterranean termite alates have been documented to swarm as early as mid-April in Mississippi, with peak activity often occurring near the latter part of May (Sun et al., 2007; Lax & Wiltz, 2010). Traps were placed on sites at the end of the first week in April, early enough to detect initial swarms. Glue boards were replaced once a week for a seven week period, with a final collection date of May 26. This sampling window allowed ample time to catch alates during their swarming periods determined by previous studies (Sun et al., 2007; Lax & Wiltz, 2010). All glue boards were dated and labeled by site when removed, wrapped in plastic wrap, and placed in a freezer for preservation. Alates were identified in the laboratory using morphological characteristics provided by Gleason & Koehler (1980).

A total of twelve alate light traps were placed on eight sites. The Lamar County sites (E, I, & J) were excluded from the alate trapping portion of this study due to travel constraints. Additionally, Lamar County had a low number of alate captures during a previous study (Sun et al., 2007) relative to all other counties sampled in Mississippi. The number of traps placed per site was determined by area, with five sites receiving one trap each (<40.5ha), two sites receiving two traps each (83 and 89.4ha respectively), and one site receiving three traps (117.4ha). Traps were placed at the approximated center of each site. On sites with multiple traps, the location was partitioned according to area, with traps located in the approximate center of each partition.

Statistical Analyses

The proportion of stumps infested with termites was compared to stump density and site using the BEINF family of the GAMLSS package (Rigby & Stasinopoulos, 2001; Stasinopoulos & Rigby, 2007; Stasinopoulos et al. 2009), in the R statistical package (R Core Team, 2012). While an arcsine transformation is commonly used for proportional data, this transformation has a number of limitations and has been superseded by a variety of alternative techniques (Warton & Hui, 2011). A zero-inflated distribution was required because the data demonstrated a bimodal distribution with an excess of zeros (Fig 4). The BEINF family of GAMLSS models the data using a beta distribution and allows for inflation of both zeros and ones. Inflation at both zero and one was chosen over zero-inflation alone because it was at least theoretically possible that all stumps in a plot could be infested with termites. This model assumes that there at least two processes that combined to produce the original data. The beta distribution models the range of proportion infestations between zero and one, while a binary process models an excess of plots with no infestations due to another process. The high number of plots with zero infestations and the potential for numerous infested trees within infested plots is potentially explained by the social/colonial nature of subterranean termites. Subterranean termite workers only feed within a finite distance of the colony. This biological process likely explains the zero-inflated distribution of our data.

Pearson's correlation coefficients (PROC CORR) were used to assess the relationship between infested pine and infested hardwood stumps in SAS 9.2 (SAS Institute, Cary, NC). Significance for all analyses was determined at $\alpha \leq 0.05$.

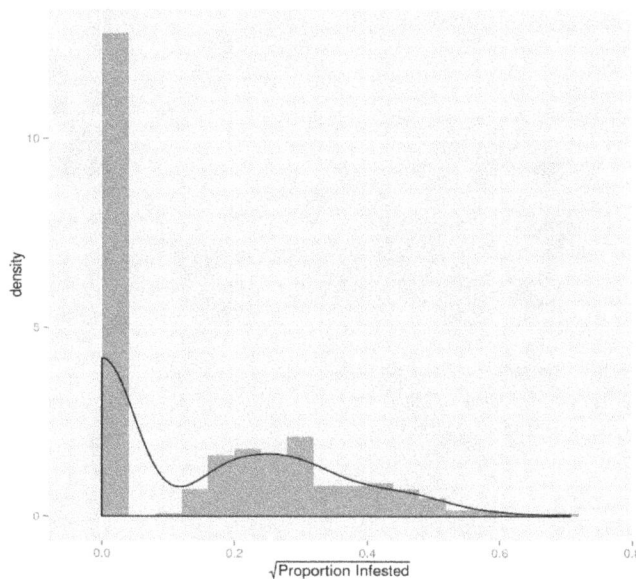

Figure 4. The distribution of the square root transformed proportion of stumps in a plot infested with termites compared to a kernel density distribution for the same variable. An excess of zeros is indicated by the large bar corresponding to zero infested stumps.

Results and Discussion

Stump Inspections

A total of 7,413 stumps were inspected, consisting of 6,072 softwoods (hereafter pines) and 1,341 hardwoods. No Formosan subterranean termites were found; however, 406 stumps containing *Reticulitermes* spp. were recorded, resulting in 5.48% of all stumps containing subterranean termites. These included occurrences in 381 pine (6.27% of total pines) and 25 hardwood stumps (1.86% of total hardwoods) (Table 2). Site F had the highest overall occurrences of *Reticulitermes* spp., with 14.97% of all stumps containing subterranean termites. Sites C and G followed, with 10.81% and 7.42% of stumps containing subterranean termites, respectively. Site I had the fewest subterranean termite occurrences among all sites with 0.94% of stumps containing termites, followed closely by sites D (1.33%) and K (1.58%).

Several sites had low occurrences of hardwood stumps. No hardwood stumps were encountered during inspections on site K, while site B had only 3 hardwood stumps in the plots. Sites D and F also had low numbers of hardwood stumps, 16 and 33 respectively. Five sites contained hardwood stumps that had no subterranean termites (B, D, H, I, and J).

For further analysis and uniformity, stumps that contained *Reticulitermes* species were converted to a per hectare basis using a conversion factor relative to the percent cruise conducted on each site. Overall, the mean number of subterranean termites per hectare across all stumps and sites was 27.41 (se=0.28, n=7,413). Mean number of pine stumps containing subterranean termites per hectare was 26.36 (se=0.30, n=6,072), with hardwood stumps at 1.05 (se=0.06, n=1,341). Subterranean termite occurrences per hectare for all stumps on a per site basis ranged from 5.06 (site A) to 79.39 (site F). Sites A, B, and I had fewer than 7.16 occurrences per hectare. Site C had the second highest occurrence of subterranean termites per hectare, 65.08, followed by site G, 38.72. Subterranean termite occurrence in hardwood stumps per hectare for site C was 7.16, which was the only site with more than 1.68 infested hardwood stumps per hectare.

Table 2. Summary of Stumps Containing Subterranean Termite Infestations by Site.

Site	Stumps Inspected			Stumps Infested			% Infestation		
	¹P	²H	Total	¹P	²H	Total	¹P	²H	Total
A	530	360	890	20	1	21	3.77	0.28	2.36
B	411	3	414	16	0	16	3.89	0.00	3.86
C	1499	268	1767	170	21	191	11.34	7.84	10.81
D	585	16	601	8	0	8	1.37	0.00	1.33
E	181	166	347	15	1	16	8.29	0.60	4.61
F	488	33	521	77	1	78	15.78	3.03	14.97
G	258	52	310	22	1	23	8.53	1.92	7.42
H	223	204	427	11	0	11	4.93	0.00	2.58
I	304	121	425	4	0	4	1.32	0.00	0.94
J	387	118	505	19	0	19	4.91	0.00	3.76
K	1206	0	1206	19	0	19	1.58	0.00	1.58
Σ	**6072**	**1341**	**7413**	**381**	**25**	**406**	**6.27**	**1.86**	**5.48**

¹P = Pine; ²H=Hardwood

Fig 5. The distribution of the proportion of stumps in a plot infested with subterranean termites by site.

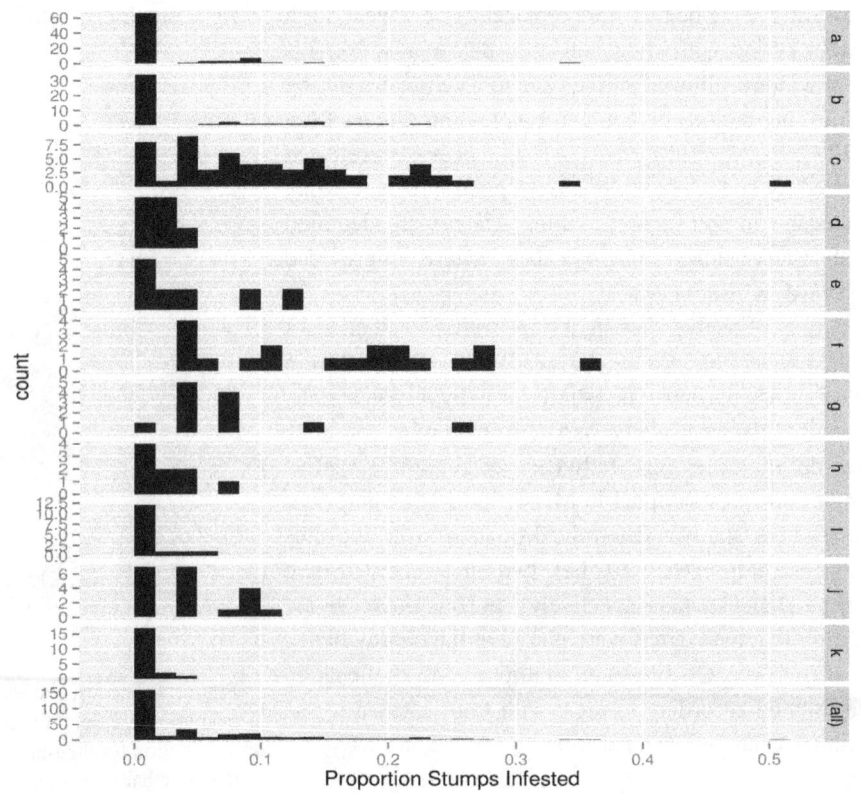

The distribution of the proportion of stumps infested varied with site (Fig. 4). In the inflated beta distribution analysis, both stump density ($\chi^2 = 8.49$, df = 1, $P = 0.0035$) and sites ($\chi^2 = 38.42$, df =10, $P = 3.21e-5$) were significant. These data suggest that across the range of overall densities experienced at various sites, native termite species infested a greater proportion of stumps in plots with lower stump densities and presumably larger stumps (Fig 6). In this agroforestry setting, lower stump densities indicate an older forest, with larger diameter trees. The BEINF family of GAMLSS fits four parameters. Two are the shape parameters for the beta distribution ($\mu = 0.1201$, $\sigma = 0.1942$), while the other two are related to the probability of the additional processes leading to zero or one values ($v = 1.047$, $\tau = 3.77e-09$). These resulted in predicted probability densities of 0.511 at zero and 1.842e-9 at one.

The frequency of subterranean termite infestations in hardwood stumps was significantly correlated with the frequency of infestations in pine stumps across all sites (Pearson $r = 0.68$, $P = 0.03$). In other words, sites with higher termite abundance had more infestations of both pine and hardwood stumps. However, the frequency of infestation was always higher in pine stumps than in hardwood stumps. Our results agree with previous studies (Osbrink et al., 1999; Guillot et al., 2010) that reported subterranean termite prefer loblolly pine over hardwoods.

Although no Formosan subterranean termites were found in stumps of recently logged forested stands in Mississippi and Louisiana during this study, native *Reticulitermes* spp. were found across all sites inspected. Subterranean termite were more prevalent in pine stumps than hardwoods. Additionally, our results confirmed findings from previous studies, which documented that the number of subterranean termite occurrences decreased with increasing wood hardness, with pines generally having higher occurrences of termites than hardwoods (Lin, 1987; Peralta et al., 2004).

Alate Survey

During the seven week trapping period a total of 14 Formosan subterranean termite alates were caught among three sites (Fig 5). Two alates were detected on the May 11 trap check of site A in Pearl River County. On May 12, one alate was present on the site D trap in Jackson County. Alates were not present again until the May 25 check, when 10 alates were caught on site A and one alate was present on site G within St. Tammany Parish, LA. Both traps on site A caught alates during this study. Since *Reticulitermes* spp. primarily swarm during daylight hours, no native termite alates were captured during this study. Due to limited alate catches, no statistical analyses were performed.

Alates of Formosan subterranean termites were captured at three of the sites sampled. Site A had multiple alate captures among two traps, while only one alate was caught on

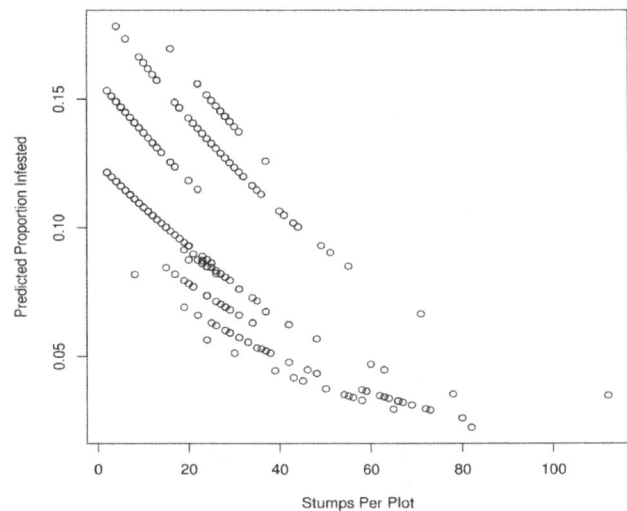

Fig 6. The predicted proportion of stumps infested in a plot based on site and stump density (the total number of stumps in the plot).

each of the other two sites. Formosan subterranean termite alates have the capability to disperse distances of nearly 900m in open areas (Messenger & Mullins, 2005); however, other studies have shown flight distances of 100m to be more common (Higa & Tamashiro, 1983). Although it is unknown if Formosan subterranean termite colonies were located on the inspection sites where alates were caught, it is likely that colonies were present within close proximity if common dispersal distances are taken into account. Additionally, since two sites only had an individual alate captured, it is possible that these alates dispersed a greater distance than the site with multiple alate captures. Multiple alates were captured on two different traps at site A, therefore it was likely that these traps were closer to parent colonies than the traps at other sites. It is also possible that Site A contained a larger colony or greater number of colonies than the other sites. Site D, which was located in a very rural location and completely surrounded by forest, only captured a single alate. The site was over 1.6km (1-mile) away from the nearest house or residential area. This increases the likelihood that the alate dispersed from an established colony in a forested area. Formosan subterranean termite populations may not currently be high enough in forested settings for easy detection, but our capture of them at secluded sites indicates that colonies may already be established in localized forested areas.

There are multiple plausible reasons as to why no Formosan subterranean termites could be located during physical stump inspections in our study. Formosan subterranean termites are known to commonly utilize living trees in urban areas (Osbrink et al., 1999). Urban environments, in general, contain far less downed woody debris than most forested areas. Less woody debris would concentrate subterranean termites into fewer resources, making them easier to detect. Soil moisture levels were also reduced during the hot summer months, which may have caused subterranean termites to travel deeper into the soil to prevent desiccation, thus avoiding detection.

Additionally, the trees on the sites inspected were mature prior to harvest, possibly limiting the import of any significant amount of infested woody debris for decades.

Given the propensity of Formosan termites for attacking living trees in urban trees in the U.S. and plantation forestry settings in other parts of the world, additional research is needed to ensure that Formosan subterranean termite populations do not pose a threat to native forests. More forested sites need to be inspected for the presence of Formosan subterranean termites, utilizing stricter site criteria such as distance to railroads and major roadways, documentation of previous alate trap catches, and proximity to urban areas. Imposing these additional requirements on inspection sites should increase the probability of locating Formosan subterranean termites in forested areas. However, due to the difficulty of locating potential sites, it may take several years to adequately complete the investigation. Continuation of alate monitoring within forested and rural areas is also very important as data collected can be useful in monitoring alate populations and range expansion. Further confirmation of Formosan subterranean termite establishment in localized forest settings would also raise the importance of determining feeding preferences and the likelihood of native subterranean termite displacement in forests, issues this study was not able to address.

We found no Formosan subterranean termites during physical stump inspections; however, only pine plantations were inspected during this study. Hardwood bottoms and older forests with less active management and higher proportions of downed woody material (DWM) may harbor Formosan subterranean termites. Furthermore, living trees might only be infested once populations in DWM are high enough and competition for woody resources is strong, a possibility that should be explored in future research. It is still debatable why more alates were caught in rural areas than in forested settings by Sun et al. (2007). Formosan subterranean termites may utilize DWM in rural forested areas. Another possibility is that light pollution may decrease alate trap efficacy in urban areas. As a consequence, traps placed in urban settings may underestimate alate numbers.

At the conclusion of this study, native *Reticulitermes* spp. were found in recently harvested timber stumps at the eleven sites inspected, but the introduced Formosan subterranean termite was not found. However, Formosan subterranean termites were located within the general area of 3 sites, as evidenced by the alate trap catches and proximity to homes. Due to lack of findings on inspected sites, feeding preference data for Formosan subterranean termites could not be collected, and native *Reticulitermes* spp. displacement is not a current problem within the study areas. Stump inspection data did reveal that *Reticulitermes* spp. occurred in a higher percentage of pine stumps than in hardwoods, but stump type did not significantly affect the number of occurrences per site. The proportion of stumps infested by native subterranean termites was negatively correlated with the number of stumps on a given plot. It is im-

portant to note that Formosan subterranean termites could be impacting forested stands in other locations, even though they were not located on the tracts inspected during this study. With rising populations, the invasive Formosan subterranean termite could still pose a major ecological threat to forested settings and native termite species, issues that need to be addressed with continued research.

Acknowledgments

The authors would like to thank Jared Seals, Kevin Chase, Brady Self, and Zach Senneff for their contributions to this research project. Their help was instrumental in achieving the objectives of this study. This research was funded by the USDA, Forest Service, Forest Health Protection and Southern Research Station, and the Mississippi Agricultural and Forestry Experiment Station. The authors would like to thank the Mississippi Forestry Commission and Weyerhaeuser for providing vital assistance in locating and accessing research sites for this study.

References

Avery, T. & Burkhart, H. (2002). Forest Measurements, 5thed. New York: McGraw-Hill.

Bess, H.A. (1970). Termites of Hawaii and the Oceanic Islands. In K. Krishna & F.M. Weesner (Eds.), Biology of Termites, vol 2. (pp. 449-475). New York: Academic Press, Inc.

Black, H.J. & Okawol, M.N. (1997). Agricultural intensification, soil biodiversity and agroecosystem function in the tropics: the role of termites. Applied Soil Ecology, 6: 37-53.

Chambers, D.M., Zungoli, P.A. & Hill H.S., Jr. (1988). Distribution and habitats of the Formosan subterranean termite (Isoptera: Rhinotermitidae) in South Carolina. Journal of Economic Entomology, 81: 1611-1619.

Crawley, M.J. (2007). The R book. Chichester: John Wiley & Sons, Ltd.

Gleason, R.W. & Koehler, P.G. (1980). Termites of the eastern and southeastern United States: pictorial keys to soldiers and winged reproductives. Bulletin 192. Institute of Food and Agricultural Sciences (IFAS), University of Florida. Gainesville, Florida.

Greaves, T., Armstrong, G.J., McInnes, R.S. & Dowse, J.E. (1967). Timber losses caused by termites, decay, and fire in two coastal forests in New South Wales. Technical Paper, C.S.I.R.O. Division of Entomology, Australia, 7: 2-18.

Guillot, F.S., Ring, D.R., Lax, A.R., Morgan, A., Brown, K., Riegel, C. & Boykin, D. (2010). Area-wide management of the Formosan subterranean termite, *Coptotermes formosanus* Shiraki (Isoptera: Rhinotermitidae), in the New Orleans French Quarter. Sociobiology, 55: 311-338.

Harris, W.V. (1966). The role of termites in tropical forestry. Insectes Sociaux, 8: 255-266.

Higa, S.Y. & Tamashiro M. (1983). Swarming of the Formosan subterranean termite, *Coptotermes formosanus* Shiraki, in Hawaii (Isoptera: Rhinotermitidae). In Proceedings of the 24th Hawaiian Entomological Society, Honolulu, Hawaii (pp. 233-238).

La Fage, J.P. & Nutting W.L. (1978). Nutrient dynamic of termites. In M. Brian (Eds.), Production Ecology of Ants and Termites (pp. 165-232). Cambridge: Cambridge University Press.

La Fage, J.P. (1987). Practical considerations of the Formosan subterranean termite in Louisiana: a 30-year old problem. In M. Tamashiro & N.-Y. Su (Eds.), The Biology and Control of the Formosan Subterranean Termite (pp. 37-42). Hawaii: College of Tropical Agriculture and Human Resources, University of Hawaii.

Lai, P.Y., Tamashiro, M., Yates, J.R., Su, N.-Y., Fujii, J.K., & Ebesu, R.H. (1983). Living plants in Hawaii attacked by *Coptotermes formosanus*. In Proceedings of the 24th Hawaiian Entomological Society, Honolulu, Hawaii (pp. 283–286).

Lax, A.R. & Wiltz, B.A. (2010). Swarming of the Formosan subterranean termite (Isoptera: Rhinotermitidae) in southern Mississippi. Mid-South Entomologist, 3: 18-25.

Lin, S.-Q. (1987). Present status of *Coptotermes formosanus* and its control in China. In M. Tamashiro & N.-Y. Su (Eds.), The Biology and Control of the Formosan Subterranean Termite (pp. 31-36). Hawaii: College of Tropical Agriculture and Human Resources, University of Hawaii.

Little, N.S. (2013). Implications for the detection, utilization, and degradation of bark beetle-attacked southern pines by subterranean termites. Dissertation, Mississippi State University, Starkville, MS.

Messenger, M.T., Su, N.-Y. & Scheffrahn, R.H. (2002). Current distribution of the Formosan subterranean termite and other termite species (Isoptera: Rhinotermitidae, Kalotermitidae) in Louisiana. Florida Entomologist, 85: 580-587.

Messenger, M.T. & Mullins, A.J. (2005). New flight distance recorded for *Coptotermes formosanus* (Isoptera: Rhinotermitidae). Florida Entomologist, 88(1): 99-100.

Morales-Ramos, J.A. & Rojas, M.G. (2001). Nutritional ecology of the Formosan subterranean termite: feeding response to commercial wood species. Journal of Economic Entomology, 94: 516-523.

Osbrink, W.L., Woodson, W.D. & Lax, A.R. (1999). Populations of Formosan subterranean termites established in living urban trees in New Orleans, Louisiana, U.S.A. In Proceedings of the 3rd International Conference on Urban Pests, New Orleans, Louisiana (pp. 341-345).

Peralta, R G., Menezes, E.B.G., Carvalho, A. & Aguiar-Mene-zes, E.L. (2004). Wood consumption rates of forest species by subterranean termites (Isoptera) under field conditions. Journal of Brazilian Forest Science, 28: 283-289.

R Core Team. (2012). R: a language and environment for statistical computing. Austria: R Foundation for Statistical Computing.

Rigby, R.A. & Stasinopoulos D.M. (2001). The GAMLSS project: a flexible approach to statistical modelling. In Proceedings of the 16th International Workshop on Statistical Modelling (pp. 337-345).

SAS Institute. (2003). SAS Version 9.1.3. Cary, NC: SAS Institute Inc.

Schultz, R.P. (1999). Loblolly - the pine for the twenty-first century. New Forests, 17: 71-88.

Smythe, R.V. & Carter, F.L. (1970). Feeding responses to sound wood by *Coptotermes formosanus*, *Reticulitermes flavipes*, and *R. virginicus* (Isoptera: Rhinotermitidae). Annals of the Entomological Society of America, 63: 841-847.

Stasinopoulos, D.M. & Rigby R.A. (2007). Generalized additive models for location scale and shape (GAMLSS) in R. Journal of Statistical Software, 23: 1-46.

Stasinopoulos, D.M., Rigby B.A., & Akantziliotou C. (2009). gamlss: Generalized Additive Models for Location Scale and Shape. R Package Version 2.0-0.

Su, N.-Y. (2003). Overview of the global distribution and control of the Formosan subterranean termite. Sociobiology, 41: 7-16.

Su, N.-Y. & Scheffrahn R.H. (1988). Foraging population and territory of the Formosan subterranean termite (Isoptera: Rhinotermitidae) in an urban environment. Sociobiology, 14: 353-359.

Su, N.-Y. & Tamashiro, M. (1986). Wood-consumption rate and survival of the Formosan subterranean termite (Isoptera: Rhinotermitidae) when fed one of six woods used commercially in Hawaii. In Proceedings of the 26th Hawaiian Entomological Society, Honolulu, Hawaii (pp. 109–113).

Sun, J.Z., Lockwood, M.E., Etheridge, J.L., Carroll, J., Holloman, C.Z., Coker, C.E.H. & Knight, P.R. (2007). Distribution of Formosan subterranean termites in Mississippi. Journal of Economic Entomology, 100: 1400-1408.

Warton, D.I. & Hui F.K. (2011). The arcsine is asinine: the analysis of proportions in ecology. Ecology, 92: 3-10.

Wang, C. & Powell J. (2001). Survey of termites in the Delta Experimental Forest of Mississippi. Florida Entomologist, 84: 222-226.

Wang, C., Powell, J.E. & Scheffrahn, R.H. (2003). Abundance and distribution of subterranean termites in southern Mississippi forests (Isoptera: Rhinotermitidae). Sociobiology, 42: 533-542.

Wood, T.G. & Sands, W.A. (1978). The role of termites in ecosystems. In M. Brian (Eds.), Production Ecology of Ants and Termites (pp. 245–292). Cambridge: Cambridge University Press.

Yamada, A., Inoue, T., Wiwatwitaya, D., Okhuma, M., Kudo, T. & Sugimoto, A. (2006). Nitrogen fixation by termites in tropical forests, Thailand. Ecosystems, 9: 75-83.

9

Two new species of *Pseudolasius* (Hymenoptera: Formicidae) from India

AA Wᴀᴄʜᴋᴏᴏ; H Bʜᴀʀᴛɪ

Department of Zoology and Environmental Sciences, Punjabi University, Patiala, India

Key words
new species, Key, Polymorphic, Taxonomy

Corresponding author
Aijaz Ahmad Wachkoo
Dept. of Zoology and Environmental Sciences
Punjabi University, Patiala, India
147002
E-Mail: aijaz_shoorida@yahoo.co.in

Abstract
Descriptions of two new species of *Pseudolasius* based on worker caste and sexuals are provided from India. With addition of these two species collected in lower Shivalik range of Northwest Himalaya, four species signify the genus *Pseudolasius* from India. An identification key to the workers of Indian *Pseudolasius* species is provided.

Introduction

Pseudolasius ants are mainly restricted to southern Asia, from India to China, stretching southwards to northern Australia and seem to be restricted to tropical localities (LaPolla et al., 2010). They are currently represented by 47 extant species, 15 subspecies and 1 fossil species across the globe (Ward, 2014). Although, *Pseudolasius* is in serious need of taxonomic revision, some important taxonomic contributions to this genus include: Emery (1911); Wheeler (1922); Menozzi (1924); Wu and Wang (1995); Xu (1997); Zhou (2001) and LaPolla (2004).

Pseudolasius are perhaps best known for possessing a polymorphic worker caste, with most species possessing complete dimorphism, characterized by major and minor workers (Holldobler & Wilson, 1990; LaPolla, 2004). Although there are clear size differences between the extremes among workers, blending of the major and minor castes occurs; this fact has caused confusion when elucidating species boundaries, and in the past has led to specimens being described several times as representing separate, distinct species (LaPolla, 2004).

Here we present descriptions of two new species, *Pseudolasius diversus* sp. n. and *Pseudolasius polymorphicus* sp. n., collected in foothills of Northwest Himalaya, the Shivalik range. Prior to this study, the genus *Pseudolasius* from India was represented by only two species, *P. familiaris* (Smith, F., 1860) and *P. machhediensis* Bharti et al., 2012 restricted to upper Himalaya.

Materials and Methods

The specimens were hand collected and their taxonomic analysis was conducted on Nikon SMZ 1500 stereo zoom microscope. For digital images, MP evolution digital camera was used on same microscope with Auto-Montage (Syncroscopy, Division of Synoptics, Ltd.) software. Later, images were cleaned with Adobe Photoshop CS5. Holotype and paratypes of both the species have been deposited in PUPAC, Punjabi University Patiala Ant Collection, Patiala. Some paratypes of both species will be deposited in BMNH, Natural History Museum, London, U.K. and California Academy of Sciences, San Francisco, United States of America. Measurements were recorded in micrometers between 80

× and 225 × with measuring accuracies of ±1 μm for small measures like eye length, such of ±2 μm for medium sized measures like head width, such of ±5 μm for larger measures like Weber's length. Morphological terminology for measurements and indices are as follows:

HL Maximum length of head in full-face view (maximum length and width of head present in the same focal level), measured in straight line from the anterior most point of the median clypeal margin to a line drawn across the posterior margin from its highest points by first measuring median length and then adding the depth of concavity.

HW Maximum width of head in full-face view (excluding the portion of eyes that extends past the lateral margins of the head) exclusive of any defined anteroposterior level.

HS Head Size, arithmetic mean of HL and HW.

EL Maximum length of eye as measured normally in oblique view of the head to show full surface of eye.

SL Maximum length of the scape excluding the basal neck and condyle.

PW Maximum width of the pronotum in dorsal view.

WL Weber's length measured from the anterior surface of the pronotum proper (excluding the collar) to the posteriormost point of the propodeal lobes.

FL Maximum length of the femur of foreleg from its margin with the trochanter to its margin with the tibia.

FW Maximum width of the femur of foreleg.

GL Maximum length of the gaster in lateral view from the anteriormost point of first gastral segment to the posteriormost point of the last segment.

TL Total length: HL + WL + GL.

CI Cephalic index: HW/HL.

SI Scape index: SL/HW.

REL Relative eye length index: EL/HL.

Results

Descriptions of new species

Pseudolasius diversus sp. n. (Figures 1-16)

Type material

Holotype (Major worker): India, Uttarakhand, Rajaji Forest Area, 30.2483°N 77.9878°E, 660m.a.s.l., 11.viii.2009, hand collecting (coll. Aijaz A. Wachkoo). Paratypes: 24 workes, 2 gynes and 6 males, with same data as holotype; 7 workers, 5.viii.2009; 55 workers 6.viii.2009; 30 workers 13.viii.2009; 62 workers 6.ix.2010, same data as holotype.

Workers (Figures 1-9)

Morphometric data of the holotype: HL 1126; HW 1108; HS 1117; EL 118; SL 681; PW 678; WL 1082; FL 743; FW 227; GL 1446; TL 3654. Indices: CI 0.98; SI 0.61; REL 0.10.

Morphometric data of the paratype workers: HL 656-1134; HW 629-1103; HS 638-1112; EL 78-118; SL 549-722; PW 454-713; WL 795-1136; FL 512-740; FW 150-234; GL 1054-1466; TL 2505-3722. Indices: CI 0.82-1.00; SI 0.61-0.88; REL 0.07-0.14 (n=29).

Head roughly heart-like in major worker, subrectangular in media and subquadrate in minor, as long as wide in major and minor workers, distinctly elongate in media workers. Posterior margin with a strong "v" shaped impression medially in major and media, weakly concave in minor; posterolateral corners rounded; sides parallel in media and minor, in major convex anteriorly, subparallel and gently converging posteriorly. Frontal carinae, nearly parallel with sides of head, not extending past posterior margin of eyes. Anterolateral corner of clypeus bluntly toothed; scape just reaches posterolateral corner in minor worker, shorter in media reaching four-fifths and even short in major worker reaching three-fourths of posterolateral corner; antennal segments 3-9 longer than wide; mandible with five teeth, fifth one mostly reduced.

In lateral view, promesonotum convex in minor and media relatively flat in major; metanotal groove strongly developed; metanotal area short but distinct. Propodeum low, nearly flat above with sides diverging basally; propodeal spiracle rounded; declivity steep. Petiole low triangular, inclined forward with posterior face longer than anterior face, dorsum mostly emarginate, transverse in some minor workers. First gastral segment with concave anterior face, receives the petiole.

Head and gaster with abundant appressed pubescence, mesosoma sparsely pubescent; head, scape, legs and gaster with abundant short erect setae, denser on gaster; relatively longer erect setae cover clypeus and mesosoma. Mandibles with short, curved setae near masticatory borders; anterior clypeal margin with a few, longer, anteriorly directed setae medially and fringe of short setae towards mandibular bases.

Full-grown workers yellowish brown, fairly microreticulate with dull and opaque cuticle; nanitic workers light yellow with superficial sculpturing and relatively shiny cuticle.

Gyne (Figures 10-12)

Morphometric data of the gynes: HL 958-987; HW 918-938; HS 938-962; EL 259-279; SL 842-861; WL 1700-1762; FL 768-782; FW 198-220; GL 2748-2791; TL 5406-5540. Indices: CI 0.95-0.96; SI 0.92; REL 0.27-0.28 (n=2).

Gyne similar to worker with usual differences indicating caste, including three ocelli, complete thoracic structure and wings. Head similar to minor worker, subquadrate with broadly emarginate posterior margin. Scape surpasses the posterior margin of head by about one-fifth its length. Propodeum indistinct. Body light yellow colored with feebler sculpture. Pubescence abundant on body including mesosoma, erect setae shorter and sparser than in conspecific worker caste.

Male (Figures 13-16)

Morphometric data of the males: HL 525-552; HW 508-531; HS 516-540; EL 228-240; SL 528-550; WL 972-1095; FL 637-661; FW 130-142; GL 1122-1181; TL 2620-2824. Indices: CI 0.95-0.96; SI 1.04-1.05; REL 0.43-0.44 (n=5).

Head oval, as long as wide excluding large compound eyes; eyes subglobulose, bulging, projecting well beyond head outline in full-face view; three prominent ocelli present. Antennae 13 segmented, filiform, scapes long, surpass posterior margin by about three-tenths their length. Anterolateral corner of clypeus bluntly toothed. Mandibles slender, armed with three teeth, apical one prominent, large and pointed, basal tooth small, blending seamlessly into inner mandibular margin.

Mesosoma modified for presence of wing; in lateral view scutum and scutellum flat; propodeum indistinct, not higher than remainder of notum with very short dorsal face and long declivitous face. Petiole as in worker; gaster elongated.

Parameres paddle-shaped, turning slightly inward toward midline of body posteriorly; long setae extending off parameres. Cuspi long and tubular, bent toward digiti; digiti weakly anvil-shaped; curved outward and covered with short peg-like teeth. Penis valves projecting.

Body mostly smooth and shiny; erect setae shorter and sparser but pubescence as in worker caste. Color light yellow, head mostly brownish.

Etymology: The species is named for its morphologically diverse worker caste.

Distribution and habitat: This species seems to be rare in Shivalik range of Northwest Himalaya and was found mostly under stones in a primary, subtropical, semi-evergreen forest with relatively high annual precipitation. It was encountered in a single locality of a reserved forest area (Rajaji Forest Area), in Uttarakhand during the intensive surveys.

Comparative notes: *Pseudolasius diversus* mostly resembles Chinese, *Pseudolasius bidenticlypeus* Xu, 1997 but can be easily separated from it by 5-toothed mandible whilst latter possess 6-toothed mandible. Scapes are longer in *P. bidenticlypeus* almost reaching posterior margin of head in major and easily surpassing posterior margin in minor and media whereas in *Pseudolasius diversus* scapes reach only up to three-fourths of posterior margin in major and never surpass the posterior margin in media and minor.

Pseudolasius polymorphicus sp. n. (Figures 17-32)

Type material

Holotype (Major worker): India, Himachal Pradesh, Andretta, 32.0744°N 76.5856°E, 940m.a.s.l., 11.vi.2010, hand collecting (coll. Aijaz A. Wachkoo). Paratypes: 45 workers, 1 gyne and 6 males, with same data as holotype.

Workers (Figures 17-25)

Morphometric data of the holotype: HL 920; HW 800; HS 860; EL 72; SL 502; PW 521; WL 864; FL 498; FW 168; GL 1282; TL 3066. Indices: CI 0.87; SI 0.63; REL 0.08.

Morphometric data of the paratype workers: HL 515-926; HW 468-804; HS 502-864; EL 27-62; SL 360-506; PW 360-532; WL 570-874; FL 350-502; FW 110-174; GL 759-1034; TL 1850-3094. Indices: CI 0.86-0.92; SI 0.61-0.80; REL 0.04-0.08 (n=24).

Head subrectangular in major worker, subquadrate in media and minor, relatively longer than wide in major as in media and minor workers. Posterior margin medially strongly impressed in major and media, only weakly so in minor; posterolateral corners rounded; sides subparallel in media, convex in minor and major, converging anteroposteriorly. Frontal carinae, divergent posteriorly, do not extend past posterior margin of eyes. Anterolateral corner of clypeus bluntly toothed; scape barely reaches posterolateral corner in minor worker, shorter in media reaching four-fifths and even short in major worker reaching two-thirds of posterolateral corner; antennal segments 3-9 wider than long; mandible with five teeth, fifth one mostly reduced.

In lateral view, promesonotum convex; metanotal groove strongly developed; metanotal area short but distinct. Propodeum low, nearly flat above with sides diverging basally; propodeal spiracle rounded; declivity steep. Petiole low triangular, inclined forward with posterior face longer than anterior face, dorsal margin weakly emarginate to transverse. First gastral segment with concave anterior face, receives the petiole.

Head and gaster covered with abundant appressed pubescence, mesosoma only sparsely pubescent; head, scape, legs and gaster with abundant short erect setae, denser on gaster; relatively longer erect setae cover clypeus and mesosoma. Mandibles with numerous, curved setae near masticatory borders; anterior clypeal margin with a few, longer, anteriorly directed setae medially and few short setae fringing margin laterally.

Full-grown workers yellow, with head and gaster usually yellowish brown, feebly microreticulate with smooth and feebly shiny cuticle; nanitic workers light yellow with superficial sculpturing and relatively shinier cuticle.

Gyne (Figures 26-28)

Morphometric data of the gyne: HL 782; HW 831; HS 806; EL 229; SL 622; WL 775; FL 661; FW 180; GL 2462; TL 4019. Indices: CI 1.06; SI 0.75; REL 0.30 (n=1).

Gyne similar to worker with usual differences indicating caste, including three ocelli, complete thoracic structure and wings. Head, trapezoidal with transverse posterior margin. Scape surpasses the posterior margin of head by about one-tenth its length. Propodeum indistinct. Head, mesosomal dorsum and gaster dark brown, antennae, legs and lateral mesosoma, brownish yellow; sculpture superficially microreticulate. Pubescence on body abundant and longer including mesosoma, erect setae shorter and sparser than in conspecific worker caste.

Male (Figures 29-32)

Morphometric data of the males: HL 436-462; HW 441-462; HS 440-464; EL 170-202; SL 350-390; WL 752-828; FL 450-512; FW 81-92; GL 841-904; TL 2027-2194. Indices: CI 1.00-1.03; SI 0.80-0.84; REL 0.38-0.44 (n=6).

Head broadly oval, as long as wide excluding large compound eyes; eyes subglobulose, bulging, projecting beyond head outline in full-face view; three prominent ocelli present. Antennae 13 segmented, filiform, scapes long, surpass posterior margin by about three-tenths their length. Anterolateral corner of clypeus bluntly toothed. Mandibles curved strap like, with acute pointed apical tooth, remainder of masticatory margin smooth, without any teeth or denticles; basal angle rounded, indistinct and seamlessly blends into inner mandibular margin; when closed their tips meet or overlap and the entire blades are tucked away under the clypeus in such a way that only their external margins show externally along the anterior clypeal border.

Mesosoma enlarged to accommodate flight muscles; in lateral view scutum and scutellum flat; propodeum indistinct, lower than remainder of notum with very short dorsal face and long declivitous face. Petiole as in worker; gaster elongated.

Parameres broad paddle-shaped, turning slightly inward toward midline of body posteriorly; long setae extending off parameres. Cuspi smaller, covered under parameres weakly paddle-shaped, slightly bent toward digiti; digiti as long as penis valves, weakly anvil-shaped; curved outward and covered with short peg-like teeth. Penis valves projecting.

Overall, body mostly smooth and shiny; erect setae shorter and sparser but pubescence as in worker caste. Head dark brown, mesosoma and gaster brown, antennae and legs yellowish brownish.

Etymology. The species is named for its polymorphic worker caste.

Distribution and habitat. Andretta, the type locality of this species falls within the Shivalik range of Northwest Himalaya and is devoid of leaf litter; surrounded on all sides by tea gardens and pine forests before they merge with the Dhauladhar ranges (a southern branch of the main Outer Himalayan chain of mountains). This species seems rare in Shivalik range of Northwest Himalaya and was encountered only once during the intensive surveys, found nesting under a stone below a shady tree.

Comparative notes. Pseudolasius polymorphicus most resembles Chinese, *Pseudolasius cibdelus* Wu & Wang, 1992 but can be easily distinguished from the latter. In *P. polymorphicus* scapes are short reaching only two-thirds of posterior margin of head in major and barely reaches poste-

rior margin in minor whereas in *P. cibdelus* scapes are longer reaching three-fourths of posterior margin of head in major and surpassing the posterior margin in minor workers. Sides of head are subparallel in major of *P. polymorphicus* with head equally broad anteroposteriorly, and minor worker has only feebly emarginate posterior margin whilst, sides of head are strongly converging anteriorly in *P. cibdelus* with head distinctly narrowing in front and posterior margin of head is fairly emarginate in minor worker.

Key to species of Pseudolasius of India based on worker caste

1. Antennal segments 3-9 longer than wide; frontal carinae parallel..2
— Antennal segments 3-9 wider than long; frontal carinae divergent posteriorly....... *P. polymorphicus* sp. n.
2. Scapes easily surpass the posterior margin of head 3
— Scapes never surpass the posterior margin of head ....*P. diversus* sp. n.
3. Mandibles armed with 8 teeth......*P. familiaris* (Smith, F., 1860)
— Mandibles armed with 6 teeth.........*P. machhediensis* Bharti *et al.,* 2012

Discussion

At times the distinction of castes among worker specimens is difficult to make. LaPolla (2004) defines, "the major as an individual worker that possesses a much wider head proportionate to the mesosoma than is otherwise observed in minor workers". In this study, with regards to shape *P. diversus* sp. n. major and minor workers can be easily differentiated from media workers by squarer head, whilst media workers distinctly have rectangular head, but in *P. polymorphicus* sp. n. minor and media workers have squarer head in comparison to rectangular head in major workers. Most of the body parts are isometric but a few are always allometric in both species i.e., they increase or decrease in relative size as total body is enlarged. Head grows wider, and the antennae become relatively shorter in major workers, whereas pronotal width etc are isometric with regard to most of the remaining body. The conclusion renders us safe to define in relativity major as worker with widest head and shortest scapes; minor as worker with narrowest head and longest scapes while media as intermediate between them.

Acknowledgments

Financial assistance rendered by Ministry of Environment and Forests (Grant No. 14/10/2007-ERS/RE), Govt. of India, New Delhi is gratefully acknowledged. We thank Xu Zheng-Hui for comparing and confirming the identity of new species. Sincere thanks are due to anonymous reviewers for helpful comments and suggestions about the manuscript.

References

Emery, C. (1911). Fragments myrmécologiques. Annales de la Société Entomologique de Belgique, 55: 213-225.

Hölldobler, B. & Wilson, E.O. (1990). The Ants. Cambridge: Harvard University Press, 732 p.

LaPolla, J.S. (2004). Taxonomic review of the ant genus *Pseudolasius* in the Afrotropical region. Journal of the New York Entomological Society, 112: 97-105.

LaPolla, J.S., Brady, S.G. & Shattuck, S.O. (2010). Phylogeny and taxonomy of the *Prenolepis* genus-group of ants. Systematic Entomology, 35: 118-131.

Menozzi, C. (1924). Alcune nuove formiche africane. Annali del Museo civico di storia naturale "Giacomo Doria", 51: 220-227.

Ward, P.S., editor (2014). Antweb: Bolton World Catalog. http://www.antweb.org/description.do?subfamily=formicinae&genus=pseudolasius&rank=genus&project=worldants (accessed date: 27 February 2014).

Wheeler, W.M. (1922). The ants of the Belgian Congo. Bulletin of the American Museum of Natural History, 45: 1-1139.

Wu, J. & Wang, C. (1995). The Ants of China. China: Forestry Publishing House, Beijing, 214 p.

Xu, Z. (1997). A taxonomic study of the ant genus *Pseudolasius* Emery in China. Zoological Research, 18: 1-6.

Zhou, S.Y. (2001). Ants of Guangxi. China: Guangxi Normal University Press, Guilin, 255 p.

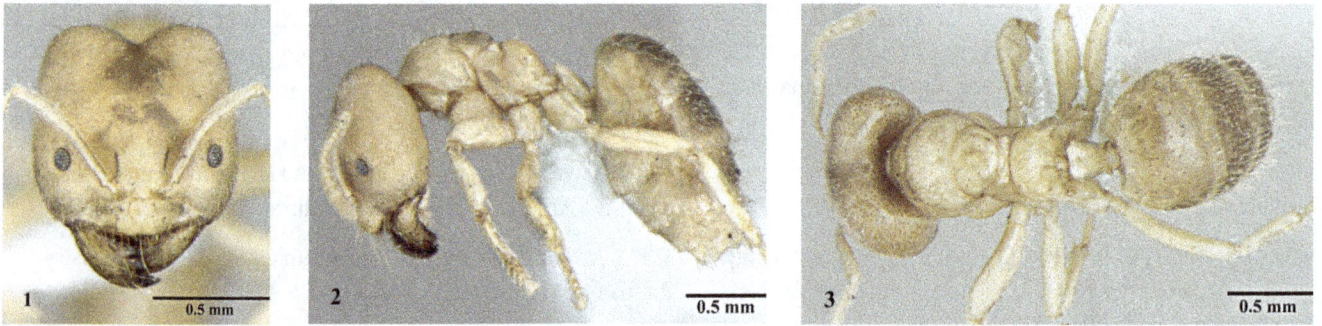

Fig. 1-3. *Pseudolasius diversus* sp. n., major worker 1) head, full face view; 2) Body, lateral view; 3) body, dorsal view.

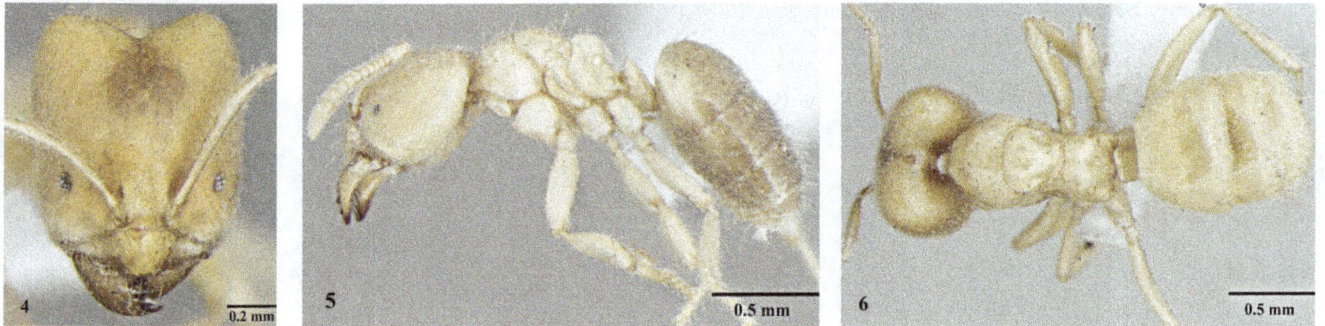

Fig. 4-6. *Pseudolasius diversus* sp. n., media worker 4) head, full face view; 5) body lateral view; 6) body, dorsal view.

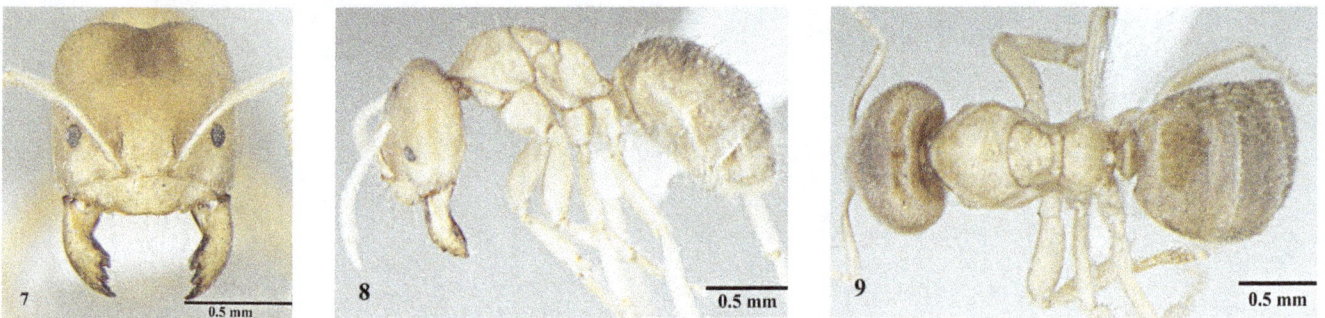

Fig. 7-9. *Pseudolasius diversus* sp. n., minor worker 7) head, full face view; 8) body, lateral view; 9) Body, dorsal view.

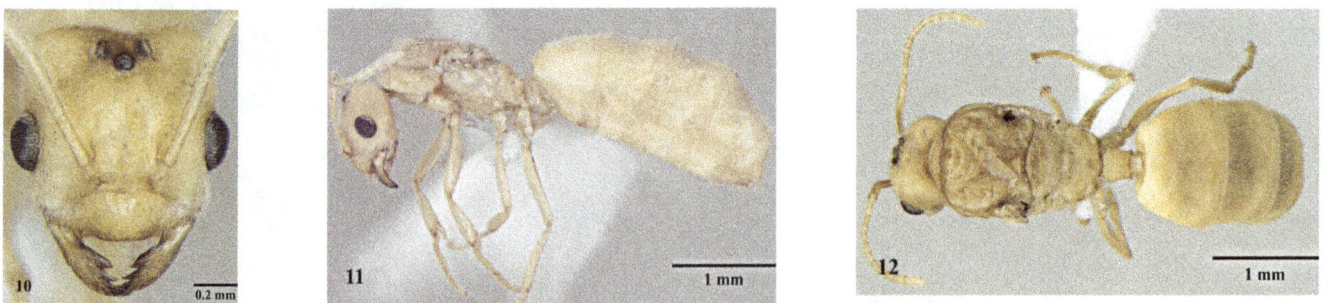

Fig. 10-12. *Pseudolasius diversus* sp. n., gyne 10) head, full face view; 11) body, lateral view; 12) body, dorsal view.

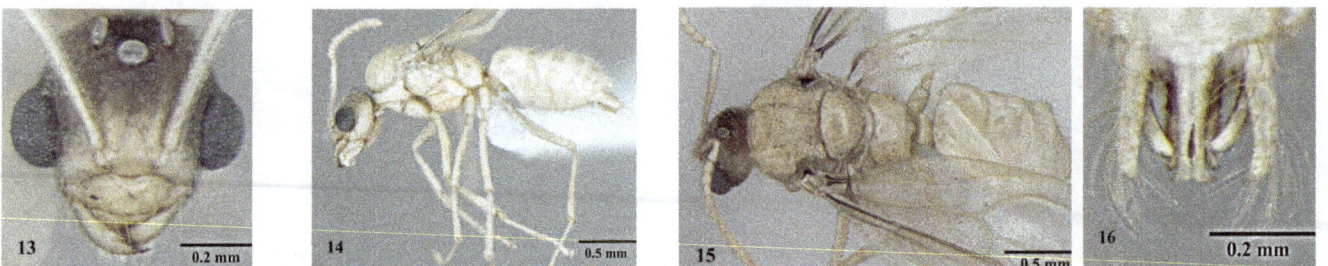

Fig. 13-16. *Pseudolasius diversus* sp. n., male 13) head, full face view; 14) body, lateral view; 15) body, dorsal view; 16) genetalia, dorsal view.

10

Assessing the Utility of a PCR Diagnostics Marker for the Identification of Africanized Honey Bee, *Apis mellifera* L., (Hymenoptera: Apidae) in the United States

AL Szalanski, AD Tripodi

University of Arkansas, Fayetteville, Arkansas, USA.

Keywords
Apis mellifera, molecular diagnostics, Africanized honey bee, mtDNA, USA

Corresponding author
Allen Szalanski
Dept. Entomology, University of Arkansas, 319 Agriculture Building
Fayetteville, AR 72701
E-Mail: aszalan@uark.edu

Abstract
An assessment of a molecular diagnostic technique for distinguishing Africanized honey bees from European honey bees in the United States was conducted. Results from multiplex PCR diagnostics of a mitochondrial DNA cyt-b marker corresponded with results based on COI-COII sequencing analysis, but differed from morphometric analysis results. We suggest utilizing both multiplex PCR and morphometric methods for Africanized honey bee diagnostics in the United States, when possible.

The Africanized honey bee was first detected in Texas in 1990 (Sugden & Williams, 1990), and USDA reports it has been established in ten states: Arizona, Arkansas, California Florida, Louisiana, Nevada, New Mexico, Oklahoma, Texas and Utah (Anonymous, 2011) Following a tragic death in 2010, Africanized honey bees were confirmed in Georgia (Berry, 2011). Hybrids between Africanized and European honey bees are morphologically similar and difficult to distinguish from one another. Two kinds of laboratory-based techniques, morphometric analysis (Rinderer et al., 1993) and molecular diagnostics (Sheppard & Smith, 2000), are commonly employed to determine Africanized status, but to our knowledge, these have not been compared.

The typical morphometric approach to Africanized honey bee detection is the Fast Africanized Bee Identification System (FABIS) method (Rinderer et al., 1987), although more precise methods such as Automatic Bee Identification System (ABIS) (Steinhage et al. 2001) are now available. Morphometric diagnostic techniques require measurements from 10 or more freshly collected specimens (Meixner et al., 2013) to assign a colony to one of four categories: 1) Africanized (AHB), 2) AHB with European (EHB) traits, 3) EHB with AHB traits and 4) EHB (Sylvester & Rinderer

1987). Molecular diagnostics rely upon mitochondrial DNA (mtDNA), typically a region of the cytochrome b gene (Pinto et al. 2003). Identification of Africanized bees with mtDNA is relatively easy, because a single worker in any condition can represent a colony's mtDNA lineage (Sheppard & Smith, 2000). Yet, since mtDNA is maternally inherited, mtDNA markers cannot determine if a European queen has mated with Africanized drones, and such colonies will remain undetected with mtDNA-based techniques. Molecular identification of Africanized bees is also conducted with an mtDNA cytochrome b marker (cyt-b) as it has a low level of intraspecific variation in honey bees (Crozier et al., 1991; Szalanski & McKern, 2007). Techniques typically used are either PCR-RFLP (Pinto et al., 2003) or multiplex PCR using a primer specific for Africanized honey bees (Szalanski & McKern, 2007). Both of these methods yield identical results. Despite the utility of molecular markers, it is unknown how well they correlate with morphometric identification and how the cyt-b marker can distinguish the A (African) lineage of honey bees relative to the O (Middle Eastern), C (Eastern Europe) and M (Western Europe) lineages (Ruttner, 1987), which are often determined through COI-COII sequencing. The objective of this study was to determine how a cyt-b multiplex technique for distinguishing Africanized from European bees compares with COI-COII DNA sequence and FABIS morphometric techniques.

A total of 968 samples from swarms, feral colonies and managed honey bee colonies collected from Arizona, Arkansas, California, Florida, Georgia, Hawaii, Kansas, Louisiana, Missouri, Mississippi, Nebraska, New Mexico, Oklahoma, Texas and Utah during 1991-2013 were analyzed. Samples were collected by various agencies and preserved in 70-100% ethanol. Samples collected from Texas from 1991 to 2008 were identified as Africanized (AHB, n = 54), AHB with evidence of introgression of European genes (AHB.E, n = 23), EHB with evidence of Africanized genes (EHB.A, n = 22) or EHB (EHB, n = 39) using FABIS following Rinderer et al. (1993).

For the multiplex PCR diagnostics, DNA was extracted from two workers from each sampled colony, and PCR of a portion of the cyt-b region was conducted per Szalanski and McKern (2007). This multiplex results in a control amplicon of 485 bp for both Africanized and European bees and a 385 bp amplicon unique to Africanized bees. Samples exhibiting a single electrophoretic band are diagnosed as European (multiplex EHB) and those with two are diagnosed as Africanized (multiplex AHB). DNA sequencing analysis of a COI-COII marker was conducted following Szalanski and Magnus (2010), and samples from that study were additionally subjected to our multiplex procedure for comparison (n = 360). DNA sequences were identified to haplotype and

assigned to lineage using GenBank BLAST searches and our own database. Africanization diagnoses from cyt-b multiplex diagnostics, COI-COII DNA sequencing and FABIS morphometric analysis were then compared with one another.

Of the 968 samples subjected to multiplex diagnostics, 318 samples were diagnosed as Africanized and 650 were diagnosed as European (Table S1). A total of 12 haplotypes from the A (African) lineage were observed, and four O (Middle Eastern), five M (Western Europe) and 11 C (Eastern Europe) haplotypes were compared with the multiplex results (Table 2). All of the multiplex PCR identified Africanized samples fell within the A lineage while the European samples were O, M or C lineages. The multiplex method reaches exactly the same diagnoses as the sequencing method but can be carried out in a single PCR step without subsequent DNA sequencing.

Africanization diagnoses in the multiplex and FABIS methods differed quite dramatically (Table 3). Of the 107 samples the exhibiting Africanized signature in the PCR-multiplex, only 84 were diagnosed as of African descent (AHB,

Table 2. Correlation of mitochondrial DNA COI-COII DNA sequence lineages with cyt-b multiplex PCR identification.

Lineage (n haplotypes)	N	Multiplex AHB	Multiplex EHB
C (11)	173	0	173
M (5)	12	0	12
O (4)	19	0	19
A (12)	156	156	0
Total	360	156	204

AHB.E and EHB.A) using FABIS. This suggests that 21% of Africanized bees in our sample were cryptically Africanized and undetectable using morphology-based methods. Similarly, of the 99 bees exhibiting morphometric characteristics of Africanization (AHB, AHB.E and EHB.A), 15% were misdiagnosed as European by the multiplex. Interestingly, our results suggest that only a small proportion (15%) of colonies that exhibit Africanized morphology were founded by European queens inseminated by Africanized drones.

COI-COII lineage data includes haplotype determinations from Szalanski and Magnus (2010).

This study verified the utility of using a cyt-b PCR-multiplex technique for identifying honey bees of the A lineage in the United States. Identification of Africanized matrilines through cytb-b diagnosis parallels classic COI-COII matriline determination through sequencing, but in a single-step, lesser cost procedure. We also found evidence that using either FABIS or mtDNA determination alone may underestimate the occurrence of Africanized populations. For greater certainty in diagnosing Africanized populations of honey bees, we suggest utilizing both morphological and molecular methods when the number of samples makes this possible.

Table 1. Sampled states and results of multiplex PCR identification of samples as Africanized or European honey bees.

State	N	Multiplex AHB	Multiplex EHB
Arizona	1	0	1
Arkansas	143	1	142
California	3	3	0
Florida	1	1	0
Georgia	2	1	1
Hawaii	124	1	123
Kansas	18	0	18
Louisiana	24	0	24
Mississippi	20	0	20
Missouri	25	0	25
Nebraska	19	0	19
New Mexico	62	45	17
Oklahoma	178	58	120
Texas	140	109	31
Utah	208	99	109
Total	968	318	650

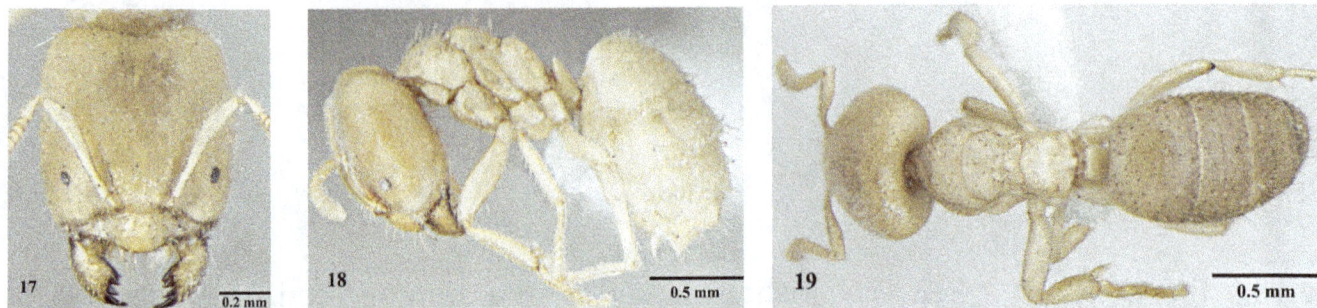

Fig. 17-19. *Pseudolasius polymorphicus* sp. n., major worker 17) head, full face view; 18) body, lateral view; 19) body, dorsal view.

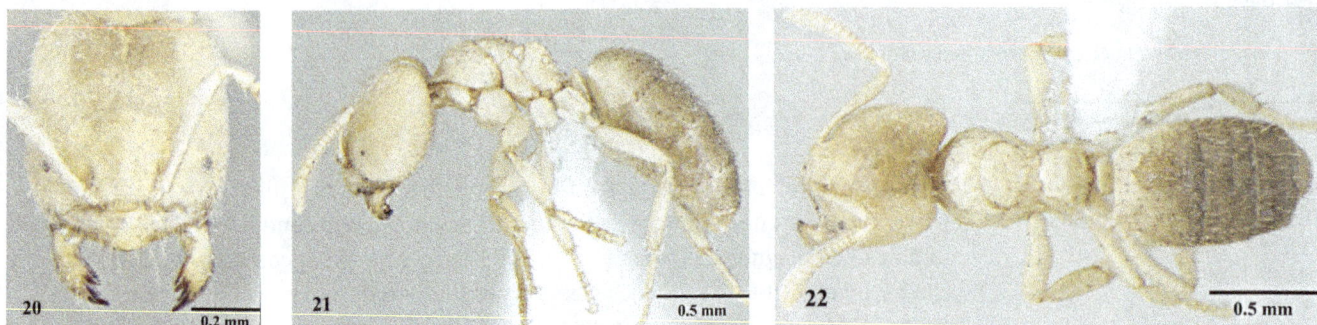

Fig. 20-22. *Pseudolasius polymorphicus* sp. n., media worker 20) head, full face view; 21) body lateral view; 22) body, dorsal view.

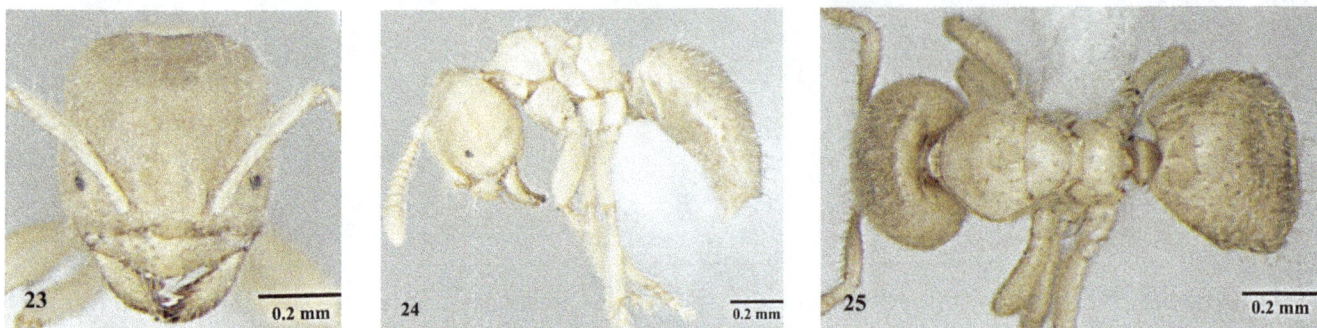

Fig. 23-25. *Pseudolasius polymorphicus* sp. n., minor worker 23) head, full face view; 23) body, lateral view; 25) body, dorsal view.

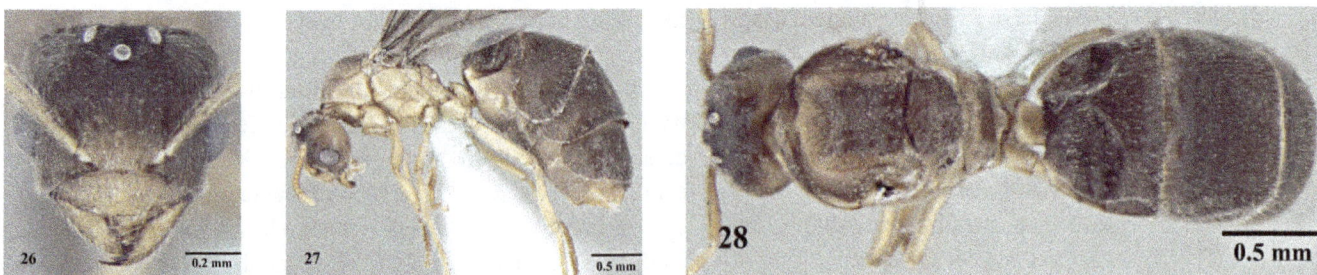

Fig. 26-28. *Pseudolasius polymorphicus* sp. n., gyne 26) head, full face view; 27) body, lateral view; 28) body, dorsal view.

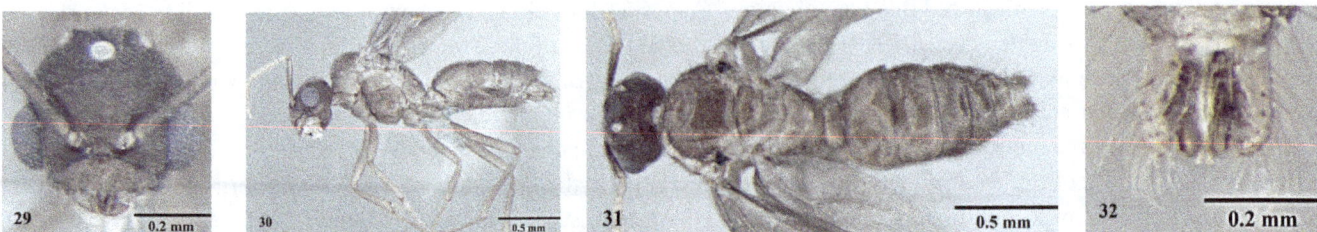

Fig. 29-32. *Pseudolasius polymorphicus* sp. n., male 29) head, full face view; 30) body, lateral view; 31) body, dorsal view; 32) genetalia, dorsal view.

Table 3 Comparison of morphometric and multiplex PCR identification of Africanized and European honey bees from Texas.

Morphometric Identification*	N	Multiplex AHB (%)	Multiplex EHB (%)
AHB	54	48 (89)	6 (11)
AHB.E	23	19 (83)	4 (17)
EHB.A	22	17 (77)	5 (23)
EHB	39	23 (59)	16 (41)
Total	138	107	31

*AHB: Africanized; AHB.E: Africanized with evidence of introgression of European genes; EHB.A: European with evidence of introgression of Africanized genes; EHB: European (Rinderer et al., 1993).

Acknowledgements

We thank numerous beekeepers and the Utah Department of Agriculture and Food, along with Ed Levi, Richard Grantham, Danielle Downey, Lisa Bradley and Clarence Collison for providing samples. We thank Lisa Bradley for doing the morphometric analysis of the Texas samples, and Clinton Trammel for assisting with the genetic analysis. This research was supported in part by the University of Arkansas, Arkansas Agricultural Experiment Station.

References

Anonymous (2011). Map of the spread of Africanized honey bee by year. http://www.ars.usda.gov/research/docs.htm?docid=11059&page=6.

Berry, J. (2011). African honey bees in Georgia. Bee Culture 139: 53-55.

Crozier, Y. C., S. Koulianos, & R. H. Crozier (1991). An improved test for Africanized honey bee mitochondrial DNA. Experientia, 47: 968-969.

Meixner M. D., M. A. Pinto, M. Bouga, P. Kryger, E. Ivanova, & S. Fuchs. (2013 Standard methods for characterizing subspecies and ecotypes of *Apis mellifera*. Journal of Apicultural Research, 52: 1-27.

Pinto, M. A., J. S. Johnson, W. L. Rubink, R. N. Coulson, J. C. Patton, & W. S. Sheppard. (2003). Identification of Africanized honey bee (Hymenoptera: Apidae) mitochondrial DNA: validation of a rapid polymerase chain reaction-based assay. Annals of the Entomological Society of America, 96: 679-684.

Rinderer, T. E., H. A. Sylvester, S. M. Buco, V. A. Lancaster, E. W. Herbert, A. M. Collins, R. L. Hellmich, G. L. Davis & D. Winfrey. (1987). Improved simple techniques for identifying Africanized and European honey bees. Apidologie, 18: 179-196.

Rinderer, T. E., S. M. Buco, W. L. Rubink, H. V. Daly, J. A. Stelzer, R. M. Riggio, & F. C. Baptista. (1993). Morphometric identification of Africanized and European honey bees using large reference populations. Apidologie, 24: 569-585.

Ruttner, F. (1987). Biogeography and taxonomy of honeybees. Springer-Verlag, Berlin. 284 pp.

Steinhage, V., T. Arbuckle, S. Schroder, A. B. Cremers, & D. Wittmann. (2001). ABIS: automated identification of bee species. BIOLOG workshop, German programme on biodiversity and global change, status report. pp. 194-195.

Sheppard, W. S., & D. R. Smith. (2000). Identification of African-derived bees in the Americas: a survey of methods. Annals of the Entomological Society of America, 93: 159-176.

Sugden, D. A., & K. R. Williams. (1990). October 15: the day the bee arrived. Gleanings in Bee Culture ,119: 18-21.

Sylvester, H. A., & T. E. Rinderer. (1987). Fast Africanized bee identification system (FABIS) manual. American Bee Journal, 127: 511-516.

Szalanski, A. L., & J. A. McKern. (2007). Multiplex PCR-RFLP diagnostics of the Africanized honey bee (Hymenoptera: Apidae). Sociobiology, 50: 939-945.

Szalanski, A. L., & R. M. Magnus. (2010). Mitochondrial DNA characterization of Africanized honey bee (*Apis mellifera* L.) populations from the USA. Journal of Apicultural Research, 49: 177-185. doi: dx.doi.org/10.3896/IBRA.1.49.2.06

11

Effect of the Habitat Alteration by Human Activity on Colony Productivity of the Social Wasp *Polistes versicolor* (Olivier) (Hymenoptera: Vespidae)

RF TORRES[1], VO TORRES[1], YR SÚAREZ[2], WF ANTONIALLI-JUNIOR[2]

1 - Universidade Federal da Grande Dourados, Dourados, Mato Grosso do Sul, Brazil.

2 - Universidade Estadual de Mato Grosso do Sul, Dourados, Mato Grosso do Sul, Brazil.

Keywords

Nest, Independent foundation, Polistinae, Reutilization of cells, Synanthropy

Corresponding author

Romario Ferreira Torres
Programa de Pós-graduação em Entomologia e Conservação da Biodiversidade
Universidade Federal da Grande Dourados
Fac. de Ciências Biológicas e Ambientais, 241
Dourados, Mato Grosso do Sul, Brazil
79804-970,
E-Mail:romario_torresmn@hotmail.com

Abstract

Currently, the main impacts on biodiversity are generated by human activities in natural environments. Monitoring the number of species of social wasps nesting attached to buildings is important to evaluate the effect of this activity on colony productivity. This study evaluated the effect of habitat alteration, particularly by human activity on the productivity of colonies of the wasp *Polistes versicolor*. We evaluated 20 abandoned nests and compared the productivity parameters: number of cells constructed, number of adults produced, nest dry mass, proportion of productive cells, number of generations, and diameter of the petiole. Most of these parameters showed higher values in the colonies nesting in the habitat less altered by human activity. Therefore, productivity was significantly higher in this habitat. In the nests, regardless of the site, the cells that were central and closer to the petiole were the most productive. Colonies in the two habitats used different strategies: in the habitat more altered by human activity, the wasps invested more in reusing cells than in enlarging the nest. However, the species continues to nest in the urban area, probably because of decreased interspecific competition, predation, and interference from climate variations.

Introduction

Eusocial wasps are represented by 29 genera in the Neotropical region, 22 of which are recorded in Brazil (Carpenter & Marques, 2001). These wasps occupy many kinds of habitats, primarily associated with human constructions (Lima et al., 2000; Prezoto et al., 2007), i.e., show a high degree of synanthropy (Fowler, 1983).

Among the numerous factors that contribute to the success of social wasps, the colony productivity stands out. Productivity depends on ecological factors including changes in temperature, prey availability, and number of founders, among others (Gamboa et al., 2005; Inagawa et al., 2001). According to Gamboa (1978), for example, colonies of *Polistes metricus* (Say) founded by association are more productive than those initiated by a single female. Tibbetts & Reeve (2003) found that *Polistes dominula* (Christ) colonies initiated by association better

defend their colonies from predators and/or conspecific wasps, and this cooperation increases the colony productivity.

Environmental factors may also be related to colony success and productivity, for example, the nesting site, as evidenced by Inagawa et al. (2001) and Nadeau & Stamp (2003) in *Polistes snelleni* (Saussure) and *Polistes fuscatus* (Fabricius), respectively. These studies showed that colonies founded in locations with higher mean temperatures had higher productivity.

For colonies under the same environmental conditions, Giannotti (1997) noted that the productivity can be influenced by intrinsic factors such as the number of cells, reuse of cells, duration of immature stages, and the mortality of the immatures in the different stages. As observed by Montagna et al. (2010), the productivity can vary within the comb; the central cells of the combs of *Mischocyttarus consimilis* (Zikán) are more productive than the peripheral cells.

Santos & Gobbi (1998), Inagawa et al. (2001) and Gamboa et al. (2005) evaluated the effect of habitat on the colony productivity of the social wasps, while others such as Penna et al. (2007), Montagna et al. (2010), Oliveira et al. (2010) and Sinzato et al. (2011) evaluated the productivity in urban environments. However, only Michelutti et al. (2013) compared the productivity of colonies between environments with high and low degrees of human interference, in *M. consimilis*.

According to Samways (2005, 2007), disturbances caused by humans in natural environments are the main factors acting to reduce biodiversity in the tropics. The human presence alters habitat quality (Raupp et al., 2010; Schowalter, 2012). The expansion of ranching and other agricultural activities as well as urbanization are the main factors modifying the natural habitats (Abensperg-Traun & Smith, 2000; New, 2005; Raupp et al., 2010) and often have forced wasps to search for new nesting environments. It is common to find wasps nesting on or in human constructions (Clapperton, 2000; Mead & Pratte, 2002). This can be explained by the abundance and quality of available sites, and the reduced interspecific competition and predation by vertebrates in these environments.

Among the Polistinae, several species of *Polistes* use human constructs as nesting substrates (Fowler, 1983; Butignol, 1992; Giannotti & Mansur, 1993). However, Gould & Jeanne (1984) and Michelutti et al. (2013) suggested that the urban environment negatively affects the productivity and development of the colonies, since it offers fewer resources than do preserved environments.

Polistes versicolor (Olivier, 1791) builds nests formed by a single, uncovered comb attached to the substratum by a single petiole (Richards, 1978). This species is facultatively synanthropic, since it is associated with human constructions (Oliveira et al., 2010) and also occurs in natural environments. Studies of predation (Butignol, 1992; Prezoto et al., 2006) and productivity (Ramos & Diniz, 1993; Oliveira et al., 2010) have been performed with this species, but none has evaluated the alteration of habitat by human activity on this aspect. Therefore, this study evaluated the effect of the habitat alteration, particularly by human activity, on the productivity of *P. versicolor.*

Material and Methods

Data collection and field procedures

To evaluate the effects of human activity on colony development, we compared the final productivity of colonies of *P. versicolor* in two habitats in the municipality of Mundo Novo in the state of Mato Grosso do Sul, Brazil.

We collected 10 abandoned nests in each habitat during April 2012 to August 2012. All these colonies were monitored weekly to determine the end of the colony cycle. We used only nests that reached the decline stage as defined by Jeanne (1972), which showed widespread presence of empty cells in the comb and nest abandonment.

Following the parameters proposed by Michelutti et al. (2013), productivity was measured from the number of cells constructed, number of adults produced, and dry mass of the nest. We also measured the diameter of the petiole, proportion of productive cells and number of generations in each cell for estimated productivity of the central and peripheral cells.

By counting the number of layers of meconium in each cell of the comb was estimated the number of adults produced. This meconium layer is formed on the floor of the cells that produced adults, since just before pupation, when the last-instar larva eliminates feces (Gobbi & Zucchi, 1985; Giannotti, 1999). Each cell of the comb was sectioned with the aid of tweezers to extract the meconium layer and for analysis of the nest dry mass, the nest was weighed on a precision balance according Michelutti et al. (2013).

Areas of study

We compared the degree of alteration of the environment by human activity, according to the parameters suggested by Michelutti et al. (2013).

Thus, the first area (23°93'77"S; 54°33'90"66W), categorized as habitat less altered by human activity, was located 8 km from the urban perimeter, and has little human activity. This area consists of rural properties with grassland, agricultural crops and forest fragments predominating, associated with rivers and streams.

The second area (23°93'79"S; 54°29'47"W), categorized as habitat more altered by human activity, contains numerous buildings of wood or brick, most of them inhabited. There is intense movement of people, and mainly areas of pavement and grass lawn adjacent to the buildings.

Statistical analyses

The t-test for independent samples was used to evaluate possible differences between the colony productivity parameters of the two populations. We applied a Pearson correlation analysis to compare the number of cells constructed with the number of adults produced, to evaluate possible differences in strategies of comb use between the two populations. We also estimated the correlation between the diameter of the petiole and the nest dry mass. For all analyses was used the Software Systat 11 and the variable was considered when the resulting regression coefficient was significant at the 0.05 level.

Results

In the habitat more altered by human activity, 60% of the nests occurred on human constructions and 40% on trees. In the habitat less altered by human activity, 100% of the nests were in trees, even if buildings were nearby.

For the 10 nests in the habitat less altered by human activity, the mean values (±SE, n=10) were: 411.30 ± 123.27

cells constructed, 493.90 ± 213.03 adults produced, nest dry mass of 9.50 ± 4.28 g, 90.20 ± 10.87% of cells were productive, 24.76 ± 14.90% of cells were reused, and 1.16 ± 0.26 adults produced per cell (Table 1). In this environment we found a significant positive correlation between the number of adults produced and the number of cells constructed (r = 0.86, p <0.01, n = 10) (Fig. 1). The correlation between the diameter of the petiole and the nest dry mass was also significantly positive (r = 0.75, P = 0.01, n = 10) (Fig. 2).

For the 10 nests in the habitat more altered by human activity, the mean values (±SE, n=10) were: 203.00 ± 44.79 cells constructed, 200.90 ± 65.69 adults produced, nest dry mass of 3.17 ± 0.99 g, 69.23 ± 17.18% of cells productive, 23.19 ± 15.95% of cells reused and 1.02 ± 0.34 adults produced per cell (Table 1). In this environment there was no significant correlation between the number of adults produced and the number of cells constructed (r = 0.30, p = 0.40, n = 10) (Fig. 1), or between the diameter of the petiole and the nest dry mass (r = 0.33, p = 0.36, n = 10) (Fig. 2).

The results of t-tests showed that the productivity in the habitat less altered was significantly higher than in the habitat more altered, with respect to the number of cells constructed (t = -5.02, df = 11.3, p <0.001), number of adults produced (t = -4.15; df = 10.7, p = 0.002), dry mass of nests (t = -5.98, df = 11.4, p <0.001) and proportion of productive cells (t = -3.29, df = 15.5, p = 0.005) (Table 1). However, the proportion of reused cells (t = -1.11, df = 17.6, p = 0.28) and the number of adults produced per cell (t = -0.62, df = 18, p = 0.53) did not differ significantly between the two habitats (Table 1).

Regarding the cell productivity in the habitat less altered, we observed that were older cells and therefore closest to the petiole produced more adults, up to 3 generations of adults (Fig. 3A), and the cells that were farther from the petiole, generally the younger cells, produced fewer individuals. In the habitat more altered a similar pattern was observed with cells producing up to 4 generations (Fig. 3B).

Discussion

In the habitat less altered, the colonies nested only on plants; whereas in the habitat more altered, most nests were sited on human constructions. This is probably related to the availability of the types of substrate in each environment, although sometimes in the habitat less altered, a few buildings were located near the nests.

Sinzato et al. (2011) found significant differences in the nesting sites of *P. versicolor*, in which 99% of the nests were located on human constructions and only 1% on vegetation. These authors found that the nests were located high on the buildings, which afforded greater protection from human interference, weather, and direct sunlight. Thus, there was a high degree of synanthropy. Butignol (1992) and Sinzato & Prezoto (2000) found similar results, demonstrating that the choice of the nesting site is based on the influence of physical climate factors such as temperature and luminosity (Butignol, 1992).

The number of cells constructed and the number of adults produced were larger in the habitat less altered (Table 1). Oliveira et al. (2010) found a positive correlation between the number of cells constructed and the total number of adults produced in colonies of *P. versicolor*; however, all these colonies were evaluated in an urban environment. These authors reported that the colonies produced on average 244 ± 89.5 cells and 171.67 ± 109.94 adults, values close to those found in this study for the habitat more altered.

The colony productivity was determined by the size of the nest in both habitats; larger nests are generally more productive. The results of t-tests (Table 1) showed that most of these parameters differed significantly between the two habitats, and nests in the habitat less altered were more productive. In the study of Michelutti et al. (2013) with *M. consimilis*, the results were similar, and nests in the habitat with less human interference are more productive; however, the proportion of productive cells showed no significant differences, as

Table 1. Comparison of the colony productivity of the social wasp *Polistes versicolor* nesting in two environments. 10 nests were used for each habitat.

Parameters	Habitat less altered by human activity		Habitat more altered by human activity				
	Mean	SE	Mean	SE	T	df	*P*
Cells constructed	411.30	123.27	203.00	44.79	- 5.02	11.3	<0.001
Adults produced	493.90	213.03	200.00	65.69	- 4.15	10.7	0.002
Nests dry mass (g)	9.50	4.28	3.17	0.99	- 5.98	11.4	<0.001
Productive cells (%)	90.20	10.87	69.23	17.18	- 3.29	15.5	0.005
Reused cells (%)	24.76	14.90	23.19	15.95	- 1.11	17.6	0.28
Adults produced/cell	1.16	0.26	1.02	0.34	- 0.62	18	0.53

observed in this study. Therefore, as previously suggested by Michelutti et al. (2013), colony productivity of the social wasps can be affected by the habitat quality, and negatively impacted by human activity.

The habitat more altered has a large concentration of buildings and predominantly grassy vegetation, which suggests a low availability of resources for the colonies. Anjos et al. (1986) and Zanuncio et al. (1991) emphasized that habitat

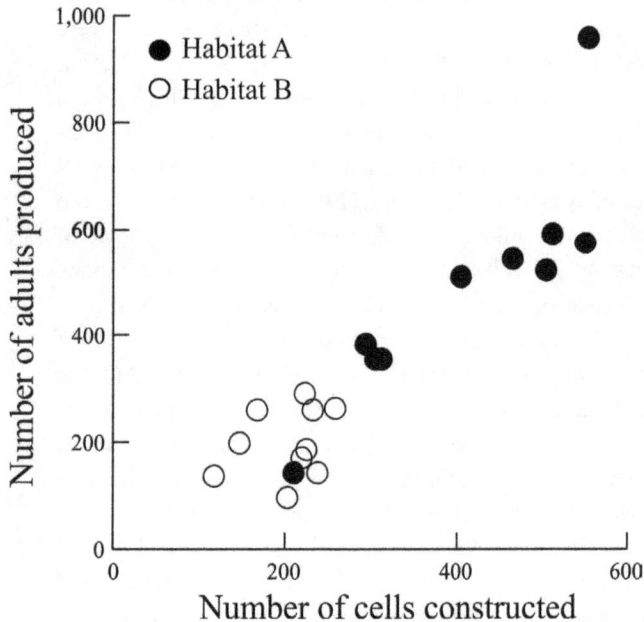

Figure 1. Linear correlation between the number of cells constructed and the number of adults produced in nests of *Polistes versicolor* in two habitats. Habitat A: less altered by human activity (r = 0.86, p <0.01, n = 10) and Habitat B: more altered by human activity (r = 0.30, p = 0.40, n = 10).

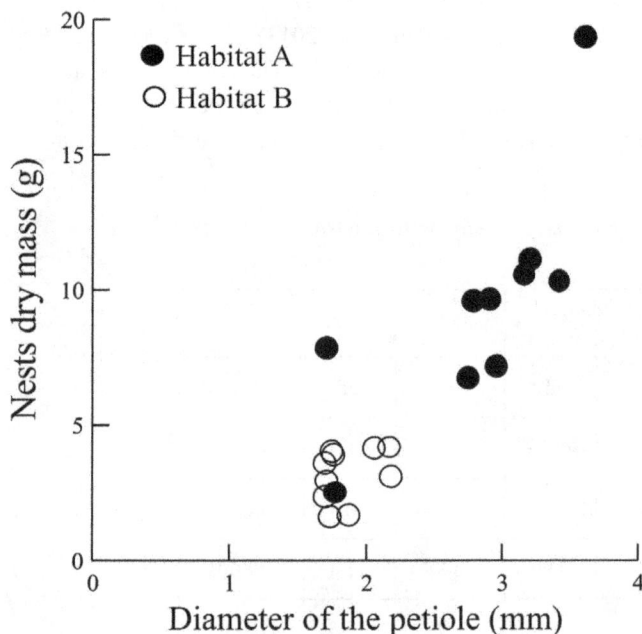

Figure 2. Linear correlation between the nest dry mass and the diameter of the petiole in nests of *Polistes versicolor* in two environments. Habitat A: less altered by human activity (r = 0.75, P <0.01, n = 10) and Habitat B: more altered by human activity (r = 0.33, p = 0.36, n = 10).

quality contributes to lower productivity in degraded environments, which may have limited resources available during unfavorable periods, especially those used in feeding immatures, such as the larvae of other insects. Mead & Pratte (2002) demonstrated that populations of *P. dominula* in situations that differed in prey availability also differed in colony growth and in the rate of production of offspring.

However, Judd (1998) and McGlynn (2012) suggested that nesting on human buildings is advantageous to reduce interspecific competition and to protect against predation, especially by vertebrates, since these factors are less intense than in preserved natural environments. Moreover, human structures can provide greater protection against variations of climatic factors (Michelutti et al., 2013).

The proportion of productive cells was lower in the habitat more altered, and the maximum number of generations was higher (Figs. 1 and 2). Ramos & Diniz (1993) and Oliveira et al. (2010) analyzed the productivity of *P. versicolor* colonies and found cells producing 4 and 6 generations, respectively, values close to those observed here. However, the difference between the 3 or 4 generations produced by the two populations is probably due to different strategies for reuse of the comb, since colonies in the habitat more altered invest more in reusing cells, while in the habitat less altered the colony productivity is ensured by construction of new cells, resulting in large nests with a higher proportion of productive cells (Figs. 1 and 2).

The existence of different strategies in colonies in the two habitats is reinforced by the significant positive correlation between the number of cells constructed and the number of adults produced, and also between the diameter of the petiole and the nest dry mass (Figs. 2 and 3) in the habitat less altered, since in these colonies the population increases concomitantly with the size of the nest. Downing & Jeanne

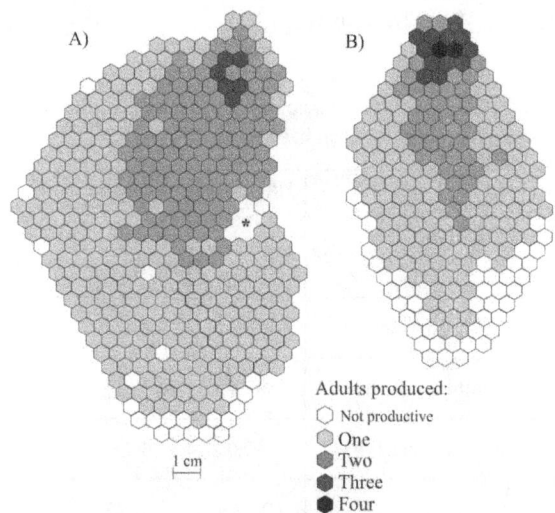

Figure 3. Productivity and number of adults produced per cell in nests of *Polistes versicolor* in two habitats. A) Habitat less altered by human activity. B) Habitat more altered by human activity. NOTE: The asterisk indicates an area of cells that were destroyed at the time of collection.

(1986) stated that in groups of independent foundation, the increase in the diameter of the petiole is associated with the construction of new cells, a relationship also observed by Montagna et al. (2010) for *M. consimilis*. According to Giannotti (1997), the petiole is reinforced with additional pulp and salivary secretions to support the weight of the nest. A nest with more cells can produce more adults.

A likely explanation for these different strategies is that in environments with more human activity, a larger nest may attract more attention and is therefore more likely to be eliminated. The insects may have adapted to this situation, preferring to reuse cells instead of increasing the size of the nest. However, Michelutti et al. (2013), comparing data for *M. consimillis* in environments with features similar to this study, did not observe differences in reuse of the comb cells, although the nests of this species in a forest environment were larger and more productive than those in an urban environment.

On the other hand, Fig. 3 shows a pattern of cell use in which the cells near the petiole are more productive, probably because these cells are older and therefore are used more often. Oliveira et al. (2010) found that cells that are central and closer to the petiole are more productive because they are older and better protected from the pressure of predators, parasites and reproductive conflicts, as also noted by Gobbi et al. (1993). Therefore, both the position and the age of the cells are important in determining the productivity of a cell.

According to Montagna et al. (2010), the central cells of the combs of *M. consimilis* are more productive than the peripheral cells, because they provide better physical conditions for the development of immature individuals; this region has a higher concentration of adults and is less exposed to attack by predators. However, in *M. consimilis* the petiole is central (Montagna et al. 2010), and therefore the more productive cells are also older.

Finally we conclude that colony productivity in *P. versicolor* is significantly higher in habitat less altered by human activity; and regardless of the habitat, the cells central and closer to the petiole are more productive. Furthermore, colonies in habitat more altered by human activity prefer to invest more in reuse of cells than to enlarge the nest structure, perhaps because a smaller nest is less conspicuous. The species probably continues to nest in this habitat because of reduced interspecific competition and greater protection against variations in climatic factors, as suggested by Michulleti et al. (2013).

Acknowledgments

The authors thank Orlando T. Silveira (Museu Paraense Emílio Goeldi) for the identification of the species, and Janet W. Reid (JWR Associates) for the revision of the English text. We are grateful to CAPES for a doctoral fellowship awarded to the second author. WFAJ acknowledges his research grants from CNPq.

References

Abensperg-Traun, M. & Smith, G.T. (2000). How small is too small for small animals? Four terrestrial arthropod species in different-sized remnant woodlands in agricultural Western Australia. Biodiversity and Conservation, 8: 709-726.

Anjos, N., Santos G.P. & Zanuncio, J.C. (1986). Pragas do eucalipto e seu controle. Informe Agropecuário, 12(141): 50-58.

Butignol, C.A. (1992). Observações sobre a bionomia da vespa predadora *Polistes versicolor* (Oliver, 1791) (Hymenoptera: Vespidae) em Florianópolis/SC. Anais da Sociedade Entomológica do Brasil, 21: 201-206.

Carpenter, J.M. & Marques O.M. (2001). Contribuição ao estudo dos vespídeos do Brasil (Insecta, Hymenoptera, Vespoidae, Vespidae). Cruz das Almas, Universidade Federal da Bahia. Publicações Digitais, 2: 147p.

Clapperton, C. (2000). Interhemispheric synchroneity of Marine Oxygen Isotope Stage 2 glacier fluctuations along the American cordilleras transect. Journal of Quaternary Science, 435-468. DOI: 10.1002/1099-1417(200005)15:4<435::AID-JQS552>3.0.CO;2-R

Downing, H.A. & Jeanne, R.L. (1986). Intra- and interspecific variation in nest architecture in the paper wasp *Polistes* (Hymenoptera, Vespidae). Insectes Sociaux, 33: 422-443.

Fowler, H.G. (1983). Human effects on nest suvivorship of urban synanthropic wasps. Urban Ecology, 7: 137-147. DOI:10.1016/0304-4009(83)90032-3

Gamboa, G. J. (1978). Intraspecific defense: advantage of social cooperation among paper wasp foundresses. Science, 199: 1463-1465. DOI:10.1126/science.199.4336.1463

Gamboa, G.J., Austin, J.A. & Monnet, K.M. (2005). Effects of different habitats on the productivity of the native paper wasp *Polistes fuscatus* and the invasive, exotic paper wasp *P. dominulus* (Hymenoptera: Vespidae). Great Lakes Entomology, 38: 170-176.

Giannotti, E. (1997). Biology of the wasp *Polistes (Epicnemius) cinerascens* Saussure (Hymenoptera, Vespidae). Anais da Sociedade Entomológica do Brasil, 26: 61-66.

Giannotti, E. (1999). Arquitetura de ninhos de *Mischocyttarus cerberus styx* Richards, 1940 (Hymenoptera, Vespidae). Revista Brasileira Zoociências, 1: 7-18.

Giannotti, E. & Mansur, C.B. (1993). Dispersion and foundation of new colonies in *Polistes versicolor* (Hymenoptera, Vespidae). Anais da Sociedade Entomologica do Brasil, 22: 307-316.

Gobbi, N. & Zucchi, R. A. (1985). On the ecology of *Polistes versicolor* (Olivier) in southern Brazil (Hymenoptera, Vespidae, Polistini). Phenological Account. Naturalia, 5: 97-104.

Gobbi, N, Fowler, H.G., Chaud-Netto, J., Nazareth, S.L.

(1993). Comparative colony productivity of *Polistes simillimus* and *Polistes versicolor* (Hymenoptera: Vespidae) and the evolution of paragyny in the Polistinae. Zoologische Jahrbucher-Abteilung Fur Allgemeine Zoologie Und Physiologie Der Tiere, 97: 1-5.

Gould, W.P. & Jeanne, R.L. (1984). Polistes wasps (Hymenoptera, Vespidae) as control agents for Lepidopterous cabbage pests. Environmental Entomology, 13: 150-156.

Inagawa, K., Kojima, J., Sayama, K & Tsuchida, K. (2001). Colony productivity of the paper wasp *Polistes snelleni*: comparison between cool-temperate and warm-temperate populations. Insectes Sociaux, 48: 259-265. doi: 10.1007/PL00001775

Jeanne, R.L. (1972). Social biology of Neotropical wasp *Mischocyttarus drewseni*. Bulletin of the Museum of Comparative Zoology - Harvard, 144(3): 63-150.

Judd, T.M. (1998). Defensive behavior of colonies of the paper wasp, *Polistes fuscatus*, against vertebrate predators over the colony cycle. Insectes Sociaux, 45: 197-208. doi: 10.1007/s000400050080

Lima, M.A.P., Lima, J.R. & Prezoto, F. (2000). Levantamento dos gêneros, flutuação das colônias e hábitos de nidificação de vespas sociais (Hymenoptera, Vespidae), no campus da UFJF, Juiz de Fora, MG. Revista Brasileira de Zoociências, 2: 69-80.

McGlynn, P.T. (2012). The ecology of nest movement in social insects. Annual Review of Entomology, 57: 291-308. doi: 10.1146/annurev-ento-120710-100708

Mead, F. & Pratte, M. (2002). Prey supplementation increases productivity in the social wasp *Polistes dominulus* Christ (Hymenoptera Vespidae). Ethology, Ecology and Evolution, 14: 111-128. doi: 10.1080/08927014.2002.9522750

Michelutti, K.B., Montagna, T.S. & Antonialli-Junior, W.F. (2013). Effect of habitat disturbance on colony productivity of the social wasp *Mischocyttarus consimilis* Zikán (Hymenoptera, Vespidae). Sociobiology, 60: 96-100. doi: 10.13102/sociobiology.v60i1

Montagna, T.S., Torres, V.O., Fernandes, W.D. & Antonialli-Junior, W.F. (2010). Nest architecture, colony productivity, and duration of immature stages in a social wasp, *Mischocyttarus consimilis*. Journal of Insect Science, 10: 1-10.

Nadeau, H. & Stamp, N. (2003). Effect of prey quantity and temperature on nest demography of social wasps. Ecological Entomology, 28: 328-339. doi: 10.1046/j.1365-2311.2003.00514.x

New, T.R. (2005). Invertebrate conservation and agricultural ecosystems. Cambridge, UK: Cambridge Univertsity Press. xiii+354p.

Oliveira, S.A., De Castro, M.M. & Prezoto, F. (2010). Foundation pattern, productivity and colony sucess of the paper wasp *Polistes versicolor*. Journal of Insect Science, 10: 125.

doi: 10.1673/031.010.12501

Penna, M.A.H., Goobi, N., Giacomini, H.C. (2007). Comparative productivity of *Mischocyttarus cerberus styx* (Richards, 1940) and *Mischocyttarus cassununga* Saussure (von Ihering, 1903) in an anthropic environment as evaluation for differences in ecological strategies. Revista Brasileira de Zoociências, 9: 205-212.

Prezoto, F., Santos-Prezoto, H.H., Machado, V.L.L., & Zanúncio, J.C. (2006). Prey Captured and Used in Polistes versicolor (Olivier) (Hymenoptera: Vespidae) Nourishment. Neotropical Entomology, 35: 707-709. doi: 10.1590/S1519-566X2006000500021

Prezoto, F., Ribeiro-Júnior, C., Oliveira-Cortes, S.A. & Elisei, T. (2007). Manejo de vespas e marimbondos em ambiente urbano, p. 123-126. In: A. S. Pinto; M. N. Rossi & E. Salmeron (eds.) Manejo de Pragas Urbanas. Piracicaba: CP 2, 208p.

Ramos, F.A. & Diniz, I.R. (1993). Seasonal cycles, survivorship and growth of colonies of *Polistes versicolor* (Hymenoptera, Vespidae) in the urban area of Brasilia, Brazil. Entomologist, 112(3-4): 191-200.

Raupp, M.J., Shrewsbury, P.M., Herms, D. A. (2010). Ecology of herbivorous arthropods in urban landscapes. Annual Review of Entomology, 55: 19-38. doi: 10.1146/annurev-ento-112408-085351

Richards, O.W. (1978). The Social Wasps of the Americas, excluding the Vespinae. London, British Museum, Natural History, 580p

Samways, M.J. (2005). Insect diversity conservation. Cambridge, UK: Cambridge University Press. 342p

Samways, M.J. (2007). Insect Conservation: A synthetic management approach. Annual Review of Entomology, 52: 465-487. doi:10.1146/annurev.ento.52.110405.091317.

Sinzato, D.M.S. & Prezoto, F. (2000). Aspectos comportamentais de fêmeas dominantes e subordinadas de *Polistes versicolor* Olivier,1791 (Hymenoptera, Vespidae) em colônias na fase de fundação. Revista de etologia, 2(2): 121-127.

Sinzato, D.M.S., Andrade, F.R., De Souza, A.R.; Del-Claro, K. & Prezoto, F. (2011). Colony cycle, foundation strategy and nesting biology of a Neotropical paper wasp. Revista Chilena de História Natural, 84: 357-363. doi: 10.4067/S0716-078X2011000300004

Santos, G.M.M. & Gobbi, N. (1998). Nesting habits and colonial productivity of *Polistes canadensis canadensis* (L.) (Hymenoptera, Vespidae) in a Caatinga area, Bahia State, Brazil. Journal of Advanced Zoology, 19: 63-69.

Schowalter, T.D. (2012). Insect responses to major landscape-level disturbance. Annual Review of Entomology, 57: 1-20. doi:10.1146/annurev-ento-120710-100610.

Tibbetts, E.A . & Reeve, H.K. (2003). Benefits of foundress associations in the paper wasp *Polistes dominulus*: increased productivity and survival, but no assurance of fitness returns. Behavioral Ecology, 14: 510-514. doi: 10.1093/beheco/arg037

Zanuncio, J.C., Batista, L.G, Zanuncio, T.V., Vilela, E.F. & Pereira, J.F. (1991). Levantamento e flutuação de lepidópteros associados à eucaliptocultura: VIII - Região de Belo Horizonte, Minas Gerais, junho de 1989 a maio de 1990. Revista Árvore, 15: 83-93.

12

New Records, Including a New Species, of Scuttle Flies (Diptera: Phoridae) Associated with Leaf Cutter Ants (Hymenoptera: Formicidae) in Brazil

R. Henry L. Disney[1] & Marcos A. L. Bragança[2]

1- University of Cambridge, Department of Zoology, Cambridge, United Kingdom
2- Universidade Federal do Tocantins, Palmas, Tocantins, Brazil

Keywords
new species, *Apterophora bragancai*
Disney new species, Taxonomy

Corresponding author
Dr Henry L. Disney
Department of Zoology,
University of Cambridge
Downing Street, CB2 3EJ, U. K.
E-Mail: rhld2@hermes.cam.ac.uk

Abstract
Among scuttle flies caught at colonies of leaf cutter ants were *Apterophora bragancai* Disney new species, and new host records for other species.

Introduction

During 2012 MALB and his colleagues Cliver Gomes, Leandro Silva, Hendria Martins and Marcos Teixeira collected Diptera attacking or hovering in the vicinity of leaf cutter ants. The scuttle flies (Phoridae) were preserved in 70% ethanol and sent to RHLD for slide mounting and identification. The ants were identified by the Brazilian team. The following species were obtained with the ant hosts indicated. The specimens are deposited in the Collection of the Museu de Zoologia, Universidade de São Paulo, Brazil (MZSP) and the University of Cambridge Museum of Zoology (UCMZ).

Genus *Allochaeta* Borgmeier

The genotype is *A. excedens* described from the female only, from Petropolis in Brazil (Borgmeier, 1924). In the same paper *A. metatarsalis* was described from the male only, from Blumenau (Santa Catharina). Later, from Petropolis, *A. longiciliata*, known only from the female and *A. propinqua* known only from the male were then added (Borgmeier, 1926). However, subsequently these two were procured *in copula*, so *A. propinqua* was then synonymised with *A. longiciliata* (Borgmeier, 1928). This served to emphasize the sexual dimorphism in this genus. Subsequently *A. senex,* known from the female only was added from São Paulo State, Campinas (Borgmeier & Prado, 1975) and *A. wallerae* Disney known from the male only was added from Mexico (Disney & Bragança, 2000) along with possible males of *A. excedens* from Viçosa, MG, Brazil. In the present study further examples of these males were obtained (see below). Borgmeier (1928) reported *A. longiciliata* soliciting food from its ant host, *Acromyrmex muticinodus.* Other reported hosts are the same species with *Acromyrmex niger* Smith (Borgmeier, 1928), *A. muticinodus* with *Ac. niger* (Borgmeier, 1928), the putative males of *A. excedens* with *Atta laevigata* Smith and *A. wallerae* with *Atta cephalotes* (Linn.) (Disney & Bragança, 2000); and *A. metatarsalis* from a colony of the termite *Nasutitermes rippertii* (Rambur) (Borgmeier, 1924). In the present study the following further records of the putative males of *A. excedens* were obtained (see below).

Material examined. 3 males, Minas Gerais State, Viçosa, vii.2012, at *Atta sexdens* (Linn.) (Marcos Bragança, TO-137). 4 males, Minas Gerais State, Florestal, x.2012, at *A. sexdens* (Cliver Gomes, TO-128, TO-129).

This species was previously reported with *Atta laevigata* (see above).

Genus *Apterophora* Brues

This genus closely resembles the large genus *Puliciphora*. The flightless females differ only in having a longer proboscis. The males differ in having modified front tibiae. However, the form of these modifications ranges from a clearly distinct excavation of the ventral margin to an apical ventral projection. In the genus *Puliciphora* modifications of the legs are not uncommon. Thus several have swollen front basitarsi or these are otherwise modified (e.g. see *Puliciphora* species B below). This genus may therefore eventually prove to be merely a subset of *Puliciphora*.

The species of *Apterophora* are discussed by Prado (1976), allowing identification of males. In our present state of knowledge, females can only be named when associated with their males.

Apterophora attophila Borgmeier

Material examined. 3 males, Minas Gerais State, Florestal, x.2012, at *Atta sexdens* (Linn.) (Cliver Gomes, TO-126, TO-127, TO-129). 1 male, Minas Gerais State, Viçosa, vii.2012, at *A. sexdens* (Marcos Bragança, TO-137).

The type series was recorded with *A. sexdens* and Prado (1976) added further records with the same ant along with records with *Eciton burchelli* (Westwood) and *Solenopsis geminata* (F.); and also with the termite *Nasutiterms sp.*

Apterophora borgmeieri Prado

Material examined. 1 male, Minas Gerais State, Viçosa, vii.2012, at *Atta sexdens* (Linn.) (Marcos Bragança, TO-136).

Apterophora bragancai Disney new species

The fore tibia closely resembles that of *A. caliginosa* Brues, but the longer anal tube and left hypandrial lobe of the hypopygium immediately distinguish the new species.

Male. Postpedicels brown, tapered, 0.16 mm long and with about a dozen subcutaneous pit sensilla. Palps brown, 0.24 mm long and 0.06 mm wide, with apical bristle 0.07 mm long. Thorax brown. Notopleuron with 3 bristles. Scutellum with and anterior pair of fine hairs and a posterior pair of bristles. Abdomen with brown tergites and paler brown venter. Hypopygium as Fig. 1. Apart from brown patches in mid coxae the legs are mainly yellow apart from the distal halves of the femora shading to brown. Front tibia as Fig. 2. Front tarsus

with posterodorsal hair palisades on segments 1 to 4. Hairs below basal half of hind femur shorter than hairs of anteroventral row of outer half. Wing 2.46 mm long. Costal thicker than vein 3. Costal index 0.65. Costal ratios 0.55 : 1. Haltere knob brown.

Material examined. Holotype male, BRAZIL, Minas Gerais State, Viçosa, vii.2012, at *Atta sexdens* (Linn.) (Marcos Bragança, TO-136) (5-162, MZSP). This fly was captured when it was flying over the ants along the foraging trail.

Genus *Mymosicarius* Borgmeier

The females are keyed by Disney, Elizalde & Folgarait (2006).

Mymosicarius grandicornis Borgmeier

Material examined. 4 females, Minas Gerais State, Viçosa, 14.iv.2012, at *Atta bisphaerica* Forel (M. Bragança, TO-132, TO-133), 1 female, same locality and ant host, 28.vi.2012 (M. Bragança, TO-135), 4 females, same locality and host ant, ix, 2012 (Hendria Martins, TO-140, TO-141); 1 female, (Jalapão) Mateiros, 3.v.2012, at *Atta sexdens* (Linn.) (Leandro Silva, TO-131). 11 females, Minas Gerais State, Florestal, x.2012, at *A. sexdens* (Cliver Gomes, TO-125-128, TO-130).

These records confirm previous records with these two ant hosts, which has also been recorded with *A. laevigata* (Disney, Elizalde & Folgarait (2006).

Genus *Puliciphora* Dahl

More than one hundred species are known in this cosmopolitan genus. Species recognition has been based on the flightless females in the first instance and for most regions keys only exist for the females, those of the Neotropical Region being keyed by Disney (2003). The boundaries of the genus are still not settled. The males of the following two species were obtained. These flies were captured when they were flying over the ants in a disturbed nest.

Puliciphora species A

Male. Frons much wider than long and with 10 bristles level with or below anterior ocellus (4 supra antennals and 2-4 bristles). Postpedicels brown, 1.0-1.1 mm wide and lacking SPS vesicles. Palps brown, 1.4 mm long, 0.3 mm wide and the longest bristles 0.9 mm long. Thorax brown, but tending to be paler on sides. Notopleuron with 3 bristles. Scutellum with and anterior pair of small hairs and a posterior pair of bristles. Abdomial tergites brown and venter gray. Hypopygium as Fig. 3, notably with a long anal tube relative to the length of the epandrium. Legs pale dusky yellow. Front tarsus with a posterodorsal hair palisade on segments 1-4. The 3 hairs below basal half of hind femur about as long as those of an-teroventral row of outer half but clearly more robust. Wings 1.1 mm long. Costa about as wide as vein 3. Costal index 0.6. Costal ratios 0.37 : 1. Haltere knob brownish gray.

Material examined.1 male, Minas Gerais State, Viçosa, vi.2012, at *Atta sexdens* (Linn.) (M. Bragança, TO-138, MZSP).

Puliciphora species B

Male. Frons brown, clearly wider than long and with 10 bristles below anterior ocellus. Subglobose postpedicels light brown, 0.1 mm wide, with numerous SPS vesicles. Palps yellowish gray, 0.1-0.12 mm long, 0.04 mm wide and longest bristle 0.06-0.07 mm long. Thorax brown, but paler at sides, with 3 bristles on notopleuron. Scutellum with an anterior pair of small hairs and a posterior pair of bristles. Abdominal terrgites brown and venter gray. Hypopygium as Fig 4. Legs with slightly dusky but pale yellow femora and tibiae but tarsi all pale. Front tarsus with posterodorsal hair palisades on segments 1-4 and basitarsus as Fig. 5. Hind femur with hairs below basal half shorter than those of anteroventral row of outer half. Wings 1.5 mm long. Costa about as wide as vein 3. Costal index 0.55. Costal ratios 0.7 : 1. Haltere with pale stem and brown knob.

Material examined.1 male, Minas Gerais State, Viçosa, vi.2012, at *Atta sexdens* (Linn.) (M. Bragança, TO-138, MZSP).

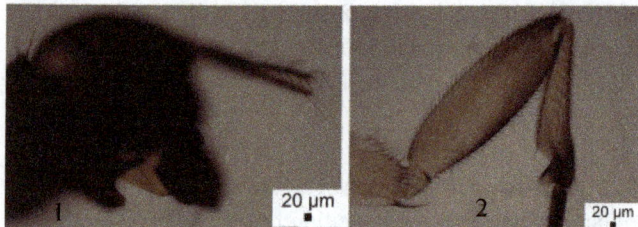

Figs 1-2. *Apterophora bragancai* male.1, left face of hypopygium; 2, front tibia.

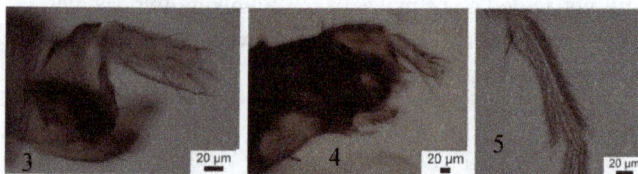

Figs 3-5. *Puliciphora* males. 3, species A, left face of hypopygium; 4-5, species B; 4, left face of hypopygium; 5, front basitarsus.

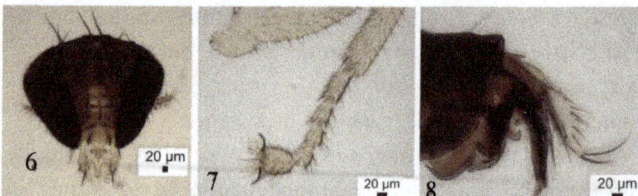

Figs 6-8. Problem damaged specimen, male. 6, frontal view of head; 7, front tarsus; 8, left face of hypopygium.

A damaged specimen

A distinctive male that has lost both its wings appears to be an undescribed species of the poorly defined genus *Macrocerides* Borgmeier or a similar genus subsequently distinguished from this genus. An intact specimen will be required to settle thee identity of this species.

Male. Frons as Fig. 6. Postpedicels yellowish gray brown, 0.14 mm long and 0.25 mm greatest breadth. Palps yellow, 0.12 mm long, 0.03 mm wide and longest bristle 0.05 mm long. Thorax brown. Notopleuron with 2 bristles. Mesopleuron bare. Scutellum with an anterior pair of small hairs and a posterior pair of bristles. Abdominal tergites and venter brown. Hypopygium as Fig. 7. Legs apart from brown on mid coxae yellow. Front tarsus as Fig. 8. Hairs below basal half of hind femur shorter and finer than those of anteroventral row of outer half.

Material examined. 1 male, Minas Gerais State, Florestal, x.2012, at *Trachymyrex* sp. (Cliver Gomes, TO-124, MZSP). This fly was captured when it was flying over the ants.

Acknowledgments

MALB thanks to CNPq and Secretaria Estadual de Desenvolvimento Econômico, Ciência, Tecnologia e Inovação do Tocantins, for the financial support to collect the phorid flies in several localities. RHLD's studies of Phoridae are currently supported by grants from the Balfour-Browne Trust Fund (University of Cambridge). We are grateful to Terezinha Della Lucia, Marco Antonio Oliveira for logistical support in localities of collect. We also thank to Cliver Gomes, Leandro Silva and Hendria Martins for helping to collect the flies.

References

Borgmeier T (1924) Novos generos e especies de Phorideos do Brasil. Boletim do Museu Nacional do Rio de Janeiro, 1: 167-202.

Borgmeier T (1926) Phorideos novos ou pouco conhecidos do Brasil. Boletim do Museu Nacional do Rio de Janeiro, 2(5), 39-52.

Borgmeier T (1928). Nota prévia sobre alguns Phorideos que parasitam formigas cortadeiras dos generos *Atta* e *Acromyrmex*. Boletim Biológico Laboratorio de Parasitologia, São Paulo, 14: 119-26.

Borgmeier T, Prado, AP do (1975) New or little-known Neotropical Phorid flies, with description of eight new genera (Diptera, Phoridae). Studia Entomologica, 18: 3-90.

Disney, R. H. L., (2003). Seven new species of New World *Puliciphora* Dahl (Diptera: Phoridae) with a new key to the Neotropical species. Zootaxa, 162: 1-22.

Disney, R. H. L. & Bragança, A. L., (2000). Two new species of Phoridae (Diptera) associated with leaf-cutter ants (Hymenoptera: Formicidae). Sociobiology, 36: 33-39.

Disney, R. H. L., Elizalde, L. & Folgarait, P. J. (2006). New species and revision of *Myrmosicarius* (Diptera: Phoridae) that parasitize leaf-cutter ants (Hymenoptera: Formicidae). Sociobiology, 47: 771-809.

Prado, AP do (1976) Records and descriptions of phorid flies, mainly of the Neotropical Region (Diptera; Phoridae). Studia Entomologica, 19: 561-609.

13

Termite infestation in historical buildings and residences in the semiarid region of Brazil

AP Mello, BG Costa, AC Silva, AMB Silva, MA Bezerra-Gusmão

Universidade Estadual da Paraíba, Campina Grande, Paraiba, Brazil.

Keywords
Infestation, Isoptera, urban environment, damage, cultural patrimony.

Corresponding author
Antonio Paulino de Mello
Laboratório de Ecologia de Térmitas
Departamento de Biologia
Universidade Estadual da Paraíba
Campina Grande, PB, Brazil
58109-753
E-Mail: antonio.pmello@hotmail.com

Abstract

This study evaluated termite infestations in historical buildings (HB) and residences (RB) in five cities in the semiarid region of Brazil (A1: Fagundes; A2: Pocinhos; A3: Alagoa Grande; A4: Areia; A5: Bananeiras). Eighty-nine percent of infestation of historical buildings and 62% of residential buildings were caused by nine species of termites, belonging to six genera and three families (Kalotermitidae, Rhinotermitidae and Termitidae). We observed greater richness in A3, A4 and A5 (*Amitermes amifer, Heterotermes tenuis, H. sulcatus, H. longiceps, Neotermes* sp. and *Nasutitermes corniger*), relating to A1 and A2 (*Nasutitermes* sp.1, *Nasutitermes* sp.2, *N. corniger* and *Microcerotermes strunckii*). *N. corniger* and dry wood termite was responsible for 66.9% and 33.1% for infestations in HB and 29.7% and 72.3% in the RB, respectively. The economic impact for these infestations was estimated at 609,956.03 USD. No correlation between the quantity of infestations and the age of the buildings was found; nor was there a correlation between the humidity and the number of infestations. However, the infestations of the RB correlated with humidity. Possibly the absence of preventative mechanisms of infestation control, along with the loss of natural habitat of these insects caused by urban expansion, explain the high indicators of registered infestations.

Introduction

Termites are recognized as insects that interact with the urban ecosystem in the most complex way (Fontes, 1995). With approximately 3,105 described species in the world (Krishna et al., 2013), of the registered 534 species in Brazil, 18 are considered urban pests (Constantino, 2005). Meanwhile, the disorganized growth of urban areas has created an impact on natural ecosystems, changing the population dynamics and resulting in the adaptation of the new species to urban conditions, which has turned them into pest species (Milano & Fontes, 2002).

Termite infestations have become increasingly more frequent inside the urban perimeter. In tropical countries and coastal regions of continents, Harris (1971) reports infestations caused mainly by *Reticulitermes* spp. (Europe and United States), *Coptotermes* spp. (Australia, Asia and part of Africa) and *Cryptotermes* spp. (Colombia, and the vast majority of South American countries). In South America the infestations are caused mainly by *Coptotermes gestroi*

(Wasmann) (Rhinotermitidae), *Heterotermes tenuis* (Hagen), *H. longiceps* (Snyder) (Rhinotermitidae), *Nasutitermes corniger* (Motschulsky) (Termitidae) and *Cryptotermes brevis* (Walker) (Kalotermitidae), the last of which is considered the main species pest termite in urban regions (Fontes, 1995; Torales et al., 1997; Bandeira et al., 1998; Eleotério & Berti-Filho, 2000; Constantino & Dianese, 2001; Vasconcellos et al., 2002; Oliveira et al., 2006).

The occurrence of termites in urban areas is seen as a complex that extrapolates the limit of economic damage to human populations. These insects can reach high levels of infestation in a short period of time, causing damage to the artistic, historic and cultural patrimony (Bandeira et al., 1989; Fontes, 1995). Currently, the annual cost in building repairs estimated in urban areas worldwide for treatment and replacements is in the order of the 40 billion USD (Rust & Su, 2012).

In Brazil, research has generated only limited discussion about the pest potential and distribution of these species in specific regions of the country, mainly the Southeast and

Center-West (Lelis, 1995; Fontes, 1999; Eleotério & Berti-Filho, 2000; Constantino & Dianese, 2001; Ferraz & Cancello, 2001; Fontes & Milano, 2002 and Costa et al., 2009). These studies revealed the potential of *N. corniger* and *C. brevis* as pest species. However, little is known about the urban fauna and termite control methods infestations in the Northeastern Region of Brazil (Bandeira et al., 1989; Bandeira et al., 1998; Matias et al., 2006; Albuquerque et al., 2012), especially in the state of Paraíba, with focus on the studies of Bandeira et al. (1998) and Vasconcellos et al. (2002).

The objective of this study was to survey the termite occurrences in historical and residential buildings in the semiarid region of the Northeastern Region, assessing the richness of species, examining the determining factors for infestation, as well as estimating the economic damage caused by infestations.

Materials and Methods

The study was developed in five cities in the State of Paraíba, in Brazil's Northeast semiarid region. The collecting occurred between 2010 and 2012 in the city of Fagundes - 11,405 inhabitants (A1) (7°20'45,56"S/35°47'51,13"W); Pocinhos - 17,032 inhabitants (A2) (7°04'36,26"S/36°03'40,22"W); Alagoa Grande - 28,479 inhabitants (A3) (07°05'20"S/35°38'36"W); Areia - 23,829 inhabitants (A4) (06° 57'46"S/35°41'31"W); and Bananeiras - 21,851 inhabitants (A5) (06° 45' 00"S/35° 37' 58"W) (IBGE, 2010).

The cities of Fagundes and Pocinhos present arid climatic characteristics, with an average temperature between 28°C and 30°C and an irregular precipitation pattern (Inmet, 2012), while the cities of Alagoa Grande, Areia and Bananeiras are located in the region of "Brejo de altitude", with milder climatic conditions (10°C to 28°C), the yearly precipitation between 900 and 1300 mm (Inmet, 2012). The municipalities contribute to the history of the state of Paraiba and of Brazil, and their houses are listed in the cultural patrimony of the cities, especially the city of Areia, which is listed in the national historical patrimony.

The general characteristics of the historical buildings present in the municipalities were analyzed, beyond 50 residences. These were chosen randomly through a raffle, placed within the 10 streets that border the central area, totaling a sample of 46 historical buildings and 250 residential buildings. The ages of the historical buildings vary from 40 to 150 years, and the residential buildings from four to 102 years.

In each building, the existence of key factors indicate interaction with the termites were analyzed, such as the presence of wooden furniture, the structural conditions of the building, coverage, beyond the existence of leaks and internal indicators the unit obtained from the thermo-hygrometer in each compartment of the buildings. The number of infestations per building was considered based on quantifying the number of pieces of wood that contained termites.

The termites were collected and conditioned in 80% alcohol, identified through specialized literature (Constantino, 1999) and confirmed by researchers Dr. Luiz Roberto Fontes and Dr. Alexandre Vasconcellos. In some cases, dry wood termites have not been identified in this study because they colonize the interior furniture and wooden structures which were not authorized to be broken down into pieces that made up the interior of the historic and residential buildings, making it difficult to collect soldiers. In these cases, we considered the presence of insect infestations (soldiers and workers) and the characteristic signs of their presence, as the existence of holes in the wooden parts for which they are expelled to be as fecal matter characteristic of dry wood termites.

The economic damage caused by infestation was calculated from an estimated market price for replacement and repair of parts and damaged structures. The samples found were deposited in the Entomological Collection of Termitology Laboratory in the Biology Department of State University of Paraíba.

Pearson correlation tests were performed using an array with information about the number of infestations by age and values of internal moisture in buildings. The frequency of infestations in wooden structures was associated with the presence of termites. The frequency of occurrence of the species in historical and residential buildings was counted according to the number of times a species was recorded. The chi-square test (χ^2) was used to check for differences in the frequencies of infestation observed and expected by age class and type of building.

Results

Infestations were diagnosed in 89% of the historical buildings and 62% of the residential buildings, caused by nine species of termites, belonging to six genera and three families (Kalotermitidae, Rhinotermitidae and Termitidae) (Table. 1).

Fagundes and Pocinhos present the least generic richness when compared with the municipalities' termite fauna in Alagoa Grande, Areia and Bananeiras (Table. 1). *N. corniger* and dry wood termites were present in real estate in the five inspected cities, with evidence such as damaged cabinets, wooden benches, tables, paintings, doors, windows and roofing (Table. 1).

The greater frequency of infestation by termites was registered in the roofing, followed by doors, windows and internal furniture in historical buildings (cabinets, secretaries, tables and painting frames) and in the residential (cabinets, tables and chairs) (Table. 2).

The economic damage caused by infestations was estimated in 609,956.02 USD, for repairs and/or replacements in damaged structures, being 555,149.59 USD for historical buildings and 54,806.43 USD for residential buildings.

Table 1 - Frequency the termite's species in historical and residential buildings in the municipalities of Fagundes (A1), Pocinhos (A2), Alagoa Grande (A3), Areia (A4) and Bananeiras (A5), located in the semiarid region of Brazil.

| Family | Species | Municipalities | | | | | Frequency (%) |
		A1	A2	A3	A4	A5	
Termitidae	*Amitermes* sp.1			x			5.64
	Nasutiternes sp.1	x					6.65
	Nasutitermes sp.2	x					1.4
	Nasutiternes corniger (Motschulsy)	x	x	x	x	x	39.4
Rhinotermidiae	*Microcerotermes strunckii* (Sörensen)		x				1.4
	Heterotermes longiceps (Snyder)				x		1.4
	Heterotermes tenuis (Hagen)		x				1.4
	Heterotermes sulcatus (Mathews)		x		x		1.4
Kalotermitidae	*Neotermes* sp.		x				1.4
	Dry wood termite *	x	x	x	x	x	40
Total	**09**	**04**	**04**	**04**	**04**	**02**	**100**

* Was not treated as a species.

Table 2 – Wooden parts and/or structures with termite's infestations in historical (HB) and residential (RB) buildings in the municipalities of de Fagundes (A1), Pocinhos (A2), Alagoa Grande (A3), Areia (A4) and Bananeiras (A5), located in the semiarid region of Brazil. (HB – historical buildings; RB – residential buildings).

| Structures and/or parts | % Infestation | |
	HB	RB
Cabinet	6.1	6.43
Chest	0.39	0.25
Wooden benches	6.84	-
Secretaries	3.33	2.23
Bed	0.59	0.74
Chair	1.18	9.65
Shelf	4.73	5.45
Stairs	0.59	6.68
Window	6.08	4.7
Table	5.1	15.1
Door	10.39	4.21
Roof lining	0.98	1.24
Framing	25.1	-
Roofing	28.6	43.32
Total	**100**	**100**

Problems related to humidity, which varied from 45% to 98%, and infiltrations in the walls and roofs were observed in 80% of the buildings and in 64.8% of the residences. Meanwhile, there was no correlation between the humidity and the number of infestations in historic buildings (r = 0.23, p = 0.12), while a correlation was found between these parameters for residences (r = 0.16, p< 0.05).

No correlation was found between the number of infestations and the age of the historic buildings (r = -0.17,

p = 0.26) and residential buildings (r = 0.12, p = 0.06). No significant difference between the observed frequency and the expected frequency of infestation was observed, concentrated by age class in both types of buildings studied (historical: χ^2 = 0.13, p = 0.99 and residential: χ^2 = 3.72 p = 0.81). However, the historic buildings up to 40 years of age represent only about 15.22% of total observed and present the greatest amount of infestation (36.5%), while in residential buildings, with ages between 21 and 40 construction years, corresponding to 33.2% of total observed, present the largest quantities of infestation (46.2%).

Discussion

The richness of termite species found in the studied areas was considered low (nine), when compared to studies with similar objectives to this study that totaled the registration of 22 species (Vasconcellos et al., 2002; Eleotério & Berti-Filho, 2000; Constantino & Dianese, 2001). Nevertheless, there is a similarity in richness of the urban fauna of termites with the observations of Albuquerque et al. (2012) in the city of Recife in the Northeast of Brazil.

The absence of ventilation mechanisms and inside temperature control (be it artificial or natural) through adequate constructive techniques, combined with precarious preservation conditions, with countless infiltrations in the walls and the roof, all favor the increase in internal humidity of real estate. These conditions combine to create a favorable environment for the nesting of new colonies and can explain the elevated infestation rate (Lelis, 1999; Hedges, 1998).

Despite the impossibility of specific identification of dry wood termites, we believe that the high number of attacks by dry wood termites verified may be related to the fact that some species of the genus *Cryptotermes* submit acclimation capacity and high level of responses to orphans, especially

Cryptotermes brevis, considered a major pest of urban Brazil (McMahan, 1962; Steward, 1983; Edwards & Mill, 1986; Bacchus, 1987; Fontes, 1995). Moreover, the low durability and natural resistance of woods used in furniture may also be contributing to this frequency in the domestic environment (Spear, 1970; Silva et al., 2004).

The elevated frequency of *N. corniger* in the study corroborates the importance of the species as one of main urban pests in Brazil (Eleotério & Berti-Filho, 2000; Vasconcellos et al., 2002; Costa et al., 2009; Albuquerque et al., 2012). Some researchers theorize that this species' success in colonizing in urban space is due to its great biological plasticity, mainly related to its reproduction mechanisms (Thorne & Noirot, 1982), its feeding habits (Bustamante & Martius, 1998; Vasconcellos & Bandeira, 2000) and aspects related to its nesting (Levings & Adams, 1984; Vasconcellos, et al., 2002).

Irregular occurrences of *Amitermes* sp., *Heterotermes longiceps, H. sulcatus, H. tenuis, Microcerotermes strunckii* and *Neotermes* sp. were also observed by Constantino and Dianese (2001); Milano and Fontes (2002); Vasconcellos et al. (2002); Costa et al. (2009); Albuquerque et al. (2012). We believe that the low frequency of these species in the urban environment is a result of lack of planning and disordered growth of urbanization that stimulate adjustment factors and the occurrence of species that previously occurred only in natural areas, as well as the product from the competition with well-established species in this environment. In 1989, Bandeira et al. highlighted the importance of *Heterotermes* as a potential pest in Brazil. These termites have been reported for being responsible for attacks on books, journals, painting frames and wooden shelves in an urban area (Pizano & Fontes, 1986; Mill, 1991; Lelis, 1995).

The infestation indicators observed in the buildings' wooden roofing and the reported attacks on doors and windows are possibly related to their state of conservation and exposure of these parts during the termites' swarm. In general, the buildings present precarious structural conditions, with roof infiltrations, and in the case of historical buildings, without performing repairs and preventative restorations over time. According to Fontes (1998), wooden sections facing the external side of buildings that are not directly exposed to weatherproofing offer good conditions for termite establishment. The presence of infiltrations in the roof promotes the elevation of environmental humidity, facilitating the proliferation of biologic agents such as fungi, organisms that facilitate the installment of termite colonies (Bandeira, 1989; Torales, 1997). However, even though the buildings present different levels of infestation, it was observed that the existence of reforms in some of the historic buildings and homes older than 50 years showed a trend toward a decrease in the number the infestations, and thus influenced the lack of correlation between age of the property and the number of reported infestations.

Importantly, even with problems of infiltration and considerable internal moisture content are responsible in the buildings do not have protocols or specific control techniques to combat and prevent infestations, they used from the indiscriminate use of insecticides of different classes resulting in failed attempts to control. No reports were found in the literature on techniques for termite control for this type of construction in Paraíba. However, for some historic buildings, an attempt to control the termites may have interfered in our findings, even in the short term, resulting in the absence of correlation between the humidity and the number of infestations.

Even though resistance and natural durability tests were not performed on the woods used for the manufacturing of the observed structures that were infested with termites, it is possible that the termite attacks on the furniture in buildings is due to the weak durability of the wood used in the furniture, the precarious structural conditions of the real estate, and the absence of repair activities and prevention by the responsible parties. These factors associated to infiltration and humidity problems are determinant in the increase in likelihood of infestations (Bandeira, 1989; Torales, 1997; Silva et al., 2004; Milano & Fontes, 2002).

Attacks on historical and residential buildings in the arid zone of Paraíba, northeastern Brazil, were caused by nine species of termites, causing severe economic consequences and damaging the historical and cultural patrimony of the inspected cities. We found that the problems of infiltration and indoor humidity coupled with few or no repairs to the structure of the buildings as well as the absence of diagnosis and prevention of termite infestation accounted for a part of the high percentage of the attacks on the buildings. However, based on the findings of the frequency of occurrence and the types of damage caused, it was discovered that only dry wood termites and *N. corniger* were present as pest species in the cities we analyzed.

Acknowledgements

We would like to thank Kátia Cristina Ferreira da Silva for her support with field work. We also thank Dr. Luiz Roberto Fontes and Dr. Alexandre Vasconcellos for the confirmation of the species. To the National Council of Scientific and Technological Development, Brazil, and the State University of Paraíba, Brazil we express our appreciation for their financial support.

References

Albuquerque, A.C., Matias, G.R.R.S., Couto, A.A.V.O., Oliveira, M.A.P., Vasconcellos, A. (2012). Urban Termites of Recife, Northeast Brazil (Isoptera). Sociobiology, 59:1-6.

Bacchus, S. (1987). A taxonomic and biometric study of the genus *Cryptotermes* (Isoptera, Kalotermitidae). Tropical Pest Bulletin, 7:1 - 91.

Bandeira, A. G., Gomes, J. I., Lisboa, P. L. B., Souza, P. C. (1989). Insetos pragas de madeira de edificações em Belém, Pará. Embrapa/cpatu. Bol. de pesquisa, 4: 1-25.

Bandeira, A. G., Miranda, C. S., Vasconcellos, A. (1998). Danos causados por cupins em João Pessoa, Paraíba - Brasil. In: L. R. Fontes, & E. Berti-Filho, (Eds.). Cupins: O desafio do conhecimento (pp.75-85). Piracicaba: FEALQ.

Bustamente, N.C.R., & Martius, C. (1998). Nutritional preferences of wood-feeding termites inhabiting floodplain forest of the Amazon river, Brazil. Acta Amazonica, 28: 301-307.

Constantino, R. (1999). Chave ilustrada para identificação dos gêneros de cupins (Insecta: Isoptera) que ocorrem no Brasil. Papéis Avulsos de Zool., 40: 387-448.

Constantino, R. & Dianese, E. C. (2001). The urban termite fauna of Brasília, Brazil. Sociobiology, 38 (3): 323-326.

Constantino, R. (2005). Padrões de diversidade e endemismo de térmitas no bioma Cerrado. In A.O. Scariot, J.C.S. Silva, & J.M. Felfili (Eds.), Cerrado: Ecologia, Biodiversidade e Conservação (pp. 319–333). Brasília: Ministério do Meio Ambiente.

Costa, D.A., Filho, K.E.S., Brandão, D. (2009). Distribution patterns of termites on urban region of Goiânia, Goiás, Brazil. Iheringia, Sér. Zoologia, 99: 364-367.

Edwards, R. & Mill, A. E. (1986).Termites in buildings: their biology and control. Rentokil Limited, 261p.

Eleotério, E. S. R., Berti Filho, E. (2000). Levantamento e identificação de cupins (Insecta: Isoptera) em área urbana de Piracicaba – SP. Ciência Florestal, 10: 125-139.

Ferraz, M. V. & Cancello, E. M. (2001). Swarming behavior of the economically most important termite, Coptotermes havilandi (Isoptera: Rhinotermitidae), in Southeastern Brazil. Sociobiology, 38: 683-693.

Fontes, L. R. (1995). Cupins em áreas urbanas. In: E. Berti Filho & L. R. Fontes (Eds.). Alguns Aspectos Atuais da Biologia e Controle de Cupins (pp.57-76). Piracicaba: FEALQ.

Fontes, L. R. & Berti-Filho, E. (1998). Cupins - O desafio do conhecimento. Piracicaba: FEALQ, 512p.

Harris, W. V. (1971). Termites: their recognition and control. England: Longman Group Ltda., 186p.

Hedges, S. (1998). Add-on for termite control. Pest Control Technology, pp. 30-35.

Instituto Brasileiro de Geografia e Estatística (IBGE). Síntese de indicadores sociais 2010: Coordenação de população e indicadores sociais. Disponível em: <http://www.ibge.gov.br>.

Inmet. (2012). Instituto Nacional de Metereologia: Boletim agroclimatológico. Secção de apoio a agricultura e recursos hídricos.

Krishna, K., Grimaldi, D. A., Hrishna, V., Engel, M. S. (2013). Treatise on the Isoptera of the world. Bulletin of American Museum of Natural History, n.377. doi: 10.1206/377.7

Lelis, A. T. (1995). Cupins urbanos: biologia e controle. In: E. Berti Filho & L. R. Fontes (Eds.). Alguns Aspectos Atuais da Biologia e Controle de Cupins (pp.77-80). FEALQ.

Levings, S.C. & Adams, E.S. (1984). Intra – interspecific territoriality in Nasutitermes (Isoptera: Termitidae) in a Panamanian mangrove forest. Journal of Animal Ecology, 53: 705-714.

Matias, G. R. R. S., Albuquerque, A. C., Matias, M. P., Silva, E. P. V., Oliveira, C. M. A. S., & Oliveira, M. A. P. (2006). Os cupins urbanos em Jardim Paulista, Paulista-PE. Diversidade e controle. O Biológico, 68: 58-61.

Silva, J.C., Lopez, A.G.C., Oliveira, J.T.S. (2004). Influência da idade na resistência natural da madeira de Eucalyptus grandis W. Hill ex. Maiden ao ataque de cupim de madeira seca (Cryptotermes brevis). Revista Árvore, 28 (4):583-587.

McMahan, E. (1962). Laboratory studies of colony establishment and development in Cryptotermes brevis (Walker) (Isoptera: Kalotermitidae). Proceedings of the Hawaiian Entomological Society, 18:145-153.

Mill, A. (1991). Termites as structural pests in Amazonia, Brazil. Sociobiology, 19 (2): 339-349.

Milano, S. & FONTES, L. R. (2002). Termite pests and their control in urban Brazil. Sociobiology, 40 (1): 63-177.

Oliveira, C.M.A.S., Matias, G.R.R.S., Silva, S.B., Moraes, F.M., Albuquerque, A.C. (2006). Diversidade de cupins no Ibura: área urbana do Recife-PE. O Biológico, 68: 264-266.

Pizano, M.A., & Fontes, L.R. (1986). Ocorrência de Heterotermes tenuis (Hagen, 1858) e H. longiceps (Snyder,1924) (Isoptera, Rhinotermitidae) atacando cana-de-açúcar no Brasil. Brasil Açucareiro, 104: 3-429.

Rust, M.K., & Su, N.Y. (2012). Managing Social Insects of Urban Importance. Annual Review of Entomol., 57: 355–75. doi: 10.1146/annurev-ento-120710-100634

Spear, P.J. (1970). Principles of termite control. In: K. Krishna & M. Weesner (Eds.). Biology of termites. Academic Press, 2: 577–604.

Steward, R. C. (1983). Microclimate and colony foundation by imago and neotenic reprodutives of drywood temite species (Cryptotermes sp.) (Isoptera: Kalotermitidae). Sociobiology, 7: 311–331.

Thorne, B., & Noirot, C. (1982). Ergatoid reproductives in Nasutitermes corniger (Motschulsy) (Isoptera: Termitidae). J. Insect Morphology and Embryology, 11: 213–226.

Torales, G. J., Laffont, E.R., Arbino, M.O., Godoy, M.C. (1997). Primeira lista faunística de los Isopteros de la

Argentina. Revista de la Sociedade Entomologica Argentina, 56: 47–53.

Vasconcellos, A., & Bandeira A.G. (2000). Avaliação do consumo de madeira porespécies de *Nasutitermes* e *Microcerotermes* (Insecta, Isoptera, Termitidae). Revista Nordestina de Biologia, 14 (1/2): 17–24.

Vasconcellos, A., Bandeira, A.G., Miranda, C.S., Silva, M.P. (2002). Termites (Isoptera) Pests in Buildings in João Pessoa, Brazil. Sociobiology, 40: 1–6.

14

Ambient Temperature Influences Geographic Changes in Nest and Colony Size of *Polistes chinensis antennalis* Pérez (Hymenoptera: Vespidae)

S Hozumi[1], K Kudô[2], H Katakura[3], S Yamane[4]

1 - Chigakukan secondary school, Mito 310-0914, Japan

2 - Niigata University, Niigata 950–2181, Japan

3 - Hokkaido University, Sapporo 060–0810, Japan

4 - Ibaraki University, Mito 310-8511, Japan

Keywords
Paper wasp, *Polistes chinensis*, geographic variation, nest architecture.

Corresponding author
Satoshi Hozumi
Chigakukan secondary school
Mito 310-0914, Japan
E-Mail: shoz@tokiwa.ac.jp

Abstract

In some *Polistes* wasps, the foundresses build huge nests during the founding phase to improve the thermal condition of these nests. This implies that *Polistes* wasps change their nesting manner in relation to ambient temperature. To test the hypothesis that nest size increases with latitude, colonies of *Polistes chinensis* were collected from 11 locations. Three nest parameters, the number of cells cell length and index of functional envelope, increased with latitude. The number of cells at the northernmost station was 60, which was 1.5 times more than in lower latitudes. Cell length increased by approximately 4 mm from low to high latitudes, indicating that extra-building in *P. chinensis* is remarkable in adding new cells. The number of first broods was not correlated with latitude, whereas the number of second brood increased with latitude because of the numerous cells built at high latitudes.

Introduction

Climate is a major physical factor that determines the geological distributions and life histories of organisms. Some climates generate different ecological traits among populations of the same species (e.g., Angilletta, 2009). Assessing adaptations to different climatic regimes helps us to understand the major selective pressures affecting organisms and to predict their reactions to global and local climatic changes in the environment (Chown, 2001).

In social insects, ambient temperature is an important physical factor that greatly affects insect activity (Heinrich, 1993). Differences in local ambient temperatures lead to differences in nesting activities (Fucini et al., 2014), rate of development of immature stages (Miyano, 1981), and duration of nesting periods (Yamane, 1969), and intra-specific changes have been reported in the genus *Polistes* (Hymenoptera, Vespidae). *Polistes* paper wasps are independent-founding social wasps. A lone foundress builds an exposed nest and nurses the first brood of larvae until worker individuals emerge (pre-emergence period; Reeve, 1991). After eclosion (post-emergence period), the adult individuals increase colony production and produce as many reproductive gynes as possible (reproduction period). The wasps build an exposed comb, in which the construction of new cells and elongation of cells are commonly related to oviposition and the development of larvae, respectively (Delaeurance's law; Delaeurance, 1957). However, species in different climates exhibit different nest-building activities and colony cycles. Yamane (1972) showed that *P. chinensis* foundresses that inhabit the cool northern areas of Japan (nesting duration 3.5–4 months) build many empty cells just before the emergence of the first adults and elongate the cells beyond the lengths of the pupae. In contrast, in the warm areas of Japan, *P. chinensis* nesting activity is consistent with Delaeurance's law, and the nesting duration is approximately 6 months.

A few previous studies have focused on the intra-specific changes in nest size, in relation to nest thermoregulation; i.e., adaptation to low temperature. Yamane & Kawamichi (1975) found that *P. chinensis* foundresses build larger nests in higher latitudinal regions and concluded that numerous and long cells function as air chambers that improve thermal conditions in the brood-rearing area of the nest, similar to the envelope of a vespine nest (functional envelope). This means that foundresses build more cells for nest thermoregulation to accelerate the development of the first batch of larvae (first brood) during the pre-emergence period. However, the observations were performed in only 2 locations, and thus further study is needed to understand the relationships between climate and the geographic variation in nest size.

In this study, we set 2 goals. The first was to verify whether nest sizes increase geographically from low to high latitudes as Yamane and Kawamichi (1975) predicted. If *P. chinensis* foundresses change their nesting behavior according to ambient temperature, the number and the length of cells should increase with latitude. We collected *P. chinensis* nests at 11 locations across Japan islands and examined the relationship between nest size and latitude. The second goal was to clarify the relationship between latitude and colony productivity. If foundresses invest more resources in nest building in higher latitudinal regions, the numbers of individuals in the first and second broods would be expected to decrease as a tradeoff. We investigated the numbers of individuals in first broods at 6 locations along a latitudinal gradient and calculated an index of functional envelope. In this study, we also discuss the difference in nesting behavior between lower and higher latitudinal areas.

Material and Methods

Nest parameters

In order to determine the latitudinal differences in nest size among the islands of Japan, a total of 168 nests of *P. chinensis antennalis* were collected at the 11 locations shown in Table 1 and Fig. 1. Two nest parameters, the number of cells and cell length, were measured for each nest. The measurement was performed as follows: number of cells, total number of cells was counted for each nest; cell length (mm), maximum value of the longest cell was measured for each nest with a vernier calliper to the nearest 0.1 mm.

Fig 1. Localities where nests of P. chinensis were collected.

Colony composition

Colony composition was investigated at 6 locations (St4, St7, St8, St9, St10, and St11) to determine latitudinal differences among colonies. The content of each cell (egg, young larva, old larva, pupa, or empty) was recorded for all nests immediately after nest collection. For this study, 1st to 3rd instar larvae were classified as young larvae, and 4th and 5th instar larvae were classified as old larvae.

To examine the hypothesis of Yamane and Kawamichi (1975), we defined the index of functional envelope (IFE) to be the nest volume per first brood, i.e., the entire nest volume per total number of individuals in the first brood. Nest volume (ml)

Table 1. Geographic positions and altitudes of localities where *Polistes chinensis* nests were collected.

Station	Locality	Latitude	Longitude	Mean annual temperature (°C)	Date collected	Collector
St1	Fukuoka	33°34′N	130°23′E	17.0	19-21 May 2000	N. Kumano
St2	Tsu	34°43′N	136°30′E	15.9	4-5 June 1997	K. Kudô
St3	Toki	35°21′N	137°11′E	15.8	7-8 June 1997	K. Kudô
St4	Itako	35°56′N	140°34′E	14.5	9-10 June 1998, 10 June 1999	S. Hozumi
St5	Tsurugi	36°27′N	136°37′E	14.3	16-17 June 1997	K. Kudô
St6	Kanazawa	36°33′N	136°39′E	14.6	19 June and 7-8 July 1997	K. Kudô
St7	Tanagura	37°01′N	140°22′E	11.5	8 June 1999	S. Hozumi
St8	Mizusawa	39°08′N	141°08′E	10.7	8 June 1998, 19 June 1999	S. Hozumi
St9	Shichinohe	40°41′N	141°08′E	10.4	4 July 1998, 3 July 1999	S. Hozumi
St10	Kanagi	40°54′N	140°27′E	10.3	4 July 1998, 3 July 1999	S. Hozumi
St11	Okushiri	42°08′N	139°29′E	9.7	21 July 1999, 18 July 2000	S. Hozumi

was estimated as follows: all of the cells were filled with small granular glass beads (diameter 0.1 mm), and the volume of beads was measured with a graduated cylinder to the nearest 0.1 ml (see Yamane et al., 1998). This measurement was repeated 3 times, and the values were averaged.

Statistics

Spearman's rank correlation analysis was used to quantify the relationship between nest and colonial parameters and latitude. All analyses were performed with EZR ver. 1.22 (Kanda, 2013) on a personal computer.

Results

Nest size variation

Table 2 shows the geographical variation in the number and length of cells. Both parameters significantly increased with latitude (Spearman's rank correlation: number of cells, $\rho = 0.6272$, $p < 0.05$; cell length, $\rho = 0.7671$, $p < 0.01$). The number and length of cells were similar (approximately 40 and 22 mm, respectively) from St1 to St8, and the values increased until St11. Cell length also increased significantly, but the range of increase was small; the maximum difference in mean value was 3.9 mm between St10 (23.9 mm) and St1 (20.0 mm).

Table 2. Mean (mean ± SD) number of cells and cell length of *Polistes chinensis* nests collected from 11 localities.

Station	N	Number of cells		Cell length (mm)	
		Mean	Median	Mean	Median
St1	5	38.8 ± 5.3	38.0	20.0 ± 1.4	19.8
St2	14	36.6 ± 5.3	38.5	22.8 ± 1.1	22.7
St3	28	33.0 ± 6.0	33.0	22.0 ± 1.6	22.1
St4	18	37.3 ± 7.6	41.0	21.6 ± 0.9	22.0
St5	10	40.5 ± 7.1	37.0	22.0 ± 1.5	22.8
St6	5	38.0 ± 4.4	38.0	22.7 ± 0.9	21.3
St7	8	39.4 ± 5.4	41.0	21.8 ± 1.7	22.2
St8	20	39.8 ± 6.6	39.0	21.4 ± 0.8	21.5
St9	19	42.9 ± 4.0	43.0	22.9 ± 1.2	23.0
St10	16	45.4 ± 6.8	43.5	23.9 ± 1.4	24.1
St11	25	60.2 ± 15.8	61.0	23.0 ± 2.1	23.5

Colony composition

Table 3 shows the colony composition. The number of individuals in the second brood and the number of empty cells increased significantly with latitude. The number of individuals in the first brood tended to decrease with latitude; however, there was no significant correlation. Nest volume and IFE increased significantly with latitude.

Table 3. Mean numbers of immature individuals (mean ± SD) of *Polistes chinensis* colonies. Index of functional envelope (IFE) was calculated by dividing the nest volume by the number of the first brood. Parentheses indicate median of each value.

	St4	St7	St8	St9	St10	St11	Spearman's rank correlation ρ	Significance
	N=19	N=8	N=19	N=18	N=13	N=24		
First brood	14.6 ± 4.4 (15.0)	14.0 ± 5.6 (15.0)	12.8 ± 5.7 (14.0)	14.2 ± 5.1 (15.0)	12.6 ± 2.3 (12.0)	12.4 ± 4.7 (14.0)	-0.6789	p=0.1381
Second brood	19.8 ± 5.6 (21.0)	15.9 ± 5.3 (18.0)	21.1 ± 4.6 (21.0)	23.3 ± 7.0 (23.5)	23.1 ± 8.8 (22.0)	38.5 ± 13.8 (40.0)	0.8407	p<0.05
Empty cells	4.1 ± 3.4 (3.5)	7.9 ± 4.9 (6.0)	6.2 ± 4.4 (5.0)	4.6 ± 3.7 (6.0)	9.1 ± 6.8 (8.0)	9.4 ± 6.1 (10.0)	0.8986	p<0.05
Nest volume	6.1 ± 1.2 (6.0)	7.0 ± 0.4 (7.1)	6.3 ± 1.3 (6.0)	7.2 ± 1.1 (7.2)	7.2 ± 0.4 (7.2)	8.7 ± 1.5 (8.5)	0.9428	p<0.05
Index of functional envelope (IFE)	0.46 ± 0.16 (0.41)	0.59 ± 0.26 (0.49)	0.65 ± 0.45 (0.49)	0.77 ± 0.80 (0.50)	0.59 ± 0.10 (0.58)	0.84 ± 0.42 (0.68)	1.000	p<0.005

Discussion

The results of this study confirmed that foundresses in higher latitudinal regions built larger nests, based on the number of cells. The increase in the IFE supports the hypothesis of Yamane and Kawamichi (1975) that the extra cells that enclose first broods may function as thermoregulatory air chambers during the pre-emergence period. Latitude is often closely correlated with ambient temperature, and low ambient temperature is believed to influence the building activities of foundresses.

The building of extra cells is known to increase cell temperatures (Hozumi & Yamane, 2001), which accelerates the rate of development of the first brood. The construction of a large number of new cells not only increases cell temperature but also enables the colonies to rear more individuals during the short nesting period. Because cell length greatly affects cell temperature, the thermal effect of the cells could further accelerate development of the immature stages. In this study, the maximum difference in cell length among the locations was only 4 mm. However, even a 4-mm elongation may increase cell temperatures by 1°C when the nest receives solar radiation (Hozumi et al., 2008).

Nest size increased with latitude, whereas the numbers of individuals in the first brood were similar among locations. The numbers of individuals in the first brood were also similar to those reported in other studies, including *P. chinensis* from other areas in Japan (Miyano, 1980; Yamane et al., 1998) and other *Polistes* species (*P. riparius*, Hozumi & Yamane, 2008; *P. biglumis*, Fucini et al., 2014). These results imply that a certain number of first brood individuals (ca. 14) are reared by foundresses during the founding phase, at least in the range of the study areas. However, the numbers of individuals from the second brood and the numbers of empty cells increased significantly with latitude. The foundresses continued to build new empty cells after laying the eggs of the first brood for nest thermoregulation and then began laying eggs to produce as many individuals as possible during the short nesting period. This indicates that *P. chinensis* foundresses at higher latitudes used more resources for nest thermoregulation and production of second broods instead of rearing more first broods.

In conclusion, *P. chinensis* foundresses can adapt to local conditions by altering nest-building activities and colony production, and ambient temperature may significantly influence nesting activities. The building of extra cells is also seen in *P. riparius*, and the building activity (elongation of cells) is influenced by the local ambient temperature (Hozumi & Kudô, 2012). This implies that elongation of cells may occur in *P. chinensis* inhabiting cooler areas. On the building activities of number and length of cells, further studies are needed to assess whether variations in *P. chinensis* populations are due to phenotypic flexibility in response to environmental conditions or to genetic differences.

Acknowledgments

We thank Dr. N. Kumano for kindly collected *P. chinensis* nests at Fukuoka. We thank Prof. S. F. Mawatari for helpful discussion and encouragement during the course of this study.

References

Angilletta, M.J. 2009. Thermal adaptation: a theoretical and empirical synthesis. Oxford Univ Press. p. 289. doi:10.1093/acprof:oso/9780198570875.001.1

Chown, S.L. 2001. Physiological variation in insects: hierarchical levels and implications. Journal of Insect Physiology, 47: 649-660.

Delaeurance, E.P. 1957. Contribution à létude biologique des *Polistes* (Hyménoptères, Vespidés). I. L'activité de construction. Annales des Sciences Naturelles Zoologie et Biologie Animale, 19: 91-222.

Fucini, S., Uboni, A. & Lorenzi, M.C. (2014). Geographic Variation in Air Temperature Leads to Intraspecific Variability in the Behavior and Productivity of a Eusocial Insect. Journal of Insect Behavior, doi: 10.1007/s10905-013-9436-y

Heinrich, B. (1993). Hot Blooded Insects. Harvard University Press, Cambridge, Massachusetts. p. 601.

Hozumi, S. & K, Kudô. (2012). Adaptive nesting tactics in a paper wasp, *Polistes riparius*, inhabiting cold climatic regions. Sociobiology, 59: 1447–1458.

Hozumi, S. & Yamane, S. (2001). Incubation ability of the functional envelope in paper wasp nests Hymenoptera, Vespidae, *Polistes*): I. Field measurements of nest temperature using paper models. Journal of Ethology, 19: 39-46. doi:10.1007/s101640170016

Hozumi, S., Yamane, S. & Katakura, H. (2008). Building of extra cells in the nests of paper wasps (Hymenoptera; Vespidae; *Polistes*) as an adaptive measure in severely cold regions. Sociobology, 51: 399-414.

Kanda, Y. (2013). Investigation of the freely available easy-to-use software 'EZR' for medical statistics. Bone Marrow Transplantation, 48: 452-458. doi: 10.1038/bmt.2012.244

Miyano S. (1980). Life tables of colonies and workers in a paper wasp, *Polistes chinensis antennalis*, in central Japan (Hymenoptera: Vespidae). Population Ecology, 22: 69-88.

Miyano, S. (1981). Brood development in *Polistes chinensis antennalis* Pérez I. Seasonal variation of duration of immature stages and an experiment on thermal response of egg development. Bulletin of the Gifu Prefectural Museum, 2: 75-83.

Reeve, H.K. (1991). *Polistes*. In: Ross, K.G. & Matthews, R.W. (eds). The Social Biology of Wasps. (pp. 191–231). Ithaca, NY: Cornell University Press.

Yamane, S. (1969). Preliminary observations on the life history of two polistine wasps, *Polistes snelleni* and *P. biglumis* in Sapporo, northern Japan. Journal of the Faculty of Sciences of Hokkaido University, Ser. VI, Zoology, 17: 78-105.

Yamane, S. (1972). Life cycle and nest architecture of *Polistes* wasps in the Okushiri island, northern Japan (Hymenoptera, Vespidae). Journal of the Faculty of Sciences of Hokkaido University, Ser. VI, Zoology, 18: 440-459.

Yamane, S., Kudô, K., Tajima, T., Nihon'yanagi, K., Shinoda, M., Saito, K. & Yamamoto, H., (1998). Comparison of investment in nest construction by the foundresses of consubgeneric *Polistes* wasps, *P. (Polistes) riparius* and *P. (P.) chinensis* (Hymenoptera, Vespidae). Journal of Ethology, 16: 97-104.

Yamane, S. & Kawamichi, T. (1975). Bionomic comparison of *Polistes biglumis* (Hymenoptera, Vespidae) at two different localities in Hokkaido, northern Japan, with reference to its probable adaptation to cold climate. Kontyû, 43: 214-232.

15

Record of Parasitoids in nests of social wasps (Hymenoptera: Vespidae: Polistinae)

A Somavilla[1,4], K Schoeninger[1], AF Carvalho[2], RST Menezes[3], MA Del Lama[2], MA Costa[3], ML Oliveira[1]

1- Instituto Nacional de Pesquisas da Amazônia, Petrópolis, Manaus, AM, Brazil
2- Universidade Federal de São Carlos, São Carlos, SP, Brazil
3- Universidade Estadual de Santa Cruz, Ilhéus, BA, Brazil

Keywords
Epiponini, eusocial, Mischocyttarini, Paper wasps, parasitoids, Polistini.

Corresponding author
Alexandre Somavilla
Coord. de Biodiversidade, Instituto
Nacional de Pesquisas da Amazônia
Av. André Araújo, 2936, Petrópolis,
69067-375, Manaus, AM, Brazil
E-Mail: alexandre.s@hotmail.com

Abstract

The aim of this study was to record the parasitoid species found in social wasps nests sampled in different localities in Brazil. We sampled nests of *Mischocyttarus cassununga*, *Mischocyttarus consimilis*, *Mischocyttarus imitator*, *Polistes canadensis*, *Polistes cinerascens*, *Polistes versicolor*, *Angiopolybia pallens*, *Leipomeles spilogastra*, *Polybia jurinei* and two indeterminate species of *Mischocyttarus*. Thus, we observed that nests of *M. cassununga*, *M. imitator* and *Mischocyttarus* (*Phi*) sp.1 were parasitized by *Toechorychus guarapuavus* (Ichneumonidae) and nests of *M. consimilis*, *M. imitator* and *Mischocyttarus* sp.2 were parasitized by *Toechorychus fluminensis* (Ichneumonidae). Nests of *P. versicolor* and *P. cinerascens* were parasitized by *Elasmus polistis* (Eulophidae) and nest of *P. canadensis* was parasitized by *Simenota depressa* (Trigonalidae); nest of *A. pallens* and *L. spilogastra*, was infested by *Brachymeria* sp.1 and *Brachymeria* sp.2 (Chalcididae), respectively. Nests of *M. cassununga* and *Polybia jurinei* were parasitized by *Megaselia scalaris* (Phoridae).

Introduction

Polistinae wasps primarily use plant materials to build their nests and can be divided into two groups according to nesting behavior and architecture: independent- and swarm-founding species (Carpenter, 1991; Wenzel, 1998). Independent-founding species (Polistini, Mischocyttarini and some Old World Ropalidiini) build unprotected combs that are small in size with few brood cells (Jeanne, 1980; Carpenter & Marques, 2001). Swarming-founding species (Epiponini and most Ropalidiini) in turn builds large nests with thousands of cells protected by an envelope (Jeanne, 1980; Carpenter & Marques, 2001).

Paper wasp (Hymenoptera: Vespidae: Polistinae) nests are frequently invaded by natural parasitoids of eggs, larvae and pupae. The nest of a social wasp is an environment rich in resources for many predators and parasitoids, which might be attracted and can cause high costs to the colony (Soares et al., 2006). Larvae and pupae are the target of many of these natural enemies including ants, birds, and parasitoid hymenopterans (Makino, 1985; Yamane, 1996; Clouse, 2001). Despite the

paucity of studies on the subject, the attack by hymenopterans parasitoids might be one of the main causes of mortality among social wasps in the early stages of development (Wenzel, 1998). In spite of that, there are few attempts to reporting parasitism in these social wasps (Makino, 1985; Yamane, 1996; Clouse, 2001). Consequently there are few records concerning natural enemies of paper wasps and how harmful such natural enemies might be for a wasp colony.

To Brazil, Soares et al. (2006) reported a species of *Pachysomoides* sp. (Ichneumonidae) and *Megaselia scalaris* (Diptera, Phoridae) parasitizing *Mischocyttarus cassununga* (Von Ihering) nests in Minas Gerais; Dorfey and Köhler (2011) reported *Elasmus polistis* Berks, 1971 parasitizing *Polistes versicolor* (Oliver, 1792) nests in Rio Grande do Sul; and Trindade et al. (2012) and Santos and Noll (2013) reported the occurrence of *Seminota marginata* (Westwood, 1874) in *Apoica flavissima* (Van der Vecht, 1973) nests in São Paulo.

The aim of this study was to record parasitoids found in social wasps nests, sampled during several years of survey in different localities in Brazil.

Material and Methods

During several collection trips, between 2009 and 2013, across Brazilian biomes as Atlantic rainforest (Gramado [RS], Sinimbu [RS], Ibirapitanga [BA], Maricá [RJ], Ubatuba [SP], Viçosa [MG]), Pantanal (Dourados [MS]), Caatinga (Jacobina [BA]) and Amazonia (Manaquiri [AM], Manaus [AM], Rorainópolis [RR]) we removed entire nests of polistine species from the field and separated all adults from the offspring. The later were put in plastic recipients covered by voile tissue and maintained in laboratory to remove individuals from the emerged social wasps.

The combs containing offspring were maintained during approximately 40 days at 28°C and relative humidity of 70% ± 5%, in a biochemical oxygen demand (BOD) incubator and monitored daily until adults emerged. After the 40 days, we verified the emergence of polistines and parasitoids as well as the nests were open for counting of cells. However, for the nest of *Angiopolybia pallens* (Olivier, 1792) and *Mischocyttarus consimilis* Zikán, 1949, this procedure could not be performed since the nest were damaged.

Most of the individuals (hosts and parasitoids) were fixed in absolute ethanol and some samples were pinned and are deposited in the Coleção Zoológica de Invertebrados of the Instituto Nacional de Pesquisas da Amazônia (INPA) and in the Laboratório de Genética Evolutiva de Himenópteros of the Universidade Federal de São Carlos (LGEH-UFSCar). Voucher nests were deposited in both collections as well. Specimens were identified using the following identification keys: Richards (1978) and Carpenter and Marques (2001) for social wasps, Burks (1960) and Boucek (1992) for parasitoid wasps and Brown (2010) for Phoridae.

Results

We sampled two nests of *Mischocyttarus cassununga* (R. von Ihering, 1903), one nest from Viçosa, in the state of Minas Gerais and one nest from Ubatuba, in the state of São Paulo; one nest of *Mischocyttarus consimilis* Zikán, 1949 from Dourados, in the state of Mato Grosso do Sul; two nests of *Mischocyttarus imitator* (Ducke, 1904) (Fig 1A, 1B), one nest from Viçosa, in the state of Minas Gerais and other from Maricá, in the

state of Rio de Janeiro; one nest of *Mischocyttarus* (*Phi*) sp.1 from Jacobina, in the state of Bahia; one abandoned nest of *Mischocyttarus* sp.2 from Manaus, in the state of Amazonas, five nests of *Polistes canadensis* (Linnaeus, 1758) (Fig 2A, 2B, 2C) from Manaquiri, Amazonas, one nest of *Polistes cinerascens* de Saussure, 1854 from Sinimbu, in the state of Rio Grande do Sul; two nests of *Polistes versicolor* (Olivier, 1791) (Fig 3A, 3B) from Gramado and Sinimbu, Rio Grande do Sul; one nest of *Angiopolybia pallens* (Lepeletier, 1836) (Fig 4A, 4B) from Ibirapitanga, Bahia; one nest of *Leipomeles spilogastra* Cameron, 1912 (Fig 5A, 5B) from Rorainópolis, in the state of Roraima and one nest of *Polybia jurinei* de Saussure, 1854 from Manaus, Amazonas (Table 1).

Fig 2. (A, B) *Polistes canadensis* nests; (C) *Polistes canadensis* habitus; (D, E) Two color forms to *Seminota depressa*.

The abovementioned nests were infected by hymenopterans or by dipterans parasitoids, as follows (see also Table 1):

(1) Mischocyttarini: *M. cassununga* was parasitized by *Toechorychus guarapuavus* Tedesco, 2013 (Ichneumonidae) and *Megaselia scalaris* (Loew, 1866) (Diptera: Phoridae), *M. consimilis* was parasitized by *Toechorychus fluminensis* Tedesco, 2013 (Ichneumonidae) and *M. imitator* was parasitized by *T. guarapuavus* (Fig 1C) and *T. fluminensis*.

(2) Polistini: Nests of *P. cinerascens* and *P. versicolor* were parasitized by *Elasmus polistis* Burks, 1971 (Fig 3C) (Eulophidae) and nests of *P. canadensis* were parasitized by *Seminota depressa* DeGeer, 1773 (Fig 2D, 2E) (Trigonalidae).

Fig 1. (A) *Mischocyttarus imitator* nest; (B) *Mischocyttarus imitator* habitus; (C) *Toechorychus guarapuavus* habitus.

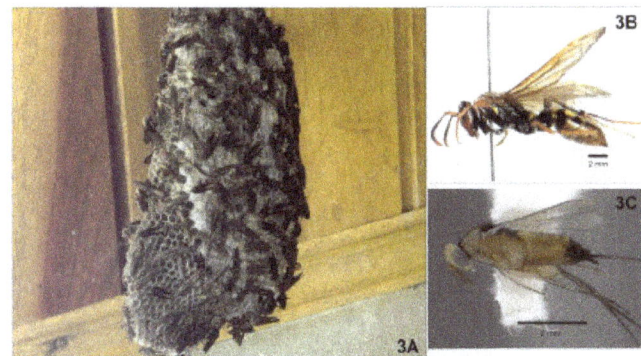

Fig 3. (A) *Polistes versicolor* nest; (B) *Polistes versicolor* habitus; (C) *Elasmus polistis* habitus.

Table 1. Records of parasitoids in nests of social wasps sampled in different localities of Brazil. Number of emerged adult social wasps from the nest (Host N); number of parasitoids emerged from the nest (Parasitoid N); Comb and number of cells from the comb (Comb / cell N).

Host	Host N	Parasitoid	Parasitoid N	Comb/cell N	Locality
Mischocyttarus cassununga (R. von Ihering, 1903)	52	*Toechorychus guarapuavus* Tedesco, 2013 (Ichneumonidae)	30 (30♂)	1/149	Viçosa, MG: UFV, (20°45'S, 42°52'W)
Mischocyttarus cassununga (R. von Ihering, 1903)	03	*Megaselia scalaris* Loew, 1866 (Phoridae)	50	1/23	Ubatuba, SP: RPPN Angelin Rainforest (23°26'S, 45°04'W)
Mischocyttarus consimilis Zikán, 1941	05	*Toechorychus fluminensis* Tedesco, 2013 (Ichneumonidae)	01 (01♂)	1/-	Dourados, MS (22°13'S, 54°51'W)
Mischocyttarus imitator (Ducke, 1792)	35	*Toechorychus guarapuavus* Tedesco, 2013 (Ichneumonidae)	04 (04♂)	1/104	Viçosa, MG: Mata do Paraíso (20°45'S, 42°52'W)
Mischocyttarus imitator (Ducke, 1792)	06	*Toechorychus fluminensis* Tedesco, 2013 (Ichneumonidae)	03 (03♂)	1/45	Maricá, RJ: Estrada do Espraiado (22°55'S, 42°47'W)
Mischocyttarus (*Phi*) sp.1	08	*Toechorychus guarapuavus* Tedesco, 2013 (Ichneumonidae)	02 (02♂)	1/124	Jacobina, BA (11°11'S, 40°29'W)
Mischocyttarus sp.2 (abandoned nest)	00	*Toechorychus fluminensis* Tedesco, 2013 (Ichneumonidae)	01 (01♂)	1/12	Manaus, AM: Ducke Reserve (02°55'S, 59°58'W)
Polistes canadensis (Linnaeus, 1758)	41	*Seminota depressa* DeGeer, 1773 (Trigonalidae)	09 (09♂)	5/131	Manaquiri, AM: BR319 (3°26'S, 60°26'W).
Polistes cinerascens de Saussure, 1854	16	*Elasmus polistis* Berks, 1971 (Eulophidae)	14 (12♀, 02♂)	1/150	Sinimbu, RS: RPPN UNISC (29°23'S, 52°32'W)
Polistes versicolor (Olivier, 1792)	121	*Elasmus polistis* Berks, 1971 (Eulophidae)	357 (240♀, 117♂)	1/696	Gramado, RS (29°22'S, 50°52'W)
Polistes versicolor (Olivier, 1792)	52	*Elasmus polistis* Berks, 1971 (Eulophidae)	439 (258♀, 181♂)	1/950	Sinimbu, RS: RPPN UNISC (29°23'S, 52°32'W)
Angiopolybia pallens (Olivier, 1792)	828	*Brachymeria* sp.1 (Chalcididae)	11 (11♀)	6/-	Ibirapitanga, BA: BR101 (14°09'S, 59°22'W)
Leipomeles spilogastra (Cameron, 1912)	15	*Brachymeria* sp.2 (Chalcididae)	04 (04♀)	3/165	Rorainópolis, RR: BR 174 (00°56'N, 60°25'W)
Polybia jurinei de Saussure, 1854	11	*Megaselia scalaris* Loew, 1866 (Phoridae)	02	1/68	Manaus, AM: Ducke Reserve (02°55'S, 59°58'W)

Fig 4. (A) *Angiopolybia pallens* nest; (B) *Angiopolybia pallens* habitus; (C) *Brachymeria* sp.1 habitus.

(3) Epiponini: Nests of *A. pallens* (Fig 4A) and *L. spilogastra* (Fig. 5A) were parasitized by *Brachymeria* sp.1 (Fig 4C) and *Brachymeria* sp.2 (Fig 5C) (Chalcididae) respectively, and in the nest of *P. jurinei* was recorded parasitism by *Megaselia*

Fig 5. (A) *Leipomeles spilogastra* nest; (B) *Leipomeles spilogastra* habitus; (C) *Brachymeria* sp.2 habitus.

scalaris (Phoridae).

Furthermore, we observed a dominance of males from species *Seminota* and *Toechorychus* which emerged from *Mischocyttarus* and *Polistes* nests, respectively, compared with *Elasmus* and *Brachymeria* which there was dominance of emerging females.

Individuals of Chalcididae started to emerge two days after collection of the nests extending to the seventh day. Individuals of Ichneumonidae started to emerge seven days after collection of the nests extending to the twelfth day. For individuals of *Seminota depressa* and *Elasmus polistis* the emergency period took place over thirty days. For *Megaselia scalaris*, we were not able to register the emergency period.

Discussion

In this study we reported parasitism in seven nests of five *Mischocyttarus* de Saussure species, eight nests of three species of *Polistes* Latreille and only in three nests of three Epiponini species. Moreover, it was possible to verify a higher number of parasitoids per nest in *Polistes* and *Mischocyttarus* when compared to Epiponini (Table 1). Social insects such as

the Polistinae wasps might be more susceptible to the attack of parasitoids when living in colonies with absence of brood keepers while foraging (Clouse, 1997; 2001). Another factor that may contribute to the attack of parasitoids is the absence of a nest-protecting envelope. *Polistes* and *Mischocyttarus* are independent-founding groups with remarkable non-enveloped nests. Epiponini species in turn shows swarm-founding behavior build nest envelopes that protect fragile brood. Strassmann (1981) suggests that swarm-founding species are less susceptible to parasitism once adults protect constantly the nest, preventing the invasion of possible enemies. Thus, the difference of parasitoidism between species with enveloped and species with non-enveloped nests and its significance must be further investigated.

Most Ichneumonidae are ecto or endoparasitoids of arthropods, usually attacking immature instars of holometabolous insects of the order Lepidoptera, Coleoptera, Diptera and Neuroptera (Clausen, 1940; Askew, 1971; Hanson & Gauld, 1995). These insects are considered important in biological control since they are parasitoids that always kill the host (Kumagai & Graf, 2000). There are few reports of Ichneumonidae parasitizing paper wasp nests and this lack of information is mostly due to the great difficult in detecting symptoms of parasitism in nests of these social insects. Until now, there are records only for *Mischocyttarus*, *Polistes* and *Dolichovespula* Rohwer. Species of *Toechorychus* Townes (Ichneumonidae) are apparently parasitoids in nests of Vespidae and pupae of Lepidoptera, but host records were previously known only for two species: *Toechorychus abactus* (Cresson, 1874) and *Toechorychus cassunungae* (Brauns, 1905) are known to attack species of *Mischocyttarus* (Brauns, 1905; Bertoni, 1911; Costa-Lima, 1962; Makino, 1985) and *Toechorychus albimaculatus* (Taschenberg, 1876) attack *P. canadensis* (Makino, 1985)

Seminota (Trigonalidae) has been reported as a parasitoid of some genera of social wasps such as *Apoica* Lepeletier, *Mischocyttarus*, *Parachartergus* R. von Ihering, *Polistes* and *Pseudopolybia* Von Dalla Torre (Makino, 1985; Weinstein & Austin, 1991; Carmean & Kimsey, 1998) but little is known on the biology of these associations being that Weinstein and Austin (1991) considered paper wasps as the secondary host of Trigonalidae. On the other hand, Santos and Noll (2013) observing the emergence of parasitoid *Seminota marginata* (Westwood, 1874) in one nest of *Apoica flavissima* Van der Vecht, 1972 suggested that social wasps may be both primary and secondary hosts of such parasitoids as they extract and chew vegetable fiber. We observed that individuals of *S. depressa* which emerged from the nests of *P. canadensis* had variations regarding the pigmentation on the integument (Fig 2D, 2E), one totally black with blackened wings and the other with black head and mesosoma, brown metasoma, and hyaline wings. The size and pattern of bristles and score were similar in both forms.

Elasmus Westwood, 1833 species are in mostly a parasitoid of Lepidoptera, although some Hymenoptera, particularly Braconidae and Ichneumonidae have also been recorded as hosts (Gibson, 1993; Graham, 1995). *E. polistis* was reported in nests of *P. versicolor* (Dorfey & Köhler, 2011). We report for the first time *E. polistis* parasitizing *Polistes cinerascens*. We could observe that a large amount of *E. polistis* (796) emerged from two differents nests of *P. versicolor* and this can be explained due to the size of referred nest that was large and contained 1.646 cells.

In nests of Epiponini we observed parasitism by two species of *Brachymeria* Westwood, 1829 (Chalcididae), which are primary or hyperparasitoids of Lepidoptera (mostly young pupae) and only one species of Diptera (mostly mature larvae) (Fernández & Sharkey, 2006). Many Neotropical *Brachymeria* are associated to social wasp nests while some parasitize moths developing in wasp nests. Social wasps known as hosts for *Brachymeria* belong to the genera *Agelaia* Lepeletier, *Angiopolybia* Araujo, *Brachygastra* Petry, *Chartergus* Lepeletier, *Metapolybia* Ducke, *Polistes*, *Polybia* Lepeletier and *Synoeca* de Saussure (Boucek, 1992). In this study we reported for the first time *Brachymeria* parasitizing *Leipomeles spilongastra*.

Megaselia scalaris (Phoridae) is a cosmopolitan insect that is primarily a detritivore and might act as facultative predator of immature hosts (Disney & Berghof, 2005). It has been recorded the presence of eggs and larvae of this dipteran in honeycombs of *Apis mellifera* Linnaeus, 1758 resulting in the nest abandonment (Ronna, 1936). This fly has already been reported in nests of *Mischocyttarus cerberus* Ducke 1918, *Protopolybia acutiscutis* (Cameron, 1907), *Polybia occidentalis* (Olivier, 1791) and *Polybia simillima* Smith, 1862, causing serious injuries to the colonies (Young, 1984; Giannotti, 1998; London & Jeanne, 1998). In this study we reported for the first time *Megaselia scalaris* parasitizing *Polybia jurinei*.

With regard to the sex ratio of parasitoids that emerged from nests of social wasps, this can be influenced by many factors, such as environmental conditions and the body size of hosts (Townes, 1958; Hamilton, 1967; Taylor & Stern, 1971; Boldt et al., 1973; Vinson, 1997). Eggs deposited in hosts with larger body size can give rise to females descendants (as observed in this study to *Elasmus* and *Brachymeria*) since they require more resources to their development and consequently eggs deposited in hosts with smaller body size can give rise to male descendants (as observed in this study to *Toechorychus* and *Seminota*) (Taylor & Stern, 1971; Boldt et al., 1973). Strassmann (1981) suggested that males typically emerge first and can wait for the females emerge to mate. Moreover, it is possible the occurrence of females egg-laying in another pupa from the same nest in which it has emerged (Macom & Landolt, 1995), which contributes to a high number of parasitoids in the same nest as observed in this study.

Concluding Remarks

In this study we reported for first time *Toechorychus guarapuavus* and *Toechorychus fluminensis* parasitizing nests of *Mischocyttarus cassununga*, *Mischocyttarus consimilis* and *Mischocyttarus imitator*; *Elasmus polistis* parasitizing *Polistes cinerascens* and *Brachymeria* sp. parasitizing a nest of *Leipomeles*

spilongastra. We also reported for the first time *Megaselia scalaris* parasitizing *Polybia jurinei*. However, further studies involving more species should bring to light about relationships between Neotropical paper wasps and their parasitoids.

Acknowledgments

We sincerely thank to Reserva Ducke (Manaus, AM), RPPN Angelin Rainforest (Ubatuba, SP) and RPPN UNISC (Sinimbu, RS) for providing sampling permissions in their areas. To Conselho Nacional de Desenvolvimento Científico e Tecnológico (CNPq), Fundação de Amparo à Pesquisa do Estado do Amazonas (FAPEAM), Fundação de Amparo à Pesquisa do Estado de São Paulo (FAPESP, process 2011/13391-2) and Fundação de Amparo à Pesquisa do Estado da Bahia (FAPESB, n° BOL0494/2011) for scholarship supports. Thanks are also to Elijah Talamas (USDA-SEL) for his suggestions on an earlier version of the manuscript. We also thank to Coleção Entomológica de Santa Cruz do Sul for dates about the parasitoids from Rio Grande do Sul. To Dr. Fernando B. Noll (UNESP) for the help in the identification of Trigonalidae and some social wasps, Dr. Alexandre Aguiar (UFES) and Bernardo Santos for the help in the identification of Ichneumonidae, Dr. Marcelo Tavares and Bruno Cancián (UFES) for the help in the identification of Chalcididae and Dayana Alves da Silva (UEMS) by providing the nest and individuals of *Mischocyttarus consimilis*.

References

Arab, A., Pietrobon, T.A.O., Britto, F.B., Rocha, T., Santos, L., Barbieri, E.F. & Fowler, H.G. (2003). Key to the nests of Brazilian Epiponini wasps (Vespidae: Polistinae). Sociobiology, 42: 425-432.

Askew, R.R. (1971). Parasitic Insects. Heinemannd Educational Books, London, 316 p.

Bertoni, A.W. (1911). Contribucion a la biologia de las avispas y abejas del Paraguay. Annales del Museo Nacional de História Natural de Buenos Aires 15: 97-143.

Boucek, Z. (1992). The New World genera of Chalcididae of Hymenoptera. Memoirs of the American Entomological Institute, 53: 49-117.

Boldt, P.E.; Marston, N. & Dickerson, W.A. (1973). Differential parasitismo of several species of lepidopteran eggs by two species of *Trichogramma*. Environmental Entomology, 2: 1121-1122.

Brauns, S. (1905). Zwei neue Mesostenus aus Brasilien (Hymenoptera). Zeitschrift für Syst Hymenop und Dipter 5: 129-131.

Brown, B.V. (2010). Phoridae. In: B.V. Brown, A. Borkent, J.M. Cumming, D.M. Wood, N.E. Woodley & M.A. Zumbado (Eds.), Manual of Central American Diptera (pp 725-761).

NRC Research Press, Ottawa, Ont. Vol. 2.

Burks, B.D. (1960). A revision of the genus *Brachymeria* Westwood in America North of Mexico (Hymenoptera: Chalcididae). Transactions of the American Entomological Society, 86: 225-273.

Carmean, D. & Kimsey, L. (1998). Phylogenetic revision of the parasitoid wasp family Trigonalidae (Hymenoptera). Systematic Entomology, 23: 35-76

Carpenter, J.M. (1991). Phylogenetic relationships and the origin of social behavior in the Vespidae. In: K.G. Ross & R.W. Matthews (Eds.), The Social Biology of Wasps (pp 7-32). Cornell University Press, Ithaca.

Carpenter, J.M. & Marques, O.M. (2001). Contribuição ao Estudo dos Vespídeos do Brasil. Universidade Federal da Bahia, Departamento de Fitotecnia. Série Publicações Digitais, v. 3, CD-ROM.

Clausen, C.P. (1940). Entomophagous Insects. McGraw Hill, New York, London, 688 p.

Clouse, R.M. (1997). Are lone paper wasp foundresses mainly the result of sister mortality? Florida Entomologist, 60: 265-274.

Clouse, R.M. (2001). Some effects of group size on the output of beginning nests of *Mischocyttarus mexicanus* (Hymenoptera: Vespidae). Florida Entomologist, 84: 418-425.

Costa-Lima, A.M. (1962). Insetos do Brasil. Rio de Janeiro, Escola Nacional de Agronomia. Série Didática, número 14, 389 p.

Disney, R.H.L. & Berghoff, S.M. (2005). New species and new records of scuttle flies (Diptera: Phoridae) associated with army ants (Hymenoptera: Formicidae) in Trinidad and Venezuela. Sociobiology, 45: 887-898.

Dorfey, C. & Köhler, A. (2011). First report of *Elasmus polistis* Burks (Hymenoptera: Eulophidae) recovered from *Polistes versicolor* (Olivier) (Hymenoptera: Vespidae) nests in Brazil. Neotropcal Entomology, 40: 515-516.

Fernández, F. & Sharkey, M.J. (2006). Introduccíon a los Hymenoptera de la Rgíon Neotropical. Sociedad Colombiana de Entomologia y Universidad Nacional de Colombia, Bogotá, 894 p.

Giannotti, E. (1998). The colony cycle of the social wasp, *Mischocyttarus cerberus styx* Richards, 1940 (Hymenoptera, Vespidae). Revista Brasileira de Entomologia, 41: 217-224.

Gibson, G.A.P. (1993). Family Elasmidae. In: H. Goulet & J.T. Huber (Eds.), Hymenoptera of the World: an identification guide to families (pp 626-668). Ottawa, Research Branch Agriculture Canada Publication.

Graham, M.V.R. de V. (1995). European *Elasmus* (Hymenoptera: Chalcidoidea: Elasmidae) with a key and descriptions of five new species. Entomologist's Monthly Magazine 131: 1-23.

Hamilton, W.D. (1967). Extraordinary sex ratios. Science, 156: 477-488.

Hanson, P.E. & Gauld, I.D. (1995). The Hymenoptera of Costa Rica. Oxford University Press, Oxford, 893 p.

Jeanne, R.L. (1980). Evolution of social behavior in the Vespidae. Annual Review of Entomology, 25: 371-396.

Kumagai, A.F. & Graf, V. (2000). Ichneumonidae (Hymenoptera) de áreas urbana e rural de Curitiba, Paraná, Brasil. Acta Biologica Paranaense, 28:153-168.

London, K.B. & Jeanne, R.L. (1998). Envelopes protect social wasps' nests from phorid infestation (Hymenoptera: Vespidae, Diptera: Phoridae). Journal of the Kansas Entomological Society, 71: 175-182.

Macom, T.E. & Landolt, P.L. (1995). *Elasmus polistis* (Hymenoptera: Eulophidae) recovered fro nests of *Polistes dorsalis* (Hymenoptera: Vespidae) in Florida. Florida Entomologist, 78: 612-614.

Makino, S. (1985). Listo f parasitoides of Polistinae wasps. Sphecos 10: 19-25.

Richards, O.W. (1978). The social wasps of the Americas (excluding the Vespinae). London: British Museum of Natural History, 580 p.

Ronna, A. (1936). Observações biologicas sobre dois dipteros parasitas de *Apis mellifica* L. (Dipt. Phoridae, Sarcophagidae). Revista de Entomolgia, 6: 1-9.

Santos, E.F. & Noll, F.B. (2013). Biological notes on the parasitism of *Apoica flavissima* Van der Vecht (Hymenoptera: Vespidae) by *Seminota marginata* (Westwood) (Hymenoptera: Trigonalidae): Are social paper wasps primary or secundary hosts of Trigonalidae? Sociobiology, 60: 123-124.

Soares, M.A., Gutierrez, C.T., Zanuncio, J.F., Bellini, L.L., Prezotto, F. & Serrão, J.E. (2006). *Pachysomoides* sp. (Hymenoptera: Ichneumonidae: Cryptinae) and *Megaselia scalaris* (Diptera: Phoridae) parasitoids of *Mischocyttarus cassununga* (Hymenoptera: Vespidae) in Viçosa, Minas Gerais State, Brazil. Sociobiology, 48: 673-680.

Somavilla, A., Oliveira, M.L. & Silveira, O.T. (2012). Guia de identificação dos ninhos de vespas sociais (Hymenoptera, Vespidae, Polistinae) na Reserva Ducke, Manaus, Amazonas, Brasil. Revista Brasileira de Entomologia, 56: 405-414.

Strassmann, J.E. (1981). Parasitoids, predators, and group size in the paper wasp, *Polistes exclamans*. Ecology, 62: 1225-1233.

Taylor, T.A. & Stern, V.M. (1971). Host preference studies with the egg parasite *Trichogramma semifumatum* (Hymenoptera: Trichogrammatidae). Annals of the Entomological Society of America, 64: 1381-1390.

Townes, H.K. (1958). Some biological characteristics of the Ichneumonidae (Hymenoptera) in relation to biological control. Journal of Economic Entomology, 51:650-652.

Vinson S.B. (1997). Comportamento de seleção hospedeira de parasitoides de ovos, com ênfase na família Trichogrammatidae.

In: J.T.P. Parra & R.A. Zuchhi (Eds.), *Trichogramma* e o controle biológico aplicado (pp 67-119). Piracicaba, FEALQ.

Yamane, S. (1996). Ecological factors influencing the colony cycle in *Polistes* wasps. In: S.E. Turillazzi & M.J. West-Eberhard (Eds.), Natural History and Evolution of Paper Wasps (pp 75-97). New York: Oxford University Press.

Young, A.M. (1984), Mechanism of pollination by Phoridae (Diptera) in some *Herrania* species (Sterculiaceae) in Costa Rica. Proceedings of the Entomological Society of Washington, 86: 503-518.

Weinstein, P. & Austin, A.D. (1991). The host-relationships of trigonalyid wasps (Hymenoptera: Trigonalyidae), with a review of their biology and catalogue to world species. Journal of Natural History, 18: 209-214.

Wenzel, J.W. (1998). A generic key to the nests of hornets, yellowjackets, and paper wasps worldwide (Vespidae: Vespinae, Polistinae). American Museum Novitates, 3224: 1-39.

16

Alternative Control of the Leaf-Cutting Ant *Atta bisphaerica* Forel (Hymenoptera: Formicidae) Via Homeopathic Baits

VM RAMOS, F CUNHA, KC KHUN, RGF LEITE, WF ROMA

1 - Universidade do Oeste Paulista, Presidente Prudente, SP, Brazil

Key words

ant control, pests control, Attini, grass-cutting ants

Corresponding author

Vânia Maria Ramos
Laboratório de Entomologia Agrícola
Faculdade de Ciências Agrárias
Universidade do Oeste Paulista
Rodovia Raposo Tavares km 572
Presidente Prudente, SP, Brazil
19067-175
E-Mail: vaniaramos@unoeste.br

Abstract

Leaf-cutting ants are pests that afflict diverse crops, and are most efficiently controlled by chemical methods that are widely utilized. Other methods have been investigated aiming to efficiently control these insects while reducing the environmental impact of applying such chemical products. Therefore, an assay was conducted to evaluate the efficiency of baits, formulated homeopathically, in nests of the leaf-cutting ant *Atta bisphaerica*, in the field. Thirty (30) colonies were chosen and divided into 10 repetitions for each of the following treatments: control (without baits), standard (8 g/m^2 of loose soil of baits based on sulfluramid 0.3%) and homeopathic (60 g/m^2 of loose soil of homeopathic baits parceled into 20g/m^2 doses applied on 3 consecutive days). At 24 hours after bait application on active foraging trails of colonies, the evaluation of of the following parameters was initiated: transport and devolution of the baits, foraging and mortality. The completed assay demonstrated that the transport of baits was greater in the standard (80%) than in the homeopathic treatment (50%). On the other hand, the devolution of baits was significantly higher in the homeopathic treatment (15%) versus the conventional, where devolution/rejection did not occur. Colony mortality was 20% under the homeopathic treatment, differing statistically from the 80% value produced by the standard treatment. Thus, the homeopathic treatment is not demonstrated to be efficient at controlling leaf-cutting ants, suggesting the need for new studies with different methodologies.

Introduction

The insects of the family Formicidae are known for their highly organized colonies that consist of millions of individuals. In terrestrial ecosystems, ants may constitute 15 to 20% of the total animal biomass (Schultz, 2000). The leaf-cutting ants belong to the tribe Attini, are present throughout tropical portions of the Americas and are the greatest consumers of vegetal mass in Brazil when compared with other insects and even with mammals. When they carry the leaves under the soil, the large quantity of organic material that is made available becomes a source of carbon and other nutrients for other organisms. These insects play an important role in ecosystems, but, despite the benefits they confer, they become pests when in proximity to agriculture and need to be controlled (Santos, 2010).

The environmental impact of anthropogenic activity directly affects terrestrial ecosystems, such as, for example, the application of agrochemicals. Before the physicochemical dispersion or biotransformation of insecticides, they can act as an important element in the disruption of trophic interactions in terrestrial ecology (Tiepo et al., 2010). Large quantities of insecticides are used for combating pests, but the introduction of xenobiotic molecules in the environment can damage the equilibrium of the ecosystem because synthetic insecticides may present high toxicity and, generally, low biodegradability, which can lead to a persistent toxic action (Tiepo et al., 2010). A large amount of chemical pesticides is utilized to meet commercial demands of industry and consumers, but given the growing public concern about the presence of their residues in water supplies or food, many producers are coming to adopt production systems that are purely organic or more ecologically integrated (Boff et al., 2008).

In order to reduce the harm caused by leaf-cutting ants, humans have pursued diverse forms of control, ranging from homemade methods to the use of advanced techniques (Souza et al., 2011). The economic and environmental aspects have led businesses to improve the operational yield of the chemical control techniques employed (baits and thermal nebulization), and to experiment with new technologies and new toxic active ingredients (Boaretto & Forti, 1997). During the long-term evolution in the methods of controlling leaf-cutting ants, one of the great concerns has been to reconcile the conflict among the three principles of efficiency, economy and safety (Cantarelli et al., 2005). As a consequence of the unfavorable aspects presented by granulated baits (deterioration of the environment, elimination of natural enemies and emergence of resistance), research lines have been generated to discover products with greater specificity and smaller environmental impact (Hebling et al., 2000).

In developed countries, intensive agriculture has not only augmented agricultural productivity, but also, due to its high dependence on great quantities of nonrenewable energy and raw materials, frequently has resulted in soil degradation, environmental pollution and damage to wildlife. For this reason, recent years have seen growing interest in agricultural methods that are both economically and environmentally sound (Betti et al., 2009). Among these, the agrohomeopathy has raised the interest of researchers. Its potential benefits are significant because homeopathic preparations, on account of their extremely high dilution, are relatively inexpensive, have little or no ecological side effect and appear to be, when considered jointly, inoffensive (Elmaz et al., 2004; Lotter, 2003).

All of these attributes make homeopathy optimally adjusted to the holistic approach of organic agriculture, and above all, biodynamic, in which the plants and their interactions with the environment are treated as a unified living organism (Carpenter-Boggs et al., 2000; Heimler et al., 2009). Furthermore, the new approach of applying homeopathic principles can also add to the improvement of nutritional properties (for example the levels of components that induce physiological benefits to human health) (Fonseca et al., 2006) and physiological and qualitative characteristics of plants, in addition to their resistance to stress factors of both biotic (insects and pathogens) and abiotic in nature (physical and chemical damage) (Betti et al., 2009).

In recent years, increasing levels of resistance to insecticides and concerns about insecticide residues in agricultural products have stimulated a growing demand for products cultivated under new strategies of control, raising, in this context, the question of whether homeopathic preparations are able to control pest species (Wiss et al., 2010).

Material and Methods

The assay was conducted on a cattle ranch located in Presidente Prudente, Sao Paulo state, Brazil, in the period from August to October of 2011. The experimental area was constituted by a pasture composed of *Brachiaria decumbens* and some spots of *Paspalum notatum*, with the presence of colonies of the leaf-cutting ants *Atta bisphaerica* and *Atta capiguara*. From the experimental area 30 adult colonies of *Atta bisphaerica* were selected and identified with numbered wooden stakes one meter in height. After this procedure, the area measurements of loose soil were estimated for each colony, with the aid of a measuring tape, namely, the greatest length and greatest width. The quantity of baits to be applied to each colony was calculated from its apparent external area.

The following treatments were applied: T1 – control (0 g of baits/m^2 of loose soil), T2 - standard (8 g of commercial bait based on 0.3% of sulfluramid per m^2 of loose soil) and T3 – homeopathic (baits formulated homeopathically and applied at the dose of 60g/m^2 of loose soil). Treatment T3 was formulated in a private commercial laboratory and the chosen dose had been recommended by the manufacturer, being that the total applied in each colony was parceled into 3 equal portions of 20g/m2 of loose soil, applied for the next 3 days. Each treatment was composed of 10 repetitions, with each repetition constituting one colony.

Prior to application in the field, the baits were weighed in the Laboratory of Entomology at Western Paulista University, according to the measures obtained in the field for each colony, with the total quantity of each repetition being individualized in a plastic sack identified and duly stored until their use. In the field, before application of the treatments, for each repetition one active foraging trail near its area of loose soil was selected to ensure the immediate transport of granules and to avoid competition with workers of other colonies. The parameters evaluated were: transport and devolution of the baits, foraging, presence of recent loose soil, intoxication of individuals and mortality, according to established methodology as a protocol for experiments of this nature (Zanuncio et al., 1997; Nagamoto et al., 1999; Ramos et al., 1999; Zanuncio et al., 2002; Forti et al., 2003; Zanetti et al., 2004).

The transport and devolution of baits were evaluated 24h, 48h and 72h after application, and attributed ratings of 0, 1, 2, 3 or 4, to represent the approximate respective equivalents of 0%, 25%, 50%, 75% and 100% of "pellets" transported and returned by the ants. The transport was evaluated by observing the trails were the baits had been applied, to verify the remaining quantity, while the devolution was evaluated by observing the mound of loose soil of each colony, in order to visualize and quantify the granules deposited on the ground, in relation to the quantity effectively applied.

At 7, 15, 30, 60 and 90 days after installation of the experiment in the field, the remaining parameters were evaluated, verifying the presence or absence of the following: a) foraging – presence of individuals foraging on trails around the nests or on their loose-soil mound, b) loose soil – presence of granules of loose dirt on the mounds of nests, indicating recent activity of excavation, and c) intoxication: presence of young individuals (with clear tegument) on the loose-soil mound, signaling intoxication inside the nests, since under normal conditions, such workers are engaged only with tasks internal to the colonies. These parameters were used to supply data for evaluating the death (or absence thereof) among the colonies.

In the final evaluation, at 90 days, as a function of the established criteria, the presence or absence of mortality was attributed to the colonies, considering dead those that did not show any signal of recent cutting or transporting of leaves and excavation of soil, with their mound of compacted loose soil and their foraging trails inactive.

From the obtained data the mean percentages were calculated, and the results transformed into arcsine $\sqrt{x + 0.5/100}$, and then submitted to analysis of variance, comparing the means by the test of Tukey at a 5% probability level, in an entirely randomized model of experimental design.

Results

The transport of baits by the ants to the interior of the nests produced an 80% value under the standard treatment versus 50% via the homeopathic approach, differing statistically. The control treatment was represented by the value 0 on account of not having received baits (Fig. 1-A).

The devolution of the pellets transported by ants occurred only in homeopathic treatment, at the rate of 15%, differing from the control and standard treatments, which did not present devolution (Fig. 1-B).

The treatments differed as to efficiency, evaluated through the mortality of the colonies, being null in the control treatment, as expected, 20% in the homeopathic treatment and 80% in the conventional treatment (Fig .1-C).

Discussion

The 80% bait transport rate under the conventional treatment had been expected since prior experiments utilizing products formulated with the same active ingredient at the same concentration also obtained a high pellet transport rate in the field (Zanuncio et al., 1997; Zanetti et al., 2004). The transport of baits is one of the parameters that define treatment efficiency since the product can only act and express its potential if it is found attractive initially, accepted after investigation and then effectively transported by workers. The baits formulated based on homeopathy were not shown to be sufficiently attractive to the workers, as reflected in the unsatisfactory pellet transport value obtained (50%),which may have negatively influenced the efficiency of the treatment. The cause of the low rate of transport for homeopathic baits was not investigated, and thus it could not be affirmed whether it is a function of homeopathic formulation or of some other random factor such as the unknown quality of citric pulp utilized in the manufacturing of baits.

The devolution of baits, as measured by the quantity of granules deposited on the mound of loose soil, was not ob-

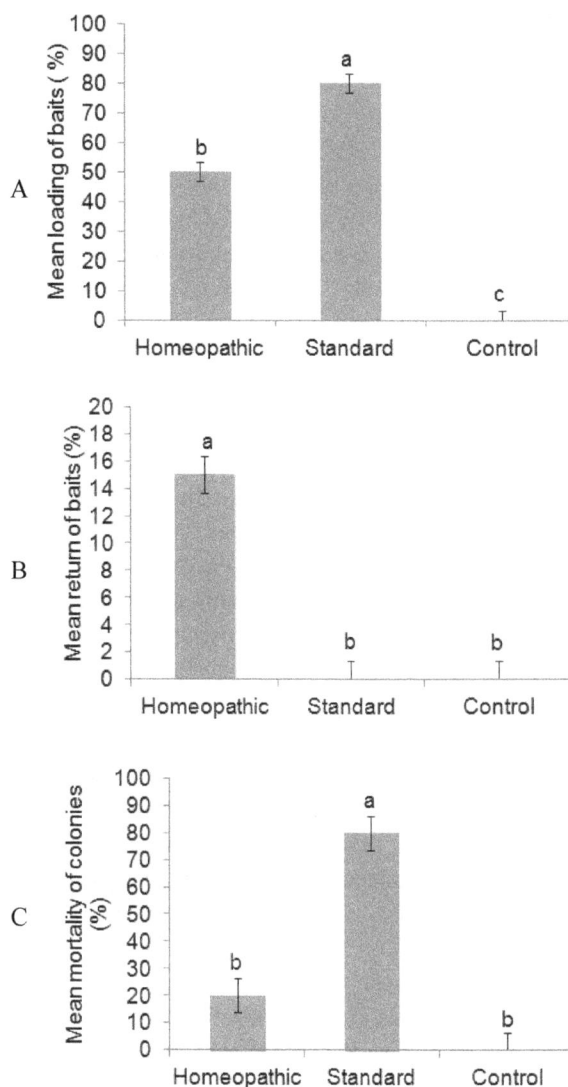

Figure 1. Transport of baits by workers (A), Devolution of baits by workers (B) and, Mortality of *Atta bisphaerica* colonies (C), 90 days after application of treatments, in the field. Presidente Prudente, SP, Brazil, 2011 (Mean percentage values ± mean standard error, obs.: means followed by the same letter do not differ from each other for $P < 0.05$ by the test of Tukey).

served in the standard treatment, a recurrent fact in field and laboratory assays that similarly evaluated sulfluramid-based baits (Zanuncio et al., 1997; Zanetti et al., 2004). In a contrary and significant manner, a portion (15%) of the homeopathic baits transported were returned by workers onto the loose-soil mound soon after transport, thus demonstrating that, for some reason, they were rejected during the process of post-selecting foraged material, which occurs in the interior of the colonies. Camargo et al. (2003) observed the occurrence of a post-selection process in colonies of *Acromyrmex subterraneus brunneus* when the colonies were offered, in the laboratory, different inert materials such as plastic, polystyrene and clay, which were rapidly differentiated and not selected by the workers for cultivation of the symbiotic fungus. The authors reported that the post-selection of foraged material constitutes strong evidence of the cognitive abilities of workers and of the colony as a whole. Devolution is an important parameter to be evaluated in experiments of this nature since it demonstrates the acceptance or rejection of a product by the ants. Thus, it becomes evident that the lethality of the active ingredient to leaf-cutting ants is necessary but not sufficient since it may be prematurely perceived and rejected by the workers when formulated into baits, a scenario that would compromise the efficiency of the product, which may have occurred in the present assay.

The efficiency of treatments, represented by the mortality of the colonies, was significantly superior in the standard treatment. The high mortality rate of colonies that received application of sulfluramid baits had been expected since there are reports in the literature of innumerable studies of the same nature that presented similar results (Zanuncio et al., 1997; Zanuncio et al., 2002; Zanetti et al., 2004). Sulfluramid is included among the active ingredients currently used for controlling leaf-cutting ants jointly with chlorpyrifos, given that chlorpyrifos is more toxic to mammals, aquatic organisms, fish and bees than sulfluramid (Tiepo et al., 2010).

The effect of the homeopathic treatment on the colonies was inconsistent, a coherent finding if considered a function of low transport of the baits presented and of the devolution rate, despite the low mortality observed. Geisel et al. (2012) found that homeopathic preparations of adults and *Acromyrmex* fungus reduced the activity of ant foraging trails in the field, from the sixth day of application (spraying), extending this effect until 20 days after last application. These results do not agree with those obtained in this test, but it is hard to compare them because the implementation and evaluation methodologies are completely different, with different goals, since in the Geisel et al. (2012) study it was not evaluated the mortality of colonies, the main focus of this study.

Similarly, in an assay conducted on potato plants sprayed with diverse homeopathic preparations, in order to verify the incidence of diseases (*Phytophthora infestans* and *Alternaria solani*) and pests (*Diabrotica speciosa*), Boof et al. (2008) found no significant difference between the homeopathic products tested and the control (water), for all parameters evaluated. In another study, aiming to evaluate the effect of homeopathic preparations on the aphid *Dysaphis plantaginea* in apple seedlings, Wyss et al. (2010) reported no significant difference between homeopathic treatments and the control, by evaluating the quantity of individuals in the seedlings, the damage to the leaves and the fresh weight of the plants.

In an assay performed by Tiepo et al. (2010) to assess a natural formicide considered a "green pesticide," the findings were promising. The product is a mixture of caffeine and common fatty acids, with citric pulp and apple as attractive agents. When its effects were observed on microbiological soil activity and the biomass of worms and plants, short-term toxicity was not detected. The authors reported that the for-

micide did not present toxic action on the experimental organisms on account of its natural composition, but that when the leaf-cutting ants ingested this product, despite not dying immediately, manifested behavioral disturbances. Field observations, according to the authors, demonstrate a disruption in the social structure of the colony, which leads to the subsequent death of the individuals, considering that the communication constitutes the basis of the ant social structure. The term "green pesticide" should be understood as a natural or synthetic pesticide produced according to the principles of "green chemistry", acting specifically and effectively on the precise target without deleterious or dangerous effects on non-target components of the ecosystem (Tiepo et al., 2010). It is expected that the toxicity of this new and necessary class of "green pesticides" would be low or nonexistent (Kahkonen & Nordstrom, 2008).

The data obtained herein do not justify a recommendation to control leaf-cutting ants by homeopathic baits. However, it is important to emphasize that, to the best of our knowledge, the literature contains no other scientific data on studies of this nature, a scenario that does not permit us to accept or reject definitively the hypothesis that homeopathy is a viable alternative for controlling these insects. In a review study on the use of homeopathic preparations in agriculture, to control pests and pathogens, Betti et al. (2009) concluded that most of the works published and consulted do not provide sufficient information to enable a clear interpretation, in particular, due to the statistical analysis being inadequate or entirely absent, the number of repetitions being unspecified and, frequently, the experimental methodology being poor. Even so, the authors concluded that the studies evaluated can serve as a starting point for future experiments that are more comprehensive and controlled. Similarly, the results presented in the present assay do not permit us to resolve the question of whether homeopathy can control leaf-cutting ants, but rather call for initiation or expansion of this research line for greater investigation, conducted within scientific premises necessary to achieve a complete understanding of the subject.

References

Betti L., Trebbi G., Majewsky V., Scherr C., Shah-Rossi D., Jäger T., Baumgartner S. (2009). Use of homeopathic preparations in phytopatological models and in field trials: a critical review. Homeopathy, 98: 244-266. doi:10.1016/j.homp.2009.09.008.

Boaretto M.A.C., Forti L.C. (1997). Perspectivas no controle de formigas cortadeiras. Série Técnica IPEF,30: 31-46.

Boff P., Madruga E., Zanelato M., Boff M.I.C. (2008). Pest and disease management of potato crops with homeopathic preparations and germplasm variability. IFOAM Organic World Congress (http://orgprints.org/view/projects/conference.html).

Camargo R.S., Forti L.C., Matos C.A.O., Lopes J.F., Andrade A.P.P., Ramos V.M. (2003). Pos-selection and devolution of foraged material by *Acromyrmex subterraneus brunneus* (Hymenoptera: Formicidae). Sociobiology, 42: 93-102.

Cantarelli E.B., Costa E.C., Oliveira L.S., Perrando E.R. (2005). Efeito de diferentes doses do formicida "Citromax" no controle de *Acromyrmex lundi* (Hymenoptera: Formicidae). Cienc. Florest., 15: 249-253.

Carpenter-Boggs L., Reganold J.P., Keneddy A.C. (2000). Biodynamic preparations: short-term efects on crops, soils and weed populations. Am. J. Alternative Agr., 15: 110-118.

Elmaz O., Cerit H., Ozcelik M., Ulas S. (2004). Impact of organic agriculture on the environment. Fresen. Environ. Bull., 13: 1072-1078.

Fonseca M.C.M., Days-Casali, V.W., Cecon P.R. (2006). Efeito de aplicação única dos preparados homeopáticos *Calcarea carbonica, Kalium phosphoricum, Magnesium carbonicum, Natrium muriaticum* e *Silicea terra* no teor de tanino em *Porophyllum ruderale* (Jacq.) Cassini. Cult. Homeopat., 14: 6-8.

Forti L.C., Nagamoto N.S., Ramos V.M., Andrade A.P.P., Santos J.F.L., Camargo R.S., Moreira A.A., Boaretto M.A.C. (2003) Eficiencia de sulfluramida, fipronil y clorpirifos como sebos en el control de Atta capiguara Gonçalves (Hymenoptera:Formicidae). Pasturas Tropicales, 25: 28-35.

Geisel A., Boff M.I.C., Boff P. (2012). The effect of homeophatic preparations on the activity level of *Acromyrmex* leaf-cutting ants. Acta Sci. Agron., 34: 445-451. doi:10.4025/actasciagron.v34i4.14418.

Hebling M.J.A., Bueno O.C., Pagnocca F.C., Da Silva, O.A., Marotti P.S. (2000). Toxic effects of *Canavalia ensiformis* L. (Legiminosae) on laboratory colonies of *Atta sexdens rubropilosa* (Hymenoptera: Formicidae). J. Appl. Entomol., 124: 33-35.

Heimler D., Isolani L., Vignolini P., Romani A. (2009). Polyphenol content and antiradical activity of *Chocorium intybus* L. from biodynamic and conventional farming. Food Chem., 114: 765-770.

Kahkonen E., Nordstrom M.K. (2008). Toward a nontoxic poison: current trends in (European Union) biocides regulation. Integr. Environ. Assess. Manag., 4: 471-477.

Lotter D.W. (2003). Organic agriculture. J. Sust. Agric., 21: 59-128.

Nagamoto N.S., Forti L.C., Ramos V.M., Garcia M.J.M. (1999). Eficiência das iscas Mirex-S Max, Blitz e Pikapau para o controle da saúva *Atta capiguara* Gonçalves (Hymenoptera: Formicidae) em condições de campo. Naturalia, 24: 277-278.

Ramos V.M., Forti L.C., Andrade A.P.P., Camargo R.S., Souza F.S. (1999). Eficiência do produto Mirex-S Max no controle de *Atta capiguara* Gonçalves (Hymenoptera: Formicidae) e análise de resíduos de sulfluramida em capim e solo. Naturalia, 24: 283-285.

Santos M.P.A. (2010). Avaliação do formicida Citromax à base de fipronil no combate às saúvas (*Atta sexdens*). Revista Controle Biológico On Line, 2: 22-27 (http://www.ib.unicamp.br/profs/eco_aplicada).

Schultz T.R. (2000). In search of ant ancestors. Proc. Natl. Acad. Sci., 97: 14028-14029.

Souza M.D., Peres Filho O., Dorval A. (2011). Efeitos de extratos naturais de folhas vegetais em *Leucoagaricus gongylophorus* (Möller) Singer, (Agaricales: Agaricaceae). Ambiencia, 7: 461-471. doi:10.5777/ambiencia. 2011.03.04.

Tiepo E.N., Corrêa A.X.R., Resgalla C., Cotelle S., Férard, J.F., Radetski C.M. (2010). Terrestrial short-term ecotoxicity of a green formicide. Ecotoxicol. Environ. Saf., 73: 939-943.

Wiss E., Tamm L., Siebenwirth J., Baumgartner S. (2010). Homeophatic preparations to control the rosy apple aphid (*Dysaphis plantaginea* Pass.). The Scient. World J., 10: 38-48.

Zanetti R., Days N., Reis M., Souza-Silva A., Moura M.A. (2004). Efficiência de iscas granuladas (sulfluramida 0.3%) no controle de *Atta sexdens rubropilosa* Forel, 1908 (Hymenoptera: Formicidae). Cienc. Agrotec., 28: 878-882.

Zanuncio J.C., Santos G.P., Firme D.J., Zanuncio T.V. (1997). Uso da isca granulada com Sulfluramida 0,3 % no controle de *Atta sexdens rubropilosa* Forel, 1908 (Hymenoptera; Formicidae). Cerne, 3: 47-54.

Zanuncio J.C., Sossai M.F., Oliveira H.N. (2002). Influência das iscas formicidas Mirex-S Max e Blitz na paralisação de corte e no controle de *Atta sexdens rubropilosa* (Hymenoptera: Formicidae). Rev. Arvore, 26: 237-242.

17

Utility of the ITS1 Region for Phylogenetic Analysis in Stingless Bees: a Case Study of the Endangered *Melipona yucatanica* Camargo, Moure and Roubik (Apidae: Meliponini)

C Ruiz[1], W de J May-Itzá[2], JJG Quezada-Euán[2], P de La Rúa[3]

1 - Universidad Técnica Particular de Loja, Loja, Ecuador

2 - Universidad Autónoma de Yucatán, Yucatán, México

3 - Universidad de Murcia, Murcia, España

Keywords

Stingless bees, *Melipona*, ITS1, phylogeny, Mesoamerica.

Corresponding author

Pilar De la Rúa
Área de Biología Animal, Departamento de Zoología y Antropología Física, Facultad de Veterinaria, Campus de Espinardo, Universidad de Murcia, 30100 Murcia, España
E-Mail: pdelarua@um.es

Abstract

The internal transcribed spacers of the ribosomal RNA gene have been recently proposed as an appropriate marker for genetic analysis of the molecular variation of stingless bees. Herein we report the characterization of the complete ITS1 region in two populations (from Mexico and Guatemala) of the endangered *Melipona yucatanica* Camargo, Moure and Roubik. Phylogenetic analyses showed low genetic variation between populations but defined a geographic structure with Mexican and Guatemalan specimens forming two well supported clades. Low ITS1 genetic variation found between populations contrasts with high genetic variation found in other markers. Phylogenetic analysis corroborates the inclusion of *M. yucatanica* within the subgenus *Melipona sensu stricto* based on previous morphological studies. The results highlight the utility of the ITS1 for the characterization of stingless bee populations.

Introduction

Molecular studies addressing the stingless bees are needed not only to support preliminary morphology-based hypotheses about phylogeny, population dynamics, species delimitation and evolution, but also to establish the basis of conservation programs for this group of bees (Arias et al., 2006). Towards these ends, a phylogeny of the stingless bee genus *Melipona* has been published (Ramírez et al., 2010) based on a multigene data set (nuclear: EF1-α, ArgK and RNA Pol-II genes; mitochondrial: cox1 and rRNA genes: *16S*) approach. In that study 35 out of the 50 *Melipona* species known-to-date were grouped in four subgenera previously described with morphological and ecological characters (Camargo & Pedro, 2007), but still some *Melipona* species remain unassigned to a specific subgenus.

Other suitable markers for phylogeny and population genetic analyses are the internal transcribed spacers (ITS1 and ITS2) of the ribosomal genes (rRNA). Concerted evolution may have led to the homogeneity of repeat motifs within the ITS regions (Hillis & Dixon, 1991). Therefore, they have been successfully applied for population and phylogenetic studies of some stingless bee species. In particular, in the tribe Meliponini, Fernandes-Salomão et al. (2005) determined the complete ITS1 sequence from three *Melipona* species and used partial sequences of eight species to infer its phylogenetic relationships. At the intraspecific level, the ITS1 region has demonstrated its usefulness for population analysis in the Mexican species *Melipona beecheii* Bennett (May-Itzá et al., 2009; 2012) and also in two Brazilian species: *Melipona subnitida* Ducke (Cruz et al., 2006) and *Melipona quinquefasciata* Lepeletier (Pereira et al., 2009).

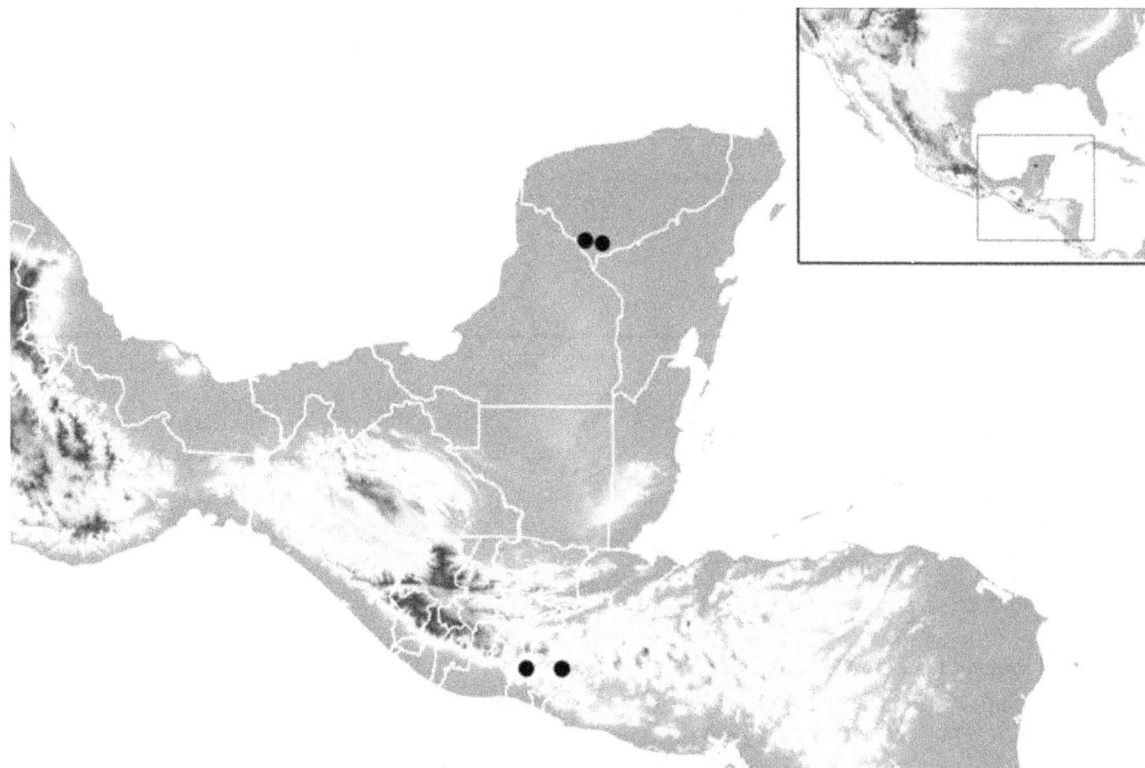

Fig 1. Study area and sampling localities of *Melipona yucatanica* in Mesoamerican region.

The Mexican stingless bee *Melipona yucatanica* Camargo, Moure and Roubik is a small bee (8 mm in average) with less than 200 worker bees per colony (Camargo et al., 1988). This species is associated with primary forest, and based on its patched distribution, various authors have suggested a recent fragmentation of *M. yucatanica* populations due to massive deforestation (Camargo et al. 1988; Ayala, 1999). Preliminary molecular studies in Mexican and Guatemalan populations of *M. yucatanica* yielded different RFLP patterns in the ITS2 region (De la Rúa et al., 2007), suggesting allopatric speciation in populations geographically separated. Furthermore, morphometric and Bayesian analyses of the mitochondrial cox1 region and microsatellite loci revealed geographic differences between Guatemalan and Mexican populations, suggesting that *M. yucatanica* from México and Guatemala could represent two distinct species (May-Itzá et al., 2010).

Because of the rapid disappearance of its habitat and the difficulty to multiply colonies domestically, conservation measures to preserve *M. yucatanica* are urgently needed (May-Itzá et al., 2010). The taxonomic position of this species based on morphological comparative studies have placed this species within the subgenus *Melipona s. str.* (Camargo & Pedro, 2007), in relation to other South-American species (Moure et al., 2007). However its taxonomic position has not been assessed on molecular grounds yet.

Herein, the entire ITS1 region of *M. yucatanica* has been sequenced and compared with available ITS1 (complete or 3' end) from other *Melipona* species as well as with previous results on *M. yucatanica* based on morphometry and molecular markers (De la Rúa et al., 2007; May-Itzá et al., 2010). We aimed to i) analyze the molecular diversity underlying this marker, and ii) test its phylogenetic signal to verify whether this species is included within the morphologically described subgenus *Melipona s. str.* as suggested by Camargo and Pedro (2007).

Material and Methods

Sampling

Feral colonies of *M. yucatanica* are difficult to find since they are restricted to preserved rain-forest, so we were only able to sample worker bees (10-20 worker bees per colony) from managed colonies located in two geographically distant Mesoamerican regions where they naturally occur: seven in México and four in Guatemala (Table 1, Fig 1). These locations showed notable climatic and geological differences: the Pacific region of Guatemala is a mountainous region, whereas the Atlantic-Caribbean region of México is a flat limestone area without mountainous zones. One specimen of each colony has been deposited in the Insect Collection of the Department of Zoology and Physical Anthropology (University of Murcia, Spain).

DNA extraction, amplification and sequencing

DNA was extracted from two worker bees per colony using the DNeasy tissue kit (QIAGEN). The ITS1 region was amplified using the primers *cas*18sf1 and *cas*5p8sB1d

Table 1. Sampling localities of *Melipona yucatanica* specimens in included in the study, geographical information and accession numbers of sequences in GenBank.

Taxon ID	Accession number	Country	State	Locality	Geographical coordinates	Altitude
MyucMer1	HQ651727	México	Yucatán	López Portillo	19.9351278 N -89.1879278 W	142 m
MyucMer2	KF564809					
MyucMer3	KF564810					
MyucMer4	KF564811					
MyucMer5	KF564812	México	Yucatán	Huntochac	19.8061861 N -89.5007389 W	80 m
MyucMer6	KF564813					
MyucMer7	KF564814					
MyucGuat1	HQ651726	Guatemala	Jutiapa	Jutiapa	14.2980778 N -89.8970972 W	898 m
MyucGuat2	KF564806					
MyucGuat4	KF564807	Guatemala	Santa Rosa	Barberena	14.3072639 N -90.3564139 W	1220 m
MyucGuat5	KF564808					

based on the conserved sequences of the 18S and 5.8S rDNA respectively (Ji et al., 2003). The amplification program consisted of 5 min. at 96 °C, 34 cycles of 45 sec. at 96 °C, 1 min. at 60 °C and 1 min. at 72 °C, and a final extension of 10 min. at 72 °C. PCR reactions were performed with PCR beads PureTaq™ Ready-To-Go™ (GE Healthcare) in a thermocycler PTC-200 (Biorad). PCR products were purified with QIAquick PCR purification kit (QIAGEN) before directly sequencing. Sequencing was performed in both directions with the same PCR primers in an ABI 3730 DNA analyzer (Applied Biosystems) at the sequencing company Secugen S. L. (Madrid, Spain). ITS1 sequences edited with the program MEGA 5 (Tamura et al., 2011) were deposited in Genbank with the accession numbers HQ651726 (*M. yucatanica* from Guatemala) and HQ651727 (*M. yucatanica* from México).

Data analyses

The ITS1 datasets used included the complete ITS1 gene for *M. yucatanica, M. beecheii* and *Melipona quadrifasciata* Lepeletier and partial sequences of ITS1 3' end for other nine *Melipona* species obtained from Genbank (Fernandes-Salomão et al., 2005; Cruz et al., 2006; Pereira et al., 2009). In order to compare ITS1 results with other markers, a multiple-gene approach was carried out. Sequence data of mitochondrial cox1 were obtained from Genbank for *M. yucatanica,* and for the species with available ITS1 sequences (namely, *M. quinquefasciata, M. beechei, Melipona compressipes* Fabricius, *M. quadrifasciata, Melipona mandacaia* Smith, *Melipona bicolor* Lepeletier, *Melipona marginata* Lepeletier, *Melipona rufiventris* Lepeletier and *Melipona scutellaris* Latreille). Separated and combined analyses were carried out.

ITS1 sequences were aligned with MAFFT software (http://mafft.cbrc.jp/alignment/server/) using E-INS-i strategy. Parameter of scoring matrix for nucleotide sequences

was set to 20PAM/ k = 2 as recommended when aligning related DNA sequences (Katoh & Toh, 2008).

ITS1 and cox1 genetic diversity in *M. yucatanica* and in other available *Melipona* sequences were measure in DNAsp (Librado & Rozas, 2009), GenAlex (Peakall & Smouse, 2012) and MEGA 5 (Tamura et al., 2011) using various variability indices: number of genotypes, genotype diversity, variable sites, parsimony-informative sites, nucleotide diversity, and intraspecific genetic distance (K2P).

Phylogenetic analysis

The best-fit evolutionary model for the Bayesian analyses was determined with jModeltest 0.1.1 (Posada, 2008) using the Akaike information criteria (AIC). Bayesian analyses were carried out using MrBayes version 3.1 (Ronquist & Huelsenbeck, 2003). Separate runs were carried out with four simultaneous Markov chains, each starting from a random tree. The analysis ran for 6 500 000 generations to allow runs to converge, and the chain was sampled every 100th generation. The first 14 000 trees were discarded as the "burn-in" before the chains converged on a stable value and the posterior probabilities of tree topology were determined from the remaining trees. Also convergence was reassessed using the potential scale reduction factor (PSRF, Ronquist & Huelsenbeck, 2003).

Results

ITS1 and cox1 sequence variation

The two *M. yucatanica* worker bees analyzed per colony showed the same ITS1 sequence, therefore only one sequence per colony was used in the subsequent analyses. The first 126 base pairs (bp) of the alignment correspond to the

Table 2. Molecular diversity of the ITS1 fragments (complete and partial 3' and fragment) and cox 1 (3' fragment) in five *Melipona* species. Data from *M. yucatanica* in bold. Data from other *Melipona* sp. obtained from GenBank.

Dataset	ITS1 complete (1671 bp)								ITS1 partial (694 bp)								cox1 (580 bp)							
Species	bp	N	h	Gd	Pi	V	Par	d	bp	N	h	Gd	Pi	V	Par	d	bp	N	h	Hd	Pi	V	Par	d
Melipona yucatanica	**1468-1472**	**11**	**3**	**0.250**	**0.0028**	**8**	**8**	**0.3**	**595-601**	**11**	**2**	**0.509**	**0.0009**	**1**	**1**	-	**580**	**7**	**4**	**0.81**	**0.0136**	**34**	**14**	**1.39**
Melipona beecheii	1519-1582	10	9	0.978	0.01239	40	20	1.21	555-558	10	5	0.844	0.01182	6	4	1.04	580	46	18	0.913	0.00744	25	16	0.75
Melipona quinquefasciata	-	-	-	-	-	-	-	-	570-609	9	9	1.00	0.0181	65	10	2.97	-	-	-	-	-	-	-	-
Melipona subnitida	-	-	-	-	-	-	-	-	590-612	13	13	1.00	0.06018	185	44	7.04	-	-	-	-	-	-	-	-
Melipona quadrifasciata	-	-	-	-	-	-	-	-	-	-	-	-	-	-	-	-	580	145	31	0.876	0.00509	55	32	0.51

bp = length of the fragment in base pairs, N = number of individuals, h = number of genotypes/haplotypes, Gd: genotype diversity, Hd: haplotype diversity, Pi: nucleotide diversity, V = variable sites, Par = parsimony-informative sites, d = intraspecific genetic distance with K2P (in %).

18S ribosomal region and the last 58 bp to the 5.8S, so the size of the ITS1 region was 1310-1312 bp in Guatemalan and 1308 bp in Mexican *M. yucatanica*. Insertion-deletion (indel) events were observed in the alignment of the ITS1 sequence of *M. yucatanica* specimens from Guatemala and México. Two indels of 4 and 8 bp corresponded to a microsatellite locus showing different number of repeat motifs. This variation in the number of motifs was not detected among the specimens within each population but between them. An additional variation was detected among Guatemalan *M. yucatanica* specimens, due to the presence of one insertion of two base pairs (TT) in the specimen 4 (*MyucGuat4*). In relation to other species, a large insertion of 80 bp was detected in *M. quadrifasciata*.

There were three genotypes, one for Mexican (n = 7, *MyucMer1-7*) and two for Guatemalan individuals (n = 1; *MyucGuat4* and n = 3 *MyucGuat1, 2* and *5*). A higher number of unique genotypes was found in *M. beechei*, *M. quinquefasciata* and *M. subnitida* populations (Table 2).

The ITS1 nucleotide diversity (Pi) was from 3 to 65 times lower for *M. yucatanica* than for the other three *Melipona* species (Table 2). This result contrasts with the nucleotide diversity and the genetic distances (K2P) values obtained for cox1, where *M. yucatanica* population had similar values or slightly higher (1.8 to 2.7 times higher) than those obtained for other available species (Table 2).

Comparisons between markers showed that in *M. yucatanica*, ITS1 nucleotide diversity (pi = 0.0028) and genetic distance (d = 0.28%) is about five times lower than in cox1 (pi = 0.0136, d = 1.39%). On the other hand, in *M. beecheii* the comparison of both markers showed a different pattern, as the diversity and distance values are about 1.6 times higher with the ITS1 than with cox1 data (Table 2).

Phylogenetic analyses

The optimal model of nucleotide substitution, following AIC criterion, was the transversional model (TVM). Bayesian analysis of the complete ITS1 region, showed that this species is divided in two well supported clades (Mexican and Guatemalan specimens, p.p. = 1.0) defining a clear phylogeographic structure (Fig 2), *M. beecheii* showed two clades with the same geographic structure, whereas *M. subnitida* and *M. quinquefasciata* showed a more complex pattern of population structure. *M. yucatanica* ITS1 sequences clustered with high support (p.p = 1.0) within the subgenus *Melipona s. str.* together with *M. quadrifasciata* and *M. subnitida*.

Analysis of available cox1 sequences resulted in the same two well supported clades of Mexican and Guatemalan populations (Fig 3). Moreover the species *M. yucatanica* was also grouped with high support within the subgenus *Melipona s. str.* The combined analysis of cox1 and ITS1 resulted in the same well supported clades (Fig S1 in Supplementary Material).

Discussion

Phylogenetic analyses showed low genetic variation in ITS1 between populations but defined a geographic structure with Mexican and Guatemalan specimens forming two well supported clades. The characterization of the ITS1 region of the stingless bee *M. yucatanica* further evidenced that this region is longer in stingless bees than in other Hymenoptera (Pilgrim & Pitts, 2006; Taylor et al., 2006). In *M. yucatanica* ITS1 is within the range of that observed in other *Melipona* species (from 1391 bp in *M. quadrifasciata* to 1940 in *M. rufiventris*, Fernandes-Salomão et al., 2005). The presence of insertions and deletions (indels) in this region is a common feature leading to sequence length variation, as it has been observed between Mexican and Guatemalan *M. yucatanica* specimens, and among Brazilian populations of the species *M. subnitida* (Cruz et al., 2006) and within *M. quinquefasciata* (Pereira et al., 2009).

The ITS1 clades separate between Mexican and Guatemalan populations showing the same phylogeographic pattern found in previous markers (ITS2, De la Rúa et al.,

2007; cox1 and microsatellites May-Itzá et al., 2010). These results are congruent with the fact that both populations are located in opposite extremes of the species distribution range (Ayala, 1999). However, an incongruent pattern of genetic variability was revealed when several markers were compared. Populations of *M. yucatanica* showed lower levels of genetic diversity and genetic distance in the ITS1 region than those reported for other markers as cox1 and microsatellite data (May-Itzá et al., 2010; Table 2). ITS1 and cox1 have different mechanisms of evolution that have led to different patterns in intraspecific variations for various taxa (e.g., Hansen et al., 2006; Carlini et al., 2009; Kornobis & Pálsson, 2011). The low variation of ITS1 could therefore be explained by intraspecific rDNA ITS1 homogenization due to the mechanisms of concerted evolution (Hillis & Dixon, 1991).

However, this explanation cannot be extended to other *Melipona* species. In *M. beecheii*, a species with wider geographic distribution, the opposite pattern was found with diversity and distance values of ITS1 higher than for cox1 data (May-Itzá et al., 2010). Moreover, when data are

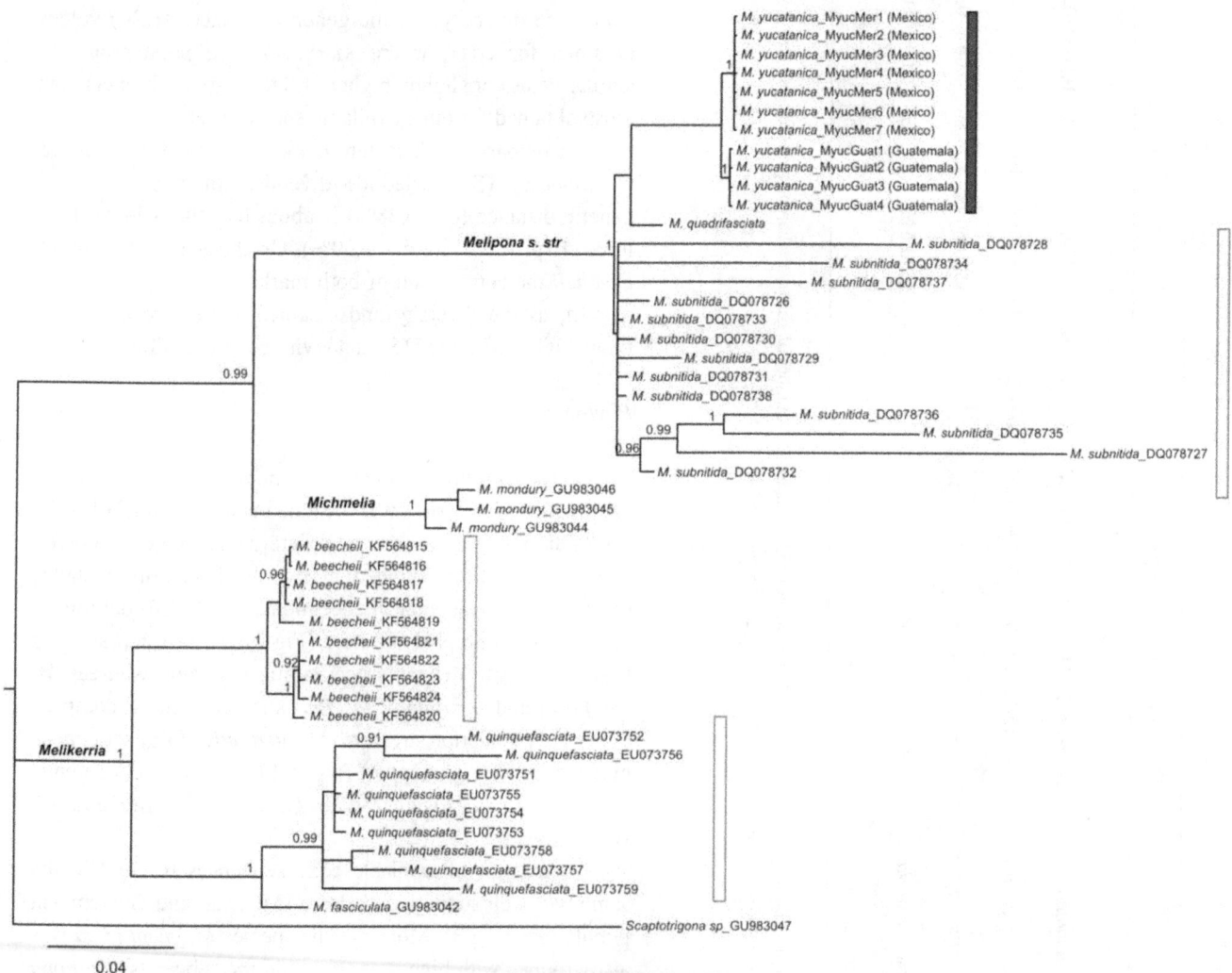

Fig 2. Bayesian analyses of the *ITS1*-complete region of several *Melipona* species. Population data of *Melipona yucatanica* are indicated with a solid bar. Available population data of several *Melipona* species are indicated with empty bars. Numbers above nodes show posterior probability ≥ 0.90.

compared between these two species, *M. beecheii* showed about four times higher divergence, genotype and nucleotide diversity of ITS1 (complete fragment) but almost two times lower values for cox1 data (Table 2). Unfortunately, there are no available data to compare this pattern with other *Melipona* species; however, *M. yucatanica* showed a remarkably lower ITS1 divergence (from 12 to 80 times) and higher cox1 divergence (from 1.8 to 2.7 times) than all other *Melipona* species analyzed (Table 2). This observed distinct pattern of divergence between nuclear and mitochondrial markers in *M. yucatanica* populations in relation to other *Melipona* species could be due to several causes which will require specific testing, among them: i) different efficiency of concerted evolution of ITS1 rDNA between *Melipona* species, as suggested for other taxa (Armbruster & Korte, 2006); ii) a complete replacement in one of the sampled populations of the original mitochondrial haplotype by introgression which led to higher levels of mitochondrial divergences. This hypothesis cannot be disregarded although until date, introgression

events have not been described in *Melipona* species yet; and iii) distinct population dynamics in relation to other the *Melipona* species analyzed. Further analyses including more markers and more *M. yucatanica* populations scattered across its distribution range are needed in order to understand the population dynamic of this endangered species.

The phylogenetic signal obtained in the ITS1 Bayesian analysis confirms the utility of this region as a potential phylogenetic marker for the genus *Melipona*. This marker seems suitable for resolving population and subgenera divergences. In this sense, the phylogenetic analysis included *M. yucatanica* in the subgenus *Melipona s. str.* together with other species as *M. quadrifasciata* and *M. mandacaia*. This molecular result confirms previous morphological studies (Moure et al., 2007) and it is in accordance with recent molecular studies of the genus *Melipona* (Ramírez et al. 2010). In the light of recent numt DNA descriptions in other *Melipona* subspecies (Cristiano et al., 2012; Ruiz et al., 2013), the ITS1 region has demonstrated its potential as an appropriate marker for genetic studies in stingless bees.

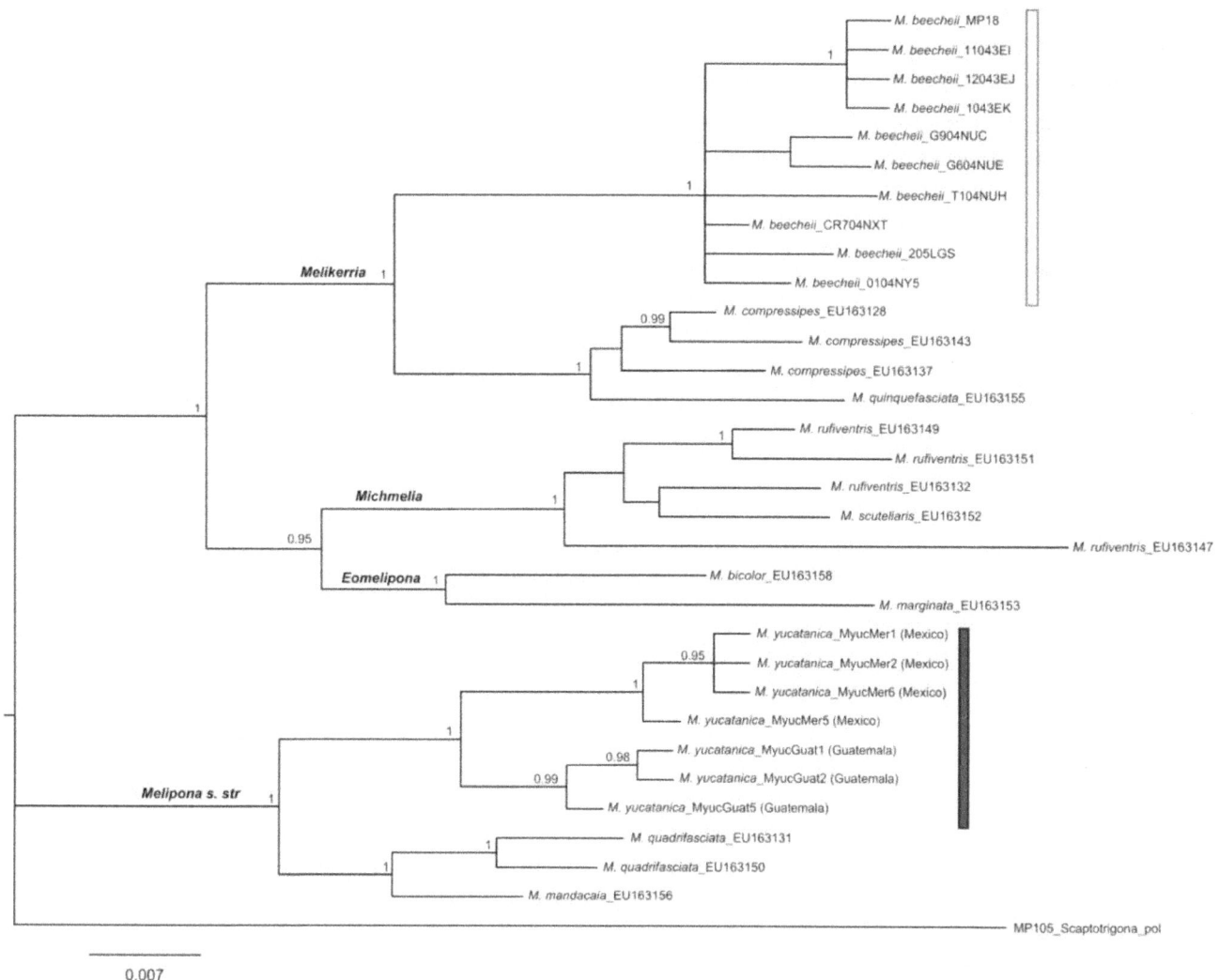

Fig 3. Bayesian analysis of the available cox1 gene from *Melipona* species. Numbers above nodes show posterior probability ≥ 0.90. Population data of *Melipona yucatanica* are indicated with a solid bar. Available population data of several *Melipona* species are indicated with empty bars.

Acknowledgements

We thank Eunice Enríquez for her help with the Guatemalan sampling and Jorge González Acereto for the Mexican samples. We also appreciate the input of the reviewers. This work has been supported by the BBVA Foundation, Fondo de Cooperación Internacional en Ciencia y Tecnología Unión Europea – México (FONCICYT) (project MUTUAL 94293), and Fondo Sectorial de Investigación para la Educación (SEP-CONACYT, project 103341).

References

Armbruster, G.F.J. & Korte, A. (2006). Genomic nucleotide variation in the ITS1 rDNA spacer of land snails. J. Moll. Stud., 72: 211–219.

Arias, M.C., Brito, R.M., Francisco, F.O., Moretto, G., de Oliveira, F.F., Silvestre, D. & Sheppard, W.S. (2006). Molecular markers as a tool for population and evolutionary studies of stingless bees. Apidologie, 37: 259–274. DOI: 10.1051/apido:2006021

Ayala, R. (1999). Revisión de las abejas sin aguijón de México (Hymenoptera: Apidae: Meliponini). Folia Entomol. Mex., 106:1 - 123.

Camargo, J.M.F., Moure, J.S. & Roubik, D.W. (1988). *Melipona yucatanica* New Species (Himenoptera:Apidae:Meliponinae): stingless bee dispersal across the Caribbean Arc and Post-Eucene Vicariance. Pan-Pacific Entomol., 64: 147–157.

Camargo, J.M.F. & Pedro, S.R.M. (2007). Meliponini Lepeletier, 1836. In Moure, J.S., Urban, D. & Melo, G.A.R. (Orgs). Catalogue of Bees (Hymenoptera, Apoidea) in the Neotropical Region - online version. Available at http://www.moure.cria.org.br/catalogue. (Accessed Oct/19/2010).

Carlini, D.B., Manning, J., Sullivan, P.G. & Fong, D.W. (2009). Molecular genetic variation and population structure in morphologically differentiated cave and surface populations of the freshwater amphipod *Gammarus minus*. Mol. Ecol., 18: 1932–1945. DOI: 10.1111/j.1365-294X.2009.04161.x

Cristiano, M.P., Fernandes-Salomão, T.M. & Yotoko, K.S.C. (2012). Nuclear mitochondrial DNA: an Achilles' heel of molecular systematics, phylogenetics, and phylogeographic studies of stingless bees. Apidologie, 43: 527-538. DOI: 10.1007/s13592-012-0122-4.

Cruz, D.O., Jorge, D.M.M., Pereira, J.O.P., Torres, D.C., Soares, C.E.A., Freitas, B.M. & Grangeiro, T.B. (2006). Intraspecific variation in the first internal transcribed spacer (ITS1) of the nuclear ribosomal DNA in *Melipona subnitida* (Hymenoptera, Apidae), an endemic stingless bee from northeastern Brazil. Apidologie, 37: 376–386. DOI: 10.1051/apido:2006003

De la Rúa, P., May-Itzá, W. de J., Serrano, J. & Quezada-Euán, J.J.G. (2007). Sequence and RFLP analysis of the ITS2 ribosomal DNA in two Neotropical social bees, *Melipona beecheii* and *Melipona yucatanica* (Apidae: Meliponini). Insectes Soc., 54: 418–423. DOI: 10.1007/s00040-007-0962-5

Fernandes-Salomão, T.M., Rocha, R.B., Campos, L.A.O. & Araújo, E.F. (2005). The first internal transcribed spacer (ITS-1) of *Melipona* species (Hymenoptera, Apidae, Meliponini): characterization and phylogenetic analysis. Insectes Soc., 52: 11–18. DOI: 10.1007/s00040-004-0767-8

Hansen, H., Martinsen, L., Bakke, T.A. & Bachmann, L. (2006). The incongruence of nuclear and mitochondrial DNA variation supports conspecificity of the monogenean parasites *Gyrodactylus salaris* and *G. thymalli*. Parasitology, 133: 639–650 DOI: 10.1017/S0031182006000655

Hillis, D.M. & Dixon, M.T. (1991). Ribosomal DNA: molecular evolution and phylogenetic inference. Q. Rev. Biol., 66: 411–453.

Ji, Y., Zhang, D. & He, L. (2003). Evolutionary conservation and versatility of a new set of primers for amplifying the ribosomal internal transcribed spacer regions in insects and other invertebrates. Mol. Ecol. Notes., 3: 581–585. DOI: 10.1046/j.1471-8286.2003.00519.x

Katoh, K. & Toh, H. (2008). Improved accuracy of multiple ncRNA alignment by incorporating structural information into a MAFFT-based framework. BMC Bioinformatics, 9: 212. DOI: 10.1186/1471-2105-9-212

Kornobis, E. & Pálsson, S. (2011). Discordance in Variation of the ITS Region and the Mitochondrial COI Gene in the Subterranean Amphipod *Crangonyx islandicus*. J. Mol. Evol., 73: 34-44. DOI: 10.1007/s00239-011-9455-2

Librado, P. & Rozas, J. (2009). DnaSP v5: A software for comprehensive analysis of DNA polymorphism data. Bioinformatics, 25: 1451-1452. DOI: 10.1093/bioinformatics/btp187

May-Itzá, W. de J., Quezada-Euán, J.J.G. & De la Rúa, P. (2009). Intraspecific variation in the stingless bee *Melipona beecheii* assessed with PCR-RFLP of the ITS1 ribosomal DNA. Apidologie, 40: 549–555. DOI: 10.1051/apido/2009036

May-Itzá, W. de J., Quezada-Euán, J.J.G., Medina Medina, L., Enríquez, E. & De la Rúa, P. (2010). Morphometric and genetic differentiation in isolated populations of the endangered Mesoamerican stingless bee *Melipona yucatanica* (Hymenoptera: Apoidea) suggest the existence of a two species complex. Conserv. Genet., 11: 2079-2084. DOI: 10.1007/s10592-010-0087-7

May-Itzá, W. de J., Quezada-Euán, J.J.G., Ayala, R., & De la Rúa, P. (2012). Morphometric and genetic analyses reveal two taxonomic units within *Melipona beecheii* (Hymenoptera: Meliponidae), a Mesoamerican endangered stingless bee. J. Insect. Conserv., 16: 723–731 DOI: 10.1007/s10841-012-9457-4

Moure, J.S., Urban, D. & Melo, G.A.R. (2007). Catalogue of Bees (Hymenoptera, Apoidea) in the Neotropical Region.

Curitiba: Sociedade Brasileira de Entomologia, 1058 p

Peakall, R. & Smouse, P.E. (2012). GenAlEx 6.5: genetic analysis in Excel. Population genetic software for teaching and research-an update. Bioinformatics, 28: 2537-2539. DOI: 10.1093/bioinformatics/bts460

Pereira, J.O.P., Freitas, B.M., Jorge, D.M.M., Torres, D.C., Soares, C.E.A. & Grangeiro, T.B. (2009). Genetic variability in *Melipona quinquefasciata* (Hymenoptera, Apidae, Meliponini) from northeastern Brazil determined using the first internal transcribed spacer (ITS1). Gen. Mol. Res., 8: 641–648.

Pilgrim, E.M. & Pitts, J.P. (2006). A molecular method for associating the dimorphic sexes of velvet ants (Hymenoptera: Mutillidae). J. Kans. Entomol. Soc., 79: 222-230.

Posada, D. (2008). jModelTest: Phylogenetic Model Averaging. Mol. Biol. Evol., 25: 1253–1256. DOI:10.1093/molbev/msn083

Ramírez, S.R., Nieh, J.C., Quental, T.B., Roubik, D.W., Imperatriz-Fonseca, V.L. & Pierce, N.E. (2010). A molecular phylogeny of the stingless bee genus *Melipona* (Hymenoptera: Apidae). Mol. Phyl. Evol., 56: 519–525. DOI: 10.1111/j.1365-3113.2006.00362.x

Ronquist, F. & Huelsenbeck, J.P. (2003). MrBayes 3: Bayesian phylogenetic inference under mixed models. Bioinformatics, 19: 1572–1574. DOI:10.1093/bioinformatics/btg180

Ruiz, C., May-Itzá, W. de J., Quezada-Euán, J.J.G. & De la Rúa, P. (2013). Nuclear copies of mitochondrial fragments (numts) and phylogenetic analysis of two related species of stingless bee genus *Melipona* (Hymenoptera: Meliponini). J. Zool. Sys. Evol. Res., 51: 107–113. DOI: 10.1111/jzs.12011

Tamura, K., Peterson, D., Peterson, N., Stecher, G., Nei, M., & Kumar, S. (2011). MEGA5: molecular evolutionary genetics analysis using maximum likelihood, evolutionary distance, and maximum parsimony methods. Mol. Biol. Evol., 28: 2731-2739. DOI:10.1093/molbev/msr121

Taylor, D.B., Moon, R., Gibson, G. & Szalanski, A. (2006). Genetic and Morphological Comparisons of new and old world populations of *Spalangia* species (Hymenoptera: Pteromalidae). Ann. Entomol. Soc. Am., 99: 799-808. DOI: 0.1603/0013-8746(2006)99[799:GAMCON]2.0.CO;2

18

Period and Time of Harvest Affects the Apitoxin Production in *Apis mellifera* Lineu (Hymenoptera: Apidae) Bees and Expression of Defensin Stress Related Gene

MS Modanesi, SM Kadri, PEM Ribolla, DP Alonso, RO Orsi

Universidade Estadual Paulista, Botucatu, SP, Brazil

Keywords
honeybees, collection, venom, welfare

Corresponding author
Dr. Ricardo de Oliveira Orsi
Center of Education, Science and Technology
in Rational Beekeeping (NECTAR)
College of Veterinary Medicine and
Animal Sciences, UNESP - University
Distrito de Rubião Jr., s/n, 18618-000,
Botucatu, São Paulo, Brazil
E-Mail: orsi@fmvz.unesp.br

Abstract
The aims of this study were to determine the period effects (morning and afternoon) and harvest time (30 and 60 minutes) in the apitoxin production as well as the management effects in the expression of a stress related gene. Five *Apis mellifera* L. beehives were used. The harvest of apitoxin and honeybees occurred three times a week (morning and afternoon at 09h00 and 14h00), according to the following treatments (period and apitoxin harvest time): T1: morning/30 minutes, T2: morning/60 minutes; T3: afternoon/30 minutes; T4: afternoon/60 minutes. The apitoxin was collected by electric collectors. The stress level was monitored by defensin gene expression, using actin gene as control. The results were evaluated by ANOVA followed by Tukey-Kramer test (p ≤ 0.05). It can be concluded that the better period and time to apitoxin harvest is in the morning for 60 minutes, associated to minor stress for honeybees.

Introduction

Apitoxin comes from a Latin word and means: *Apis* - bee and *toxikon* – venom, being produced by bees as a way of defending and protecting the hive. The *Apis mellifera* apitoxin composition varies according to the subspecies development stage and habits. The primary findings are variations in the concentrations of apitoxin proteins throughout the seasons of the year, showing an influence of the environment (Abreu, 2010).

The apitoxin injected into a sting contains about 50 mg of dry matter. The major proteins present are melittin; 50% of the dry weight of apitoxin (DWV), phospholipase A2 (DWV 12%), mast cell factor degranulation (3% DWV), hyaluronidase (3% DWV) and apamin (2% DWV). Moreover, biogenic amines are present, including histamine (1% DWV), dopamine (0.5% DWV) and norepinephrine (0.5% DWV) (Cruz-Landim et al., 2002). Many volatile acetates which presumably stimulate the defensive behavior of other bees are also found in apitoxin (Abreu et al., 2010).

The apitoxin has beneficial effects to the humans, apitherapy (the use of bee products for the benefit of humans) has been used for many health problems, such as, eczema, tropical ulcers, infections such as laryngitis and mastitis, rheumatologic problems, cardiovascular, pulmonary and orthopedic, and help in the inhibition of ovarian cancer, multiple sclerosis and have an antibiotic effect (Leite et al., 2005).

The apitoxin harvest can be performed by electrical collectors placed at the entrance of the hive containing filaments conducting an electric current, thereby promoting accessory muscle contraction release of the apitoxina, without killing the bees (Leite & Rocha, 2005).

However, even when not promoting the death of bees, there is still no knowledge of the influence of management on the hive. It is assumed that it can promote behavioral changes, interfering with routine activities of the colony, causing an acute or chronic stress.

The main goal of this study was to investigate the effects of period (morning and afternoon) and harvest time (30 and 60 minutes) in apitoxin production by Africanized honeybees and the influence of this management on the defensin gene expression.

Material and methods

The experiment was conducted by the Center for Education, Science and Technology in Rational Beekeeping - NECTAR, the Apicultural Sector, from College of Veterinary Medicine and Animal Science, Experimental Farm Lageado, UNESP, Botucatu, Brazil; with the following coordinates: 22° 49'' S, 48°24'' W and 623 meters high.

In order to perform the experiment 10 Africanized beehives in wooden Langstroth hives model (five as control and five per treatment), externally oil painted light green, kept in 50cm racks numbered for easy identification were used. The selected colonies were standardized to the number of frames of brood and food. During the experimental period, each beehive received sugar syrup (50% water + 50% sugar) weekly through Bordman feeder.

To measure the defensive behavior of the colonies, we performed the defensiveness test, according to the methodology described by Stort (1975) and Brandenburg and Gonçalves (1990). The test consisted of a black suede ball swinging in front of the entrance, attached to a string, for a minute. The first sting time (FST) were recorded and the number of stings left in the black suede ball (SNB). The tests were repeated in triplicate. Values above 36 stings designate defensive colonies.

The bee and apitoxin harvesting to evaluate expression of genes related to stress occurred three times a week, from November 2011 to January 2012 (twelve weeks), and five beehives per treatment were assessed in the same day using the same treatment according to: T1: Harvesting apitoxin in the morning, lasting 30 minutes; T2: Harvest apitoxin in the morning, lasting 60 minutes; Treatment 3: Harvest apitoxin in the afternoon, lasting 30 minutes; Treatment 4: Harvesting apitoxin in the afternoon, lasting 60 minutes. Thus, each treatment was realized in nine replicates

The harvest took place in the morning, always starting at 09:00 and in the afternoon at 14:00, by electric collectors consisted of wire filaments with six volts of electric current, and a glass plate for the fall of the released apitoxin (Leite & Rocha, 2007). At the end of each harvest, the collectors and the glass plates were removed and sent to the Laboratory of Apiculture Sector, kept at room temperature and protected

from light, until the evaporation of the volatile phase. After, the apitoxin was scraped with a spatula, mixed (pool of five beehives), weighed and stored at -20°C.

Climatic variables (mean temperature, wind speed, rainfall and relative humidity) were provided by the Department of Environmental Sciences, College of Agricultural Sciences, UNESP, Botucatu.

The climatic data for the period of the study were as follows: minimum temperature of 17.2 ± 1.9 °C, maximum temperature of 30.6 ± 2.2, precipitation of 5.0 ± 6.6 mm, relative humidity $60.13 \pm 10.4\%$, solar radiation 477.1 ± 80.0 cal / cm², insolation 8.1 ± 3.4 hours decimal and wind speed 99.7 ± 39.8 km/day.

The data for the analysis of stress-related genes were processed at Laboratory for Genetics Research and Analysis, Parasitology Department, Biosciences Institute, São Paulo State University – UNESP, Botucatu Campus.

For the experiments of *defensin* expression, a stress related gene, the harvest of the after the placement of apitoxin collectors. Ten internal worker bees and ten foragers were used. After harvest, the bees were immediately stored at - 80°C for RNA extraction. Bees were collected also of five beehives control, without apitoxin collect, in the different treatments.

For RNA extraction, the head of each worker bee was separated from the body with the aid of a disposable scalpel (Scharlaken et al., 2008). Each sample consisted of a "pool" of five heads and RNA extraction was performed by using the TRIzol method, using for each sample 500 ul of TRIzol (GIBCO BRL) to disrupt the cells and release their contents. The extraction product was visualized on a 1% agarose gel and quantified using the NanoDrop instrument (ND-1000 Spectrophotometer). Then all samples were stored at -80°C until ready to use.

After, samples were treated with DNase, and cDNA synthesis reaction was set as follows: a mix of oligodT solution (N = 18) 0.75 mM; random oligonucleotides (N = 8) 0.15 mM; 0.75 mM dNTP and 11 µl of RNA treated with DNAse in the previous step , was prepared and incubated at 65°C for 5 minutes and then placed on ice for 1 minute. To this preparation, 0.5mM DTT, 40U of RNase out and 100U of Super Script III were added. The reaction was then incubated at 50°C for 1 hour and then at 70°C for 15 minutes.

The stress level of the bees was monitored by changes in *defensin* gene expression (Scharlaken et al., 2008). As internal control for quantitative PCR reactions, we used the *actin* gene (Scharlaken et al., 2008).

The determination of gene expression was performed by real time polymerase chain reaction in triplicate, on a Real Time ABI 7300 instrument (Applied Byosystems) using the SYBR Green PCR Master Mix kit (Applied Biosystems, Foster City, CA) under the following conditions: one cycle at 50°C for 2 minutes, a cycle at 94°C for 10 minutes, followed by 40 cycles of 94°C for 15 seconds and 60°C for 1 minute. The dissociation curve was obtained as follows: 95°C for 15 seconds, 60°C for 30 seconds and 95°C for 15 seconds.

The oligonucleotides sequences used and details are shown in the Table 1:

Table 1. Genes and oligonucleotide sequences useoto avaluate the defensin expression.

Gene	Gene Bank Number Accession	Oligonucleotides sequences 5'-3'	Amplification (pb)	Ta (°C)[a]	EE (%)[b]
Actina	AB023025	TGCCAACACT GTCCTTTCTG AGAATTGAC CCACCAATCCA	155	61	99,11
Defensina	U15955.1	GTCGGCCTTCT CTTCATGGT GACCTCCAGCT TTACCCAAA	200	61	91,70

[a]Ta optimal annealing temperature specific to each oligonucleotide

[b]E Measuring the efficiency of the reaction Real Time PCR (calculated trough the standard curve)

To calculate the efficiency for the oligonucleotides used, four dilutions of cDNA samples were made: 1:5, 1:25, 1:125 and 1:625. The efficiency (E) was calculated using the formula E = 10 (-1/slope). Relative quantification (R) was determined according to Pfaffl (2001).

Data were calculated using bees taken before placing electric collectors as controls in relation to the other five harvests after the electric collectors were removed.

The analysis of the data obtained was compared by ANOVA followed by Tukey test to check for differences between means. Results were considered statistically different when $p < 0.05$ (Zar, 1996).

Results and Discussion

Do not observed difference in the first sting time between the beehives of treatments and control (1.3 ± 0.7 to 1.4 ± 0.8 seconds) and number of stings left in the black suede ball (50.3 ± 17.2 to 51.8 ± 14.4).

There was a higher apitoxin production in the treatment using 60 minutes in the morning (T2), which differed significantly from the other treatments (Table 2).

Table 2. Mean values and standard deviation for apitoxin production (milligrams) of Africanized honeybees according the day time and time of production.

Production Period (min)	Morning (minutes)	Afternoon (minutes)
30	33.7 ± 13.0 aA	24.6 ± 10.0 aA
60	49.1 ± 13.0 bB	24.9 ± 6.0 aA

Different small letters in the same column indicate statistical differences between averages ($p < 0.05$).
Different capital letters in the same line indicate statistical differences between averages ($p < 0.05$).

The relative quantification (R) results from Defensin gene, using Actin as the endogenous gene in the samples from the different treatments and control are shows in Table 3.

Table 3. Mean and standard deviation of defensin gene relative quantification of foragers workers (FW) and internal workers (IW), in different treatments with Afr canized honeybees.

Treatment	FW	FW - Control	IW	IW - Control
1	2.02 ± 1.6aAα	0.007 ± 0.0aBα	1.84 ± 1.7abAα	0.005 ± 0.0aBα
2	1.23 ± 0.1aAα	0.003 ± 0.0aBα	0.16 ± 0.2aAα	0.06 ± 0.01aAα
3	1.35 ± 1.4aAα	0.002 ± 0.0aBα	2.97 ± 1.6bAα	0.07 ± 0.0aBα
4	0.59 ± 0.4aAα	0.025 ± 0.0bBα	0.56 ± 0.6abAα	0.003 ± 0.0aBβ

Different small letters in the same column indicate statistical differences between averages ($p < 0.05$).
Different capital letters in the same line, to group of foraging or internal bees (treatment vs control), indicate statistical differences between averages ($p < 0.05$).
Different greek letters in the same line, to foraging vs internal, indicate statistical differences between averages ($p < 0.05$)
Treatment 1: Apitoxin production in the morning, during 30 minutes; Treatment 2: Apitoxin production in the morning, during 60 minutes; Treatment 3: Apitoxin production in the afternoon, during 30 minutes , Treatment 4: Apitoxin production in the afternoon, during 60 minutes.

Relative quantification of the *defensin* gene in foragers bees showed no significant difference between treatments and treatments versus control. In control worker bees was observed that treatment using 60 minutes in the afternoon (T4) showed significant difference between others treatments.

Internal workers showed significant difference between treatments 2 and 3 throughout the experimental period. The control beehives do not observed differences. It was observed significant difference between internal works to treatment 4 and the control.

No difference was observed between the foragers and internal workers for all treatments and was observed difference between foragers and internal worker bees from T4 and control.

The colonies used in the experiment can be classified as defensive, regardless the period of the day, during apitoxin production, according to the Stort (1975) classification, which the bees are classified with high defensiveness when they sting 36 more times the black suede ball. We could observe that there was no influence of climatic variables on colonies defensiveness. Previous work, in the same place, found no significant correlations between time and number of first sting in the black suede ball with climatic variables (Lomele et al., 2010).

The higher apitoxin production occurred in the morning period during 60 minutes. A probable cause for this result would be the foraging behavior of bees. Malerbo and Souza (2011) observed in the end of spring and in the beginning of summer, that nectar were collected by the bees throughout the day, with preference for the hottest hours of the day, between 10 and 14h, when the temperature was between 15 and 30°C. Thus, a greater flow of bees results in greater amount of foragers through the collector and consequently a significant increase on the release of apitoxin through electrical stimulation.

However, the apitoxin harvest could promote alterations in all beehives, affecting the grooming, brood care and hygienic behavior, besides could affect the immunity sistem. As the honeybees could be exposed to field pathogens, *defensin* gene related to stressor stimulus like changes in immune system was quantified (Ursic-Bedoya & Löwenberger, 2007).

There was significant increase in the *defensin* gene expression in the internal worker bees in the treatment 2 (morning/60 minutes) compared to the treatment 3 (late/30 minutes). This higher *defensin* gene expression in the internal worker bees could be related to the increase of alarm pheromones released at the time of apitoxin release by the electrical stimulation (isopentilacetate and 2 heptanone) promoting alertness in other worker bees to protect the hive, increasing discomfort and possibly higher stress in the afternoon period than morning.

It was observed that the apitoxin harvest promote a highest defesin expression in treatments when compared as control, with exception for internal worker bees from treatment 2 that do not differ as control. Its result could be explained by the worker bees alarm pheromone adaptation in the collector. This is important because apitoxin harvest would not promote stress in the bees, associated with highest production, and can be used by beekeepers.

It can conclude that the better period and time to apitoxin harvest is in the morning for 60 minutes, associated the minor stress for honeybees.

Acknowledgments

We would like to thank (CAPES), for the master's degree scholarship and FAPESP by financial support (process number 2012/23466-2).

References

Abreu, R.M., Moraes, R.L.M.S. & Mathias, M.I.C. (2010) Biochemical and cytochemical studies of the enzymatic activity of the venom glands of workers of honey bee *Apis mellifera* L. (Hymenoptera, Apidae). Micron, 41: 172-175. doi: 10.1016/j.micron.2009.09.003.

Brandeburgo, M.A.M. & Gonçalves, L.S. (1990) Environmental influence on the aggressive (defense) behaviour and colony development of africanized bees (*Apis mellifera*). Ciência e Cultura 42: 759-771. http://www.cabdirect.org/abstracts/19920510324.html (Accessed date: 28 June, 2014).

Lomele, R.L., Evangelista, A., Ito, M.M., Ito, E.H., Gomes, S.M.A. & Orsi R.O. (2010) Natural products on the defensive behavior of *Apis mellifera* L. Acta Scientarium Animal Science, 32: 285-291. doi: 10.4025/actascianimsci.v32i3.8486.

Funari, S.R.C., Rocha, H.C., Sforcin, J.M. & Orsi, R.O. (2004) Influence of smoke and lemon grass (*Cymbopogon citratus*) on the defensive behavior of africanized bees and their hybrid European (*Apis mellifera* L). Boletim de Indústria Animal, 61: 121-125. http://revistas.bvs-vet.org.br/bia/article/view/8704. (Accessed date: 28 June, 2014).

Leite, G.L.D. & Rocha, S.L. (2005) - Apitoxin. Revista UniMontes Científica, 7: 115-125. http://www.ruc.unimontes.br/index.php/unicientifica/article/viewArticle/145. (Accessed date: 28 June, 2014).

Malerbo-Souza, D.T. & Silva, F.A.S. (2011) Comportamento forrageiro da abelha africanizada *Apis mellifera* L. no decorrer do ano. Acta Scientarium Animal Sciences, 33: 183-190. doi: 10.4025/actascianimsci.v33i2.9252.

Pfaffl, M. (2001) A new mathematical model for relative quantification in real-time RT–PCR. Nucleic Acids Research, 29: e45.

Scharlaken, B., Graaf, D.C., Goossens, K., Peelman, L.J. & Jacobs F.J. (2008) Differential gene expression in the honeybee head after a bacterial challenge. Developmental and Comparative Immunology, 32: 883-889. doi: 10.1016/j.dci.2008.01.010.

Stort, A.C. (1975) Genetic study of the aggressiveness of two subspecies of *Apis mellifera* in Brazil. Behavior Genetics, 3: 269-274. http://link.springer.com/article/10.1007/BF01066178. (Accessed date: 28 June, 2014).

Ursic-Bedoya, R.J. & Lowenberger, C.A. (2007) *Rhodnius prolixus*: Identification of immune-related genes up-regulated in response to pathogens and parasites using suppressive subtractive hybridization. Developmental and Comparative Immunology, 31: 109-120. doi: 10.1016/j.dci.2006.05.008.

Zar, J. H. (1996) Biostatistical analisys. 4th ed. New Jersey: Prentice Hall 663 p.

19

Nesting sites, nest density and spatial distribution of *Melipona colimana* Ayala (Hymenoptera: Apidae: Meliponini) in two highland zones of western, Mexico

JO Macías-Macías[1], JJG Quezada-Euán[2], JM Tapia-Gonzalez[1], F Contreras-Escareño[3]

1 - Universidad de Guadalajara. Centro Universitario del Sur, México

2 - Universidad Autónoma de Yucatán, Mexico

3 - Universidad de Guadalajara. Centro Universitario de la Costa Sur, México

Keywords

Melipona colimana, nest density, spatial distribution, *Quercus laurina*, Jalisco, México.

Corresponding author

José Octavio Macias-Macias. Universidad de Guadalajara. Centro Universitario del Sur. Departamento de Desarrollo Regional. Enrique Arreola Silva Av. 883. Cd. Guzmán, Jalisco. Mexico. CP. 49000
E-Mail: joseoc@cusur.udg.mx

Abstract

Melipona colimana Ayala is endemic to the temperate forests of western Mexico and may be in conservation risk due to forest exploitation. Differences between the density of nests, nesting sites and spatial distribution in two places with different levels of human disturbance were established. A preserved (P) and a disturbed area (D) were identified: the forest had not been exploited for more than 18 years in the P zone, while there had been recent forest exploitation of D zone in less than two years. It was determined that nesting sites, nest density and the number of potential nest sites were predominant in the P zone. In total, 27 of 30 colonies were found on oak trees *(Quercus laurina)* with a diameter at breast height of 183.4 ± 34.21 cm which shows a close relationship of this bee species with this type of tree. A positive correlation between the DBH of the nesting sites in relation to the trees with nests and the presence of cavities was found. The nests are distributed in the form of aggregates in P and D zones (R = 0.31 and 0.39) with a density of 0.17 ha^{-1} and 0.04 ha^{-1} colonies respectively. Forestry exploitation seems to be affecting wild populations since the trees that bees use as nesting sites are destroyed in D zone.

Introduction

Stingless bees or Meliponini is a group of bees with great biological and morphological diversity distributed mainly in the tropical and subtropical regions of the world (Michener, 2000) and 250 species have been described in South and Central America (Camargo & Pedro, 1992; Nogueira-Neto, 1997; Michener, 2000). In Mexico, 46 species of stingless bees have been identified; two of them are classified as endemic to the mountain ranges of western Mexico: *Melipona colimana* Ayala and *Melipona fasciata* Latreille (Ayala, 1999). *M. colimana* endemism is linked to the mesophyll mountain forest in elevations over 1000 meters above sea level, in the geographical zones that correspond to the Manantlan mountain range, the National Park Nevado of Colima and the Tigre mountain range (Ayala, 1999). Usually

stingless bees use diverse lodgings to establish their nests, as gaps and cavities in trees, electricity poles, under house's roofs and abandoned termites and wasps' nests. (Nogueira-Neto, 1997, Roubik, 2006). For that reason, it is hard to acknowledge which are the places *M. colimana* sets its nest, therefore there is a scientific interest in studying and get to know the places it uses for this purpose in its origin habitat, which will provide us new information about the nesting sites of stingless bees in temperate weather.

On the other hand, the conservation of stingless bees is important because they take part in the ecological interactions that contribute to the biodiversity maintenance, acting as pollinators of different plant species, and because they can act as indicators of environmental disturbances (Brown & Albrecht, 2001; Imperatriz-Fonseca, 2002; Slaa et al., 2006). Forest exploitation has negatively affected the presence of stingless bee nests and their density, often leading to their disappearance (Cannon et al., 1994; Brown & Albrecht, 2001; Venturieri, 2002; Eltz et al., 2003; Samejima et al., 2004). In this regard, *M. colimana* might be in conservation risk since the species original habitat is a zone where there is commercial exploitation of forest resources, especially oak trees (*Quercus* spp) and pines (*Pinus* spp*)* (Comision Nacional Forestal [CONAFOR], 2010). The purpose of this work is to know the type of tree that *M.colimana* prefers to nest, and evaluate nest density, nesting sites and spatial distribution in two zones with different level of human disturbance; to infer if deforestation could affect the presence of this species.

Material and Methods

Study site

The observations were carried out in the Halo mountain range in southern Jalisco, Mexico (18° 58' 00" N, 102° 59' 45" W, 1600 meters above sea level), in the surroundings of San Isidro village. The territory of this zone is 132, 644 km², in which 79,055 km² are used for forestry. Vegetation is mainly mountain mesophyll forest, made up of various useful timber-yielding species such as oak and pine trees and other species such as *Abies religiosa, Lysoma apaculcensis, Betula pendula, Cornus disciflora* and *Litzea glucesens* (Instituto Nacional de Estadistica Geografia e Informatica [INEGI]; 2005, Carranza, 2008).

The weather is characterized by sub-humid temperate with summer rain (Relative Humidity 69%). The average temperature in the last 10 years was 69.08 °F (min. 52.23 °F and max. 79.88 °F) with an average rainfall of 930mm (min. 120 mm and max. 1230 mm) (Comision Nacional del Agua [CONAGUA], 2011). *M. colimana* is a 9.5 mm long bee, with black integument, yellow marks and orange pubescence; it is a species morphologically close to *M. fasciata* with the difference that *M. colimana* has black terga and yellow apical segments (Ayala, 1999). Two wild populations of *M. colimana* colonies were found in two contiguous areas with different degree of human perturbation. The first area was a disturbed zone (D) there had been recent forest exploitation in less than two years. In the second area, the forest had not been exploited for more than 18 years (P).

Nesting sites and nest density

In order to find wild nests, fourteen 1000-meter-long transects were established with a distance of 200m between each other. A 100 x 100 meter quadrant was traced around every 100 meters of each transect; in this area, wild colonies were located and recorded. The geographical position of each nest and the scientific name of the tree, where the nest was located, were recorded. The total study area and the nesting density per hectare were calculated with the geographical

position data and the ArcView 8.3 software. In order to obtain numerical data from the oak trees, which the bees used as a nesting site, a 17.84-meter-long string-stick was placed every 100 meters along the transects. The string was spun around in circles making a circumference that covered a measuring area of 1000 m².

One hectare per measured transect was covered with this method (Instituto Nacional de Investigaciones Forestales Agricolas y Pecuarias [INIFAP], 2006). The number of oak trees, the diameter at breast height (DBH), and the number of oak trees with holes were counted at each measuring station. Afterwards, there was a comparison of this data in both areas P and D by using a t-test for the diameter at breast height (DBH) of the nesting sites; and a Chi-square goodness-of-fit test (X^2) for the number of nests, the trees with cavities (potential nest sites) and trees without cavities. The relationship between the DBH of trees with nests and the ones with cavities was evaluated with a correlation. To make a comparison with other tree types that exist in the region and the ones *M. colimana* prefers, a record of other tree species was made for every 100 oak trees found in each zone. The statistical software Statgraphics Plus® (1999) was also used for this statistical analysis.

Spatial distribution

The nearest neighbour distance (Clark & Evans, 1954) was used in order to determine the distribution pattern in the nesting sites. The comparison of average distances between the closest neighbour nest that was observed (XrA) and the average distances in the closest neighbour nest that was expected (Xre), which would be obtained if the nests were placed randomly. Clark and Evan's index is calculated as follows (Krebs, 1998): R= XrA/XrE. If the R value >1, the spatial distribution pattern is regular or uniform, if R<1 the organisms are forming aggregations, and if R=1 the distribution is random.

Results

The total study area was 594 hectares from which 280 were registered in the search of nests. Thirty wild nests were located: 24 in the P area and six in the D area. 27 nests were found in oak trees *Q. laurina;* the other three nests were found in other tree species: *L. apaculcensis, C. disciflora* and *L. glucesens* in the P area. The average diameter at breast height of trees with nests was 169.20 ± 42.74 in the P area and 197.16 ± 56.27 in the D area. There were statistical differences between the two zones in the number of nests (X^2 = 10.8; p<0.05, DF= 1, 29; 0.05 = 3.84), the number of *Q. laurina* trees (X^2 = 568; p<0.05, DF= 1, 1286; 0.05 = 3.84), the oak trees DBH (T=35.14, DF=1, 1285, p<0.05) and the number of oak trees with cavities (X^2 = 23.52; p<0.05, DF= 1, 67; 0.05 = 3.84).The P area had the highest values in almost

Table 1. Number of oak trees without nests, diameter at breast height, number of trees with cavities and density of the nests in the zone preserved (P) and disturbed (D).

Study zone	Number of oak trees	Diameter at breast height (cm)	Number of trees with cavities	Density of nests (ha⁻¹)
Zone P	216 a	116.22 ± 40.91 a	54 a	0.17 a
Zone D	1071 b	26.24 ± 32.83 b	14 b	0.04 b

Different letters denote statistic differences P<0.05

all parameters, except that in the D area the number of oak trees was higher, but with a DBH lower than the P area. The number of oak trees without nests, their diameter at breast height, the number of trees with cavities and the density of nests in every zone are shown in table 1.

For every 100 trees of *Q. laurina* there is an average 8.5 of other trees in the P area. In the D area there were 2.6 other trees for every 100 *Q. laurina* that had no cavities in their trunks. In both areas there is a relationship between the nesting sites DBH (oak trees), the nests in trees and tree holes availability in them. (Ji²= 62.244, DF=1, 58 P<0.01 y Ji²= 252.456, DF=1, 282 P<0.01). Nests presented aggregated distribution in both areas (P= 0.39 y D=0.31). The number of nests and their spatial distribution can be seen in Fig. 1.

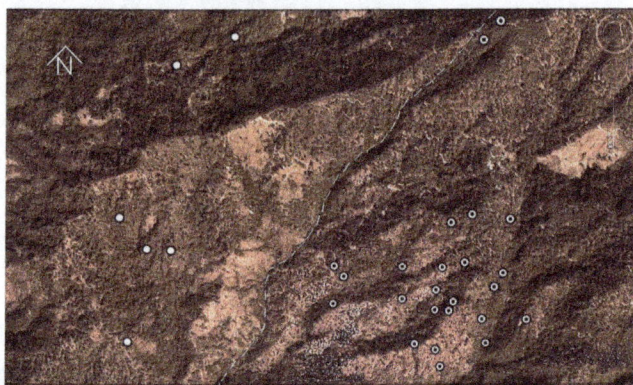

Fig 1. The spatial distribution of *Melipona colimana* wild nests in the study area. Every O mark indicates the position of a nest in the disturbed (D) area (n=6) and the ⊙ mark indicates a nest in preserved (P) area (n=24).

Discussion

Many stingless bee species are opportunistic and take advantage of several trees' cavities for nesting (Hubbell & Johnson 1977; Roubik, 1989). *M. colimana,* should be closely related to the *Q. laurina* trees with cavities, since it is one of the main species in this region and it is very common to have natural trunk cavities in old trees (Cuevas et al., 2004). Meliponini species have a natural tendency of taking advantage of the arboreal species predominant in their place of origin, such as *Melipona subnitida* Ducke*, Melipona asilvai* Moure, and *Melipona quadrifasciata* Lepeletier in Brazil, which nesting in *Commiphora lepthophoeos,*

Caesalpinia pyramidalis and *Caryocar brasiliense* trees (Martins et al., 2001; Antonini & Martins, 2003). If there is a strong preference for a predominant tree to nesting in the distribution area, it would leave the bees in a risky situation as the absence of these tree species would directly affect bee populations. Unfortunately, for *M. colimana*, the oldest *Q. laurina* trees having higher biomass are the most useful ones for companies dedicated to vegetal coal production (Reyes, 2012), so the activities arising from this industry may be decreasing the nesting site availability. The activity of the forest exploitation of zone D seems to be also affecting the presence of different trees to *Q. laurina*, given that the low number of trees that are not oak trees, reflects the negative impact that has been held in the zone.

It is observed that the P zone was the one with the highest number of nests, DBH and *Q. laurina* trees with cavities that might be a consequence of having no recent forest exploitation in the zone, which has allowed the subsistence of thicker oak trees. In forests habitats stingless bees have been found making their nests mainly in trees over 60 cm in diameter (Brown & Albrecht 2001; Antonini & Martins 2003; Eltz et al., 2003; Fierro et al., 2012) thus, a higher DBH and quantity of oak trees with available holes might have been a factor for the P zone to have a higher number of nests of *M. colimana*. Despite the higher number of oak trees found in D zone compared to the P zone, these are not useful for bees, since they are young trees which do not have any cavities in their structure. The number of nests found per hectare in both zones was low compared to what was reported by other authors. Roubik (2006) estimates that the usual colonies of *Melipona* and *Trigona* quantity per hectare is from 2 to 6, while in works that were carried out in rainy and broken up forest areas, the general density of stingless bees wild nests was 8.4 and 6.7 nests per ha⁻¹ (Batista et al., 2001; Eltz et al., 2002). As the same way, Antonini & Martins (2003), found a density of 3 nests per ha⁻¹ of *M. quadrifasciata* in a study conducted in a Brazilian savanna, which is a high nest density compared with the density nests found in *M. colimana*.

Although this low nest density found in *M. colimana* is the first report about stingless bees in a temperate climate, the results suggest that wild populations of this species might be endangered by forest exploitation. The *M. colimana* nests were found forming aggregations in both study zones, which coincides with the report about other stingless bees species such as *Scaptotrigona. pectoralis* Dalla Torre*, Frieseomelitta nigra* Cresson*, Trigona. fulviventris* Guérin*, Tetragonisca angustula* Latreille and *Nannotrigona testaceicornis* Lepeletier (Slaa, 2002; Santos, 2006). Since stingless bees nest on tree's hollow spaces, its density and spatial distribution depends on the presence and distribution of the trees that are used as nesting site (Batista et al, 2001; Santos, 2006), in this case the nest distribution of *M. colimana* depended on the distribution of *Q. laurina* trees.

Despite not having replicas of these observations, this information can give us sign that forest exploitation could be affecting wild nesting of *M. colimana*, resulting in a lower number of trees which bees can use as nesting sites. It has been found that the anthropogenic activity, directly or indirectly, affects density, diversity and spatial distribution of stingless bee communities (Brown & Albrecht, 2001, Hsiang et al., 2001; Moreno & Cardoso 2002; Samejima et al., 2004), therefore, it would be important to do another study with more replicas to get data that show us in a conclusive way that the forest exploitation of *Q. laurina* can put in risk the presence of *M. colimana* nests in its original habitat, which is why it is important to emphasize its protection.

Acknowledgements

Especially to Angelica F.S. This article is part of J. O. Macias Macias research as an Agriculture and Livestock Science PhD scholarship student in CCBA-UADY sponsored by CONACYT-PROMEP. Thanks to Bernardo Soto, to the community of San Isidro and Gustavo Alcázar Oceguera for his invaluable help and knowledge about the stingless bees' location. To the CUSUR Research Coordination (University of Guadalajara) and to the Apiculture students for their help in the fieldwork.

References

Antonini, Y. & Martins P.R. (2003). The value of tree species (*Caryocar brasiliense*) for a stingless bee *Melipona quadrifasciata quadrifasciata*. J. Insect Cons. 7:167-174. DOI 10.1023/A: 1027378306119.

Ayala, R. (1999). Revisión de las abejas sin Aguijón de México (Hymenoptera: Apidae: Meliponini). Fol. Entom. Mex. 106: 1-123.

Batista, M.A., Ramalho, S. & Soares A.E.E. (2001). Nest sites and diversity of stingless bees in the Atlantic rainforest, Bahia, Brazil. II Mexican Seminar on stingless bees. Autonomous University of Yucatan. F.M.V.Z. U.A.D.Y. Mérida, Yucatán, México.

Brown, C.J. & Albrecht C. (2001). The effect of tropical deforestation on stingless bees of the genus *Melipona* (Insecta: Hymenoptera: Apidae: Meliponini) in central Rondonia. Brazil. J. Biol. 28: 623-634.

Camargo, J.M.F. & Pedro, S.R.M. (1992). Systematics, phylogeny and biogeography of the Meliponinae (Hymenoptera, Apidae): a mini review. Apidologie 23: 293–314.

Cannon, C.H., Peart, D.R., Leighton, M. & Kartawinata, K. (1994). The structure of lowland rainforest after selective loggin in West Kalimantan, Indonesia. For. Ecol. Man. 69: 49-68.

Clark, P.J. & Evans, F.C. (1954). Distance to nearest neighbour as a measure of spatial relationship in populations. Ecology 35: 445-453.

CONAFOR (2010). Forestry National Program. Comisión Nacional Forestal. Diario Oficial de la Federación. February 2. Secretaria del Medio Ambiente Recursos Naturales y Pesca, México, D. F.

CONAGUA (2011). Regional Meteorological Observatory, Southern Jalisco files, Comisión Nacional del Agua. Cd. Guzmán, Zapotlán el Grande. Jalisco, México.

Cuevas, G.R., Koch, S., Garcia, M.E., Nuñez, L.N.M. & Jardel, P.E.J. (2004). Flora vascular de la estación científica las Joyas, in: Cuevas G.R, Jardel P.E.J. (eds) Flora y Vegetación de la Estación Científica las Joyas, University of Guadalajara, México, pp 119-176.

Eltz, T., Bruhl, A.C.,Van Der, K.S. & Linsenmal, E.K. (2002). Determinants of stingless bee density in lowland dipterocarp forest of Sabah, Malaysia. Oecologia 131: 27-34. DOI 10.1007/s00442-001-0848-6

Eltz, T., Carsten, A., Imiyabir, Z. & Linsenmal, E.K. (2003). Nesting and nest trees of stingless bees (Apidae: Meliponini) in lowland dipterocarp forest of Sabah, Malaysia, with implications for forest management. For. Ecol. Manag. 172: 301-313. PII: S0378-1127(01)00792-7

Fierro, M.M., Cruz-Lopez, L., Sanchez, D., Villanueva-Gutierrez, & Vandame R. (2012). Effect of biotic factors on the spatial distribution of stingless bees (Hymenoptera: Apidae: Meliponini) in fragmented neotropical habitats. Neotrop. Entomol. 41:95-104. DOI 10.1007/s13744-011-0009-5.

Hsiang, L.L., Sodhi, S.N. & Elmqvist, T. (2001). Bee diversity along disturbance gradient in tropical lowland forest of southeast Asia. J. App. Ecol. 38: 180-192. DOI 10.1046/j.1365-2664.2001.00582.x

Hubbell, S.P. & Johnson, L.K. (1977). Competition and nest spacing in a tropical stingless bee community. Ecology 58: 949-963.

Imperatriz-Fonseca, V.L. (2002). Best management practices in agriculture for sustainable use and conservation of pollinators. Agreement FAO-Sao Paulo University.

INEGI (2005). Anuario estadístico del Estado de Jalisco. Instituto Nacional de Estadística Geografía e Historia, Guadalajara, Jalisco, México.

INIFAP (2006). Forestry resources evaluation course. Instituto Nacional de Investigaciones Forestales Agrícolas y Pecuarias, 11-15 November. CIPAC-INIFAP Guadalajara, Jalisco, México.

Krebs, C.J. (1998). Ecological Methodology. Benjamin Cummings, New York.

Martins, C.F., Cortopassi-Laurino, M., Koedam. D. & Imperatriz-Fonseca, V.L. (2001) The use of trees for nesting by stingless bees in Brazilian Caatinga. Procedings XXXVI International Apicultural Congress. Durban.

Michener, C.D. (2000). The bees of the World. The Johns Hopkins Univesity Press, Baltimore, USA.

Moreno, E.F.A. & Cardozo, A. (2002). Parámetros biométricos y estados de colonias de abejas sin aguijón (Meliponinae) en restos de árboles después de la explotación maderera en el Estado de Portuguesa-Venezuela. Investigation report. Investigation deanship. Department of Agronomic Engineering, National University of Tachira, Venezuela.

Nogueira-Neto, P. (1997).Vida e criacao de abelhas indigenas sem ferrao. Ed. Nogueirapis. São Paulo. Brazil. 385 p.

Carranza, G.E. (2008). Flora de Jalisco y áreas colindantes. Centro Universitario de Ciencias Biologicas y Agropecuarias. Universidad de Guadalajara. 22:34-37.

Reyes, F. (2012). Es carbón vegetal fuente de ingresos para familias colimenses. México Forestal Magazine. Comisión Nacional Forestal, No. 70, October-November. Mexico.

Roubik, D.W. (1989). Ecology and natural history of tropical bees. Cambridge University Press, London.

Roubik, D.W. (2006). Stingless bee nesting biology. Apidologie 37: 124-143. DOI: 10.1051/apido: 2006026

Samejima, H., Marzuki, M., Nagamitsu, T. & Nakashizuka, T. (2004). The effects of human disturbance on a stingless bee community in a tropical rainforest. J. Biol. Cons. 120: 577-587. DOI: 10.1016/j.biocon.2004.03.030

Santos, L.A.C. (2006). Distribución espacial de los sitios de anidación de abejas eusociales (Hymenoptera: Apidae: Meliponini y Apini) en Sudzal, Yucatan, Mexico, Dissertation, Autonomous University of Yucatan, Mexico.

Slaa, E.J. (2002). Community ecology of stingless bees in forested and deforested habitats in Guanacaste, Costa Rica, In: Universidad de Utrech (Ed) Foraging Ecology of stingless bees: from individual behaviour to community ecology. The Netherlands.

Slaa, E.J., Sanchez-Chavez, L.A., Malodi-Braga, K.S. & Hofstede, F.E. (2006). Stingless bees in applied pollination: practice and perspectives. Apidologie 37: 293-315. DOI 10.1051/apido: 2006022

Statgraphics Plus (1999). For Windows 4.1 Copyrigth©, By Statistical Graphics Corp, All Rights Reserved.

Venturieri, C.G. (2002). Exploração florestal e impacto sobre abelhas indigenas sem ferrão. In: Imperatriz-Fonseca VL (Ed). Best management practices in agricultura for sustainable use and conservation of pollinators. Agreement FAO Sao Paulo University.

20

First records of the myrmecophilous fungus *Laboulbenia camponoti* Batra (Ascomycota: Laboulbeniales) from the Carpathian Basin

F BÁTHORI[1], WP PFLIEGLER[2] & A TARTALLY[1]

1 - *Department of Evolutionary Zoology and Human Biology, University of Debrecen, Debrecen, Hungary*
2 - *Department of Genetics and Applied Microbiology, University of Debrecen, Debrecen, Hungary*

Keywords
Austria, *Camponotus aethiops*, Central-Europe, mycology, Romania, social parasite

Corresponding author
András Tartally
Department of Evolutionary Zoology and Human Biology, University of Debrecen
Egyetem Tér 1, H-4032 Debrecen, Hungary
E-mail: tartally.andras@science.unideb.hu

Abstract
Laboulbenia camponoti Batra, 1963 (Ascomycota: Laboulbeniales), has been found on *Camponotus aethiops* (Latreille, 1798) (Hymenoptera: Formicidae) workers in the Carpathian Basin: in Baziaş, Caraş-Severin (Romania), and Vienna (Austria). Vienna is the northernmost known locality of this fungus (48°12′ N). These new observations expand the area of *L. camponoti* from regions with Mediterranean and subtropical climatic influences to the common borders of the Continental and Pannonian regions. These results show that *Camponotus* samples from other climatic regions should be examined more closely for this fungal parasite.

The order Laboulbeniales comprises more than 2000 species in about 140 genera (Santamaria, 2001; Weir & Blackwell, 2005; Kirk et al., 2008). They are obligate ectoparasites of arthropods, and approximately 80% of the described Laboulbeniales species parasitize Coleoptera species (Santamaria, 2001; Henk et al., 2003; Weir & Blackwell, 2005).

In the order Hymenoptera, only ants are known to be hosts of certain species of Laboulbeniales (Espadaler & Santamaria, 2003). Thus far, four species of these fungi have been reported to be associated with ants in Europe: *Rickia wasmannii* Cavara, 1899, is found in 15 countries on eight *Myrmica* species; *Laboulbenia formicarium* Thaxt, 1908, in France, Portugal and Spain on two *Lasius* species; *Laboulbenia camponoti* Batra, 1963, in Bulgaria and Spain on five *Camponotus* species; and *Rickia lenoirii* Santamaria and Espadaler, 2014, in Greece and France on two *Messor* species (Herraiz & Espadaler, 2007; Lapeva-Gjonova & Santamaria, 2011; Espadaler & Santamaria, 2012; Santamaria & Espadaler, 2014).

The effect of these ant parasitic fungi on their hosts is rather understudied except for the work of Csata et al. (2014).

They found that under laboratorial conditions the lifespan of *Myrmica scabrinodis* Nylander, 1846 individuals infected with *R. wasmannii* was significantly reduced in comparison with the lifespan of uninfected ants. Moreover auto- and allogrooming increased in infected nests. These facts support the parasitic character of ant-associated Laboulbeniales fungi.

Only *R. wasmannii* has been reported among these four species in the Carpathian Basin (Tartally et al., 2007). As *Camponotus aethiops* (Latreille, 1798) is a relatively common species in this region (Csősz & al., 2011; pers. observ.), which is one of the known hosts of *L. camponoti* (Espadaler & Santamaria, 2012), we suspected the possibility to record *L. camponoti* from the Carpathian Basin. Our aim was therefore to prove the presence of *L. camponoti* within this region by checking museum specimens of *C. aethiops*. Though the other known (Espadaler & Santamaria, 2012) host ants (*C. universitatis* Forel, 1890; *C. pilicornis* (Roger, 1859); *C. sylvaticus* (Olivier, 1792)) are not known from this region (Csősz & al., 2011), we aimed to search for individuals among museum specimens from the Carpathian Basin.

Finding *L. camponoti* for a new region may call the attention of myrmecologists and mycologists to check *Camponotus* specimens more intensively for the presence of this small and understudied fungus.

Materials and Methods

To reveal the presence of *L. camponoti*, all the specimens of *Camponotus aethiops* (Hymenoptera: Formicidae) (workers, males, and queens) in the Hymenoptera Collection of the Hungarian Natural History Museum were examined under an Olympus SZX9 stereomicroscope at magnifications of 12.6x-114x. No *C. universitatis*, *C. pilicornis* or *C. sylvaticus* specimens were found in this collection from the Carpathian Basin.

Pinned specimens of the host that were found to be infested were soaked in 70% ethanol for 5-12 hours and examined using transmitted light under a binocular microscope at 10x magnification. Thalli were removed with an insect pin and cleared in lactic acid (12 hours) before being mounted in a PVA-glycerol medium and photographed with an Olympus digital camera through an Olympus BX-40 microscope equipped with 40x and 100x lenses. Measurements were taken with the manufacturer's image acquisition software (DP Controller).

Specimens are deposited in the Fungi Collection of the Hungarian Natural History Museum on slides (inventory numbers: BP 105023, BP 105024).

Results and Discussion

More than 200 *C. aethiops* specimens were examined, originating from 34 parts of the Carpathian Basin (sites in Hungary, Romania, Slovakia, Austria, and Serbia). Only three specimens (less than 1.5% of the investigated samples) of *C. aethiops* workers were found to be parasitized by *L. camponoti*: two workers from Vienna, Austria (48°12' N, 16°22' E, 180 m a.s.l.), and one from Baziaş, Romania (44°48' N, 21°23' E, 85 m a.s.l.). The fungus grew from the cuticle of different body parts of the workers, mainly on the head and the legs (Fig. 1-2). No infested queens or males were found. However, the numbers of queens and males in the museum collection were small.

The number of thalli observed on infected *Camponotus* specimens was relatively small. A dozen (mostly immature) thalli were found in two groups on an antenna of one specimen from Vienna, while the other worker from the same location had only two immature thalli with developing perithecia (the spore-producing fruiting body of the fungus) on one leg. A single, mature thallus with visible spores was found on the head of the Romanian specimen collected at Baziaş (Fig. 1). Variation in the length and number of the sterile appendages was observable, as also noted in the species' original description (Batra, 1963), where explanations of life stages and morphology are also available.

The ectoparasitic fungus *L. camponoti* was found for the first time in Romania and Austria (see: Espadaler and Santamaria, 2012 and references therein). The number of countries this fungus is recorded in is now increased from four to six: it has previously been found only in Spain, Bulgaria, Turkey (for a review: Espadaler and Santamaria, 2012 and references therein) and India (Batra, 1963). In its prior known localities, the Mediterranean or subtropical climatic influence is strongly expressed. This may have led myrmecologists and mycologists to consider *L. camponoti* to be distributed solely in such climatic areas. However, the two newly recorded localities are in the common borders of the Continental and Pannonian regions (see: EEA, 2011), and the new locality at Vienna is the northernmost (48°12' N) known latitude of *L. camponoti* in the world. These facts give a new picture of the potential distribution of this fungus.

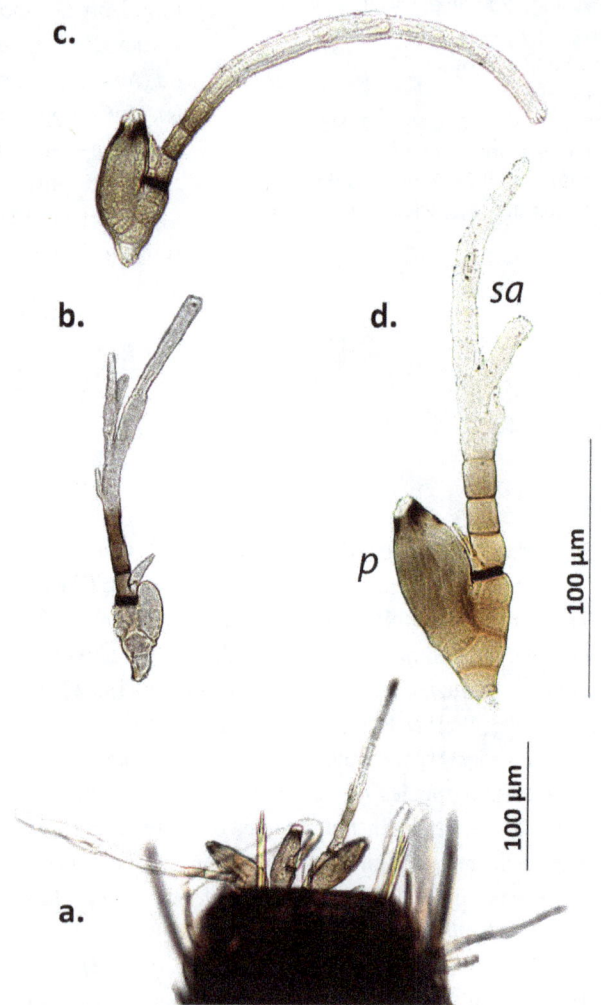

Fig. 1. *Laboulbenia camponoti*. a. Group of thalli on antenna (Vienna). b. Young immature thallus (Vienna). c. Immature thallus with developing perithecium (Vienna). d. Mature thallus (Baziaş). Legend: p - perithecium; sa – sterile appendages (their numbers show individual differences).

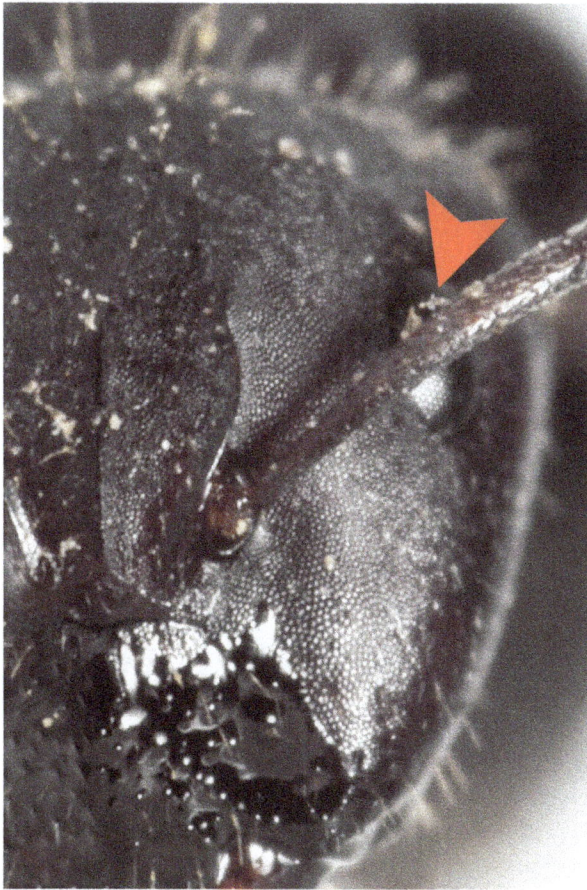

Fig. 2. A *Laboulbenia camponoti* individual on the scapus of a *Camponotus aethiops* worker (Vienna), the figure illustrates how meticulous it is to find this small fungus on a large *Camponotus* individual, especially when dust on the host prevents easy recognition

The inconspicuous nature of *L. camponoti* has undoubtedly contributed to the scarcity of its distribution records. As illustrated by Fig. 2., the thalli are very hard to locate, especially on older museum specimens with dust particles. Determination of the fungus must be validated by light microscopy. Because European *Camponotus* species are usually large (see e.g. Seifert, 2007), and therefore usually easily observed with the naked eye, myrmecologists rarely examine them by microscopy. However, these results demonstrate that a thorough examination of *Camponotus* specimens from other climatic regions may reveal the presence of this little-known parasitic fungus.

Acknowledgments

We would like to thank the numerous collectors whose work provided the samples that we examined. Zoltán Vas, curator, helped us in our work in the Hymenoptera Collection of the Hungarian Natural History Museum. BF and AT were supported by the 'AntLab' Marie Curie Career Integration Grant (of AT) within the 7th European Community Framework Programme. AT was supported by a 'Bolyai János' scholarship of the Hungarian Academy of Sciences (MTA).

References

Batra, S. W. T. (1963). Some Laboulbeniaceae (Ascomycetes) on insects from India and Indonesia. American Journal of Botany, 50: 986-992.

Csata, E., Erős, K. & Markó, B. (2014). Effects of the ectoparasitic fungus *Rickia wasmannii* on its ant host *Myrmica scabrinodis*: changes in host mortality and behavior. Insectes Sociaux, 61: 247-252. doi: 10.1007/s00040-014-0349-3

Csősz, S., Markó, B. & Gallé, L. (2011). The myrmecofauna (Hymenoptera: Formicidae) of Hungary: an updated checklist. North-Western Journal of Zoology, 7: 55-62.

European Environment Agency (EEA) (2011). Biogeographic regions in Europe.

Espadaler, X. & Santamaria, S. (2012). Ecto- and endoparasitic fungi on ants from the Holarctic Region. Psyche, 2012 (168478): 1-10. doi: 10.1155/2012/168478

Henk, D.A., Weir, A. & Blackwell, M. (2003). *Laboulbeniopsis termitarius*, an ectoparasite of termites newly recognized as a member of the Laboulbeniomycetes. Mycologia, 95: 561-564.

Herraiz, J.A. & Espadaler, X. (2007). *Laboulbenia formicarum* (Ascomycota, Laboulbeniales) reaches the Mediterranean. Sociobiology, 50: 449-455.

Lapeva-Gjonova A. & Santamaria, S. (2011). First record of Laboulbeniales (Ascomycota) on ants (Hymenoptera: Formicidae) in Bulgaria. Zoonotes, 22: 1-6.

Kirk, P.M., Cannon, P.F., Minter, D.W. & Stalpers, J.A. (eds) (2008). Ainsworth and Bisby's Dictionary of the Fungi (10th Edition). CABI Europe-UK, Cromwell Press, Trowbridge, 771 p

Santamaria, S. (2001). Los Laboulbeniales, un grupo enigmático de hongos parásitos de insectos. Lazaroa, 22: 3-19.

Santamaria, S., & Espadaler, X. (2014). *Rickia lenoirii*, a new ectoparasitic species, with comments on world Laboulbeniales associated with ants. Mycoscience (in press) doi: 10.1016/j.myc.2014.06.006

Seifert, B. (2007). Die Ameisen Mittel- und Nordeuropas. Görlitz/Tauer: Lutra Verlags- und Vertriebsgesellschaft, 368 p

Tartally, A., Szűcs, B. & Ebsen, J.R. (2007). The first records of *Rickia wasmannii* Cavara, 1899, a myrmecophilous fungus, and its *Myrmica* Latreile, 1804 host ants in Hungary and Romania (Ascomycetes: Laboulbeniales, Hymenoptera: Formicidae) Myrmecological News, 10: 123.

Weir, A., & Blackwell, M. (2005). Fungal biotrophic parasites of insects and other Arthropods. In: F.E. Vega & M. Blackwell (Eds.), Insect-Fungal Associations: ecology and evolution (pp. 119-145). Oxford: Oxford University Press.

21

Geographic Distribution, Key Challenges, and Prospects for the Conservation of Threatened Stingless Bee *Melipona capixaba* Moure and Camargo (Hymenoptera: Apidae: Meliponini)

HC Resende[1,2], TM Fernandes-Salomão[1], MG Tavares [1], LAO Campos[1]

1 - Universidade Federal de Viçosa, Viçosa, MG, Brazil

2 - Universidade Federal de Viçosa, Campus Florestal, MG, Brazil

Keywords

Meliponini, Endangered Species, Atlantic Forest, Restricted Distribution, Mountain Habitats.

Corresponding author

Helder Canto Resende
Universidade Federal de Viçosa - campus UFV Florestal; Laboratório de Genética e Conservação da Biodiversidade, Florestal, 35690-000, Minas Gerais, Brazil
E-Mail: helder.resende@ufv.br

Abstract

The stingless bee *Melipona capixaba* Moure and Camargo is endemic to the Brazilian Atlantic Forest. Its occurrence is restricted to highlands in the Espírito Santo State, and it has possibly the smallest known geographic distribution among the cataloged stingless bees. It is therefore considered to be an endangered species. Perhaps because of its small area of occurrence, or because it was only identified two decades ago, little is known about the biology of this species, its current geographic distribution, or its actual preservation status. Here, we present the results from the largest sampling of *M. capixaba* conducted in its natural habitat. We developed a distribution map by using a geographic information system. Our data indicate that *M. capixaba* is found in the municipalities of Espírito Santo State at altitudes between 800 m and 1,200 m; with annual average temperatures around 18–23°C; precipitation more than 1,200 mm per year; and vegetation cover-type Mountain Dense Ombrophylous Forest, restricted to an area of approximately 3,450 km². We observed colonies both in their natural habitat and under conditions of *ex situ* maintenance, and identified the key challenges and prospects for the conservation of this endangered bee.

Introduction

The stingless bee *Melipona capixaba* Moure and Camargo is endemic to the Brazilian Atlantic Forest. The species is popularly known as "uruçu-preta" (black uruçu) because of its dark coloring and as "uruçu-capixaba" in reference to its endemism. It is the only social bee included in the List of Species of Brazilian Fauna Threatened with Extinction (Brasil, 2003) and classified in the category Vulnerable - VU-B1ab (iii): taxon facing high risk of extinction in the wild, geographical distribution whose extent of estimated occurrence is less than 20,000 km², severely fragmented, or with known distribution in not more than ten localities. A continuous decline was observed, inferred, or projected in its area of occupation, extension, and/or quality of habitat (Machado et al., 2008). The main threat for this species is massive deforestation in its habitat, which fragments and isolates populations, reduces food sources and nesting areas, and hinders the survival and reproduction of colonies (Silveira et al., 2008).

The factors that threaten *M. capixaba* are local examples of the factors that affect bees globally. There has been a worldwide decline in the population of these pollinators, and this is mainly due to habitat loss and fragmentation, habitat agrochemicals, pathogens, alien species, climate change, and the interactions among these factors (Freitas et al., 2009; Potts et al., 2010). Thus, monitoring the remaining populations of *M. capixaba* might provide information about their long-term population trends and offer interesting perspectives for bee conservation actions on a larger scale.

The currently known geographic range of *M. capixaba* is restricted to the mountainous regions of the state of Espírito Santo in Brazil. Melo (1996) registered the first occurrence of the species and found that it occurred only at higher altitudes (900-1,000 m) and suggested that this is possibly the smallest known distribution among the cataloged stingless bees. Serra et al. (2012) identified native colonies of *M. capixaba*, all found in regions at altitudes between 900 m and 1,200 m. Nogueira et al. (2014) observed the occurrence of *M. capixaba* colonies in regions at altitudes between 900 m and 1,100 m.

The remaining population of *M. capixaba* is drastically reducing, mainly because of loss and fragmentation of its habitat and activities of the local population, which includes removing the hives from the natural environment for beekeeping. A direct consequence of population reduction is the reduction in genetic diversity, which was verified by Nogueira et al. (2014) through assessing microsatellite markers, inter simple sequence repeat (ISSR) markers, and mitochondrial haplotypes. The loss in genetic diversity reduces the evolutionary potential of the species to adapt to environmental changes and endangers its long-term survival.

Considering the risk of extinction of *M. capixaba*, it is essential to gain some insight into the area of its occurrence, conservation status, state of maintenance of colonies, and the risks and threats faced by the bees. Serra et al. (2012) modeled the potential geographic distribution based on a sample of 14 *M. capixaba* wild nests. However, extensive sampling of natural and beekeeper nests inside its natural habitat would provide an accurate idea of the current geographic distribution of the species and of potential areas for its management and conservation status. Here, we present the results of the largest sampling of *M. capixaba* conducted in its natural habitat. Our goal was to confirm the current geographic distribution area of *M. capixaba*, identify potential areas for management and conservation of the remaining populations, and suggest conservation actions for this bee threatened with extinction.

Materials and Methods

Between 2007 and 2014, field surveys were conducted to collect bee specimens. The field surveys had two objectives: (i) identification of wild colonies, i.e., those located in native forests, and (ii) identification of colonies managed by humans that are close to their natural habitats. The points of occurrence were recorded using Global Positioning System (GPS), and spatial analyses were performed using ArcGIS 9.3 ESRI®.

For spatial analyses, free-access Shapefiles were used. The following programs were used: Shapefiles of the digital elevation of the ASTER GDEM program (NASA, 2009) Shapefiles of the physical limits of municipalities and states of the Brazilian federation, IBGE (http://www.ibge.gov.br/home/download/geociencias.shtm); Shapefiles of the Atlantic Forest vegetation and conservation areas MMA - Ministry of Environment (http://mapas.mma.gov.br/i3geo/datadownload.htm); and IBAMA - Brazilian Institute of Environment (http://siscom.ibama.gov.br/shapes/); and Shapefiles of Forest Remnants of the Atlantic Forest provided by SOS Atlantic Forest Foundation and INPE - National Institute for Space Research (http://mapas.sosma.org.br/dados/).

Contour maps and area delineation considering elevation were generated using the *Surface Analysis* method of ArcGIS 9.3 ESRI®. Temperature maps were generated by interpolation of 30-year historical series data (1977–2006) obtained from 110 weather stations located in the Espírito Santo State and neighboring states. The Kriging spherical method of ArcGIS 9.3ESRI® was used to map the variable temperature, considering the range of annual average temperatures in the historical series.

Contour maps, of temperature and vegetation, were compared, and the intersection areas of the three variables were taken into account in the delimitation of the area estimated for the occurrence of *M. capixaba* and in the delimitation of potential areas for reintroduction and management of the species. The extent of the delimited areas were deduced by calculating the irregular polygon areas with the geometric calculation tool of the ARCGis ESRI® 9.3 mapping platform, using the projected coordinate systems SAD′1969 UTM Zone 24S with km² as the area unit.

Results and Discussion

We recorded the occurrence of 194 *M. capixaba* colonies, 27 wild nests and 167 beekeeper nests, at 63 sampling points in 12 municipalities of Espírito Santo State, Brazil (Fig 1). In 10 municipalities, the species is found in native forests or is maintained in areas near the site from which they were collected: Afonso Claudio (AF), Alfredo Chaves (AC),

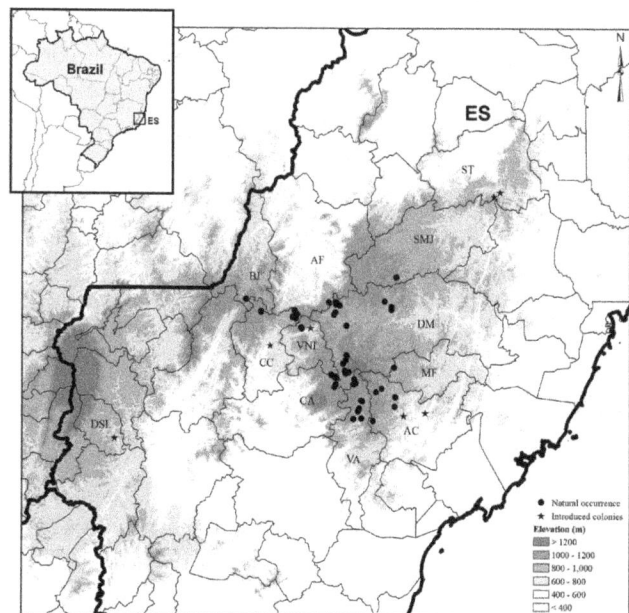

Fig 1. Map of 63 sampling points from 194 colonies of *Melipona capixaba* in the area of current occurrence. Circles on the map represent native or managed colonies close to their natural habitat. Stars on the map represent introduced colonies that were derived from other locations. Legend of municipalities: SMJ - Santa Maria de Jetibá, ST - Santa Teresa, DM - Domingos Martins, MF - Marechal Floriano, AC - Alfredo Chaves, VA - Vargem Alta, CA - Castelo, VNI - Venda Nova do Imigrante, CC - Conceição do Castelo, BJ - Brejetuba, AF - Afonso Cláudio, DSL - Divino de São Lourenço.

Brejetuba (BJ), Castelo (CA), Conceição do Castelo (CC), Domingos Martins (DM), Marechal Floriano (MF), Vargem Alta (VA), Venda Nova do Imigrante (VNI), and Santa Maria de Jetibá (SMJ).

We recorded the occurrence of colonies that were introduced in the municipalities of Divino de São Lourenço (DSL, 4 nests) and Santa Teresa (ST, 8 nests), originated from VNI and SMJ, respectively. Although there are no records of natural colonies in these two locations, these colonies have been successfully maintained for at least 5 years, indicating potential areas of management for conservation near the areas of occurrence.

Geographic distribution

Native colonies of *M. capixaba* were observed at a minimum altitude of 882 m and a maximum altitude of 1,168 m, and no data from naturally occurring colonies or managed colonies were found in regions above 1,200 m and below 600 m (Table 1, Fig 2A).

Although natural occurrence has been observed only between 800 m and 1,200 m, colonies of *M. capixaba* are managed with relative success at altitudes between 700 m and 800 m in colder regions and near native forests. At 3 sites, we verified the maintenance of colonies between 600 m and 700 m (minimum altitude, 612 m), but in these cases, management requires special care with food supplementation. To the best of our knowledge, *M. capixaba* colonies do not remain at altitudes lower than 600 m, especially in regions with warmer temperatures, even when managed properly. Although Serra et al. (2012) suggested a wider distribution area for *M. capixaba* species, obtained by potential distribution modeling, our sampling efforts reported no occurrence *M. capixaba* in regions north of Santa Tereza (ST), east of Marechal Floriano (MF), south of Vargem Alta (VA), and west of Brejetuba (BJ) (Fig 1).

Climatological stations located in areas where we commonly observed *M. capixaba* had recorded average annual temperatures (Aat) below 19°C in the 30-year historical series (1977–2006). We observed neither the occurrence nor the management of *M. capixaba* colonies in regions of Espírito Santo where the Aat in the historical series was greater than 23°C. Thus, it appears that regions with Aats exceeding 23°C are not favorable for the occurrence or management of this

species. Even in regions between 800 m and 1,200 m, but with an Aat of 20–21°C, occurrence of *M. capixaba* has not been verified or it was restricted; it is more common to verify its occurrence in regions at temperatures between 18°C and 20°C (Fig 2B). The observed precipitation in the region of occurrence, using the 30-year historical series, was estimated to be more than 1,200 mm per year, which is higher than the amount of evapotranspiration, thus, the region of occurrence is characterized as a moist environment. According to the climatic classification of Köppen-Geiger, updated by Peel et al. (2007), the prevailing climate in the area estimated for the occurrence of *M. capixaba* is the Tropical Monsoon Climate class "Am." According to this classification, tropical climate regions are characterized by the coldest month of the year with an average temperature greater than or equal to 18°C; absence of seasons with low temperatures (winter seasons); and annual rainfall greater than the annual potential evapotranspiration, with high rates of annual precipitation and precipitation of less than 60 mm in the driest month.

Local vegetation in areas where we sampled *M. capixaba* is almost exclusively Dense Ombrophylous Forest, with some occurrence of Open Ombrophylous Forest, according to the Map of Law Application Area N° 11,428 from 2006 (Atlantic Forest

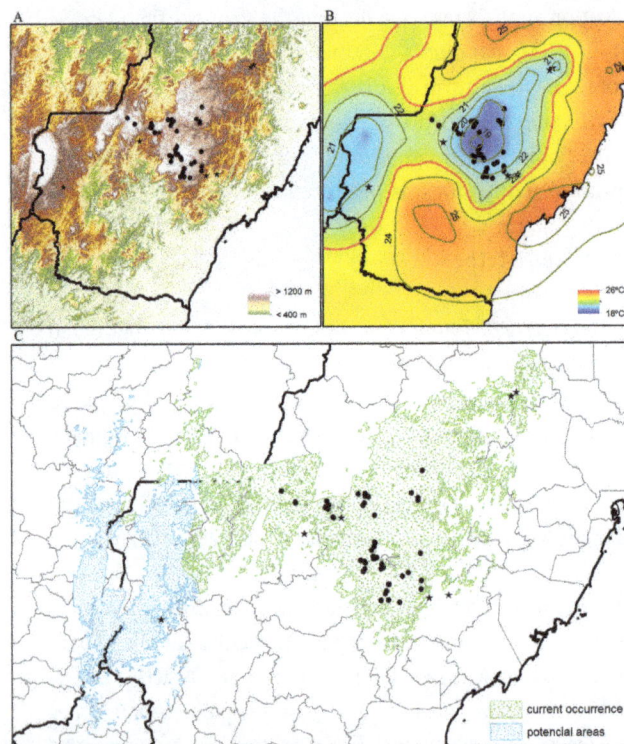

Fig 2. A - Altitude map and occurrence area of *Melipona capixaba*, indicating sampling points and terrain elevation according to the Global Digital Elevation Model - GDEM ASTER (NASA, 2009). B - Spatial distribution map of mean annual temperatures (Tmax), 30-year (1977–2006) time series obtained by interpolation of data from 110 weather stations by using the Kriging spherical model. C - Map of the estimated area of current geographic occurrence (in green) and potential areas (in blue) for species management and conservation, considering altitude, temperature, precipitation, vegetation, native occurrence, and management of species.

Table 1. The number of samples and percentage of colonies sampled at each altitude and the minimum and maximum altitudes at which native and managed colonies of *Melipona capixaba* were sampled.

Altitude (m)	Number of colonies	%		Minimum altitude	Maximum altitude
600–700	11	5.67	Native colonies	882 m	1168 m
700–800	16	8.25			
800–900	24	12.37	Managed colonies	612 m	1180 m
900–1000	77	39.69			
1000–1100	61	31.44			
1100–1200	5	2.58			
	194	100			

Law - IBGE, 2008). In areas where we observed the presence of *M. capixaba*, the Dense Ombrophylous Forest can be classified as Mountain Dense Ombrophylous Forest. The mountain formation is located in high plateaus and/or at ranges of 500 m to 1,500 m, between 16°S and 24°S (IBGE, 1992).

Current estimated area of occurrence for M. capixaba and potential area for the management and conservation of the species

By comparing the sites where the presence of *M. capixaba* was verified with elevation maps of the spatial distribution of vegetation and annual temperature averages and precipitation, it is possible to delimit an Estimated Area of Natural Occurrence for *M. capixaba*, comprising regions at altitudes between 800 m and 1,200 m, annual temperature averages between 18°C and 23°C, precipitation more than 1,200 mm per year, and with vegetation cover-type Mountain Dense Ombrophylous Forest. The regions with these characteristics are restricted to an area of 3,453 km² (Fig 2C, green area).

A potential area for maintenance and occurrence of *M. capixaba* is around the Caparaó National Park, a calculated area of 1,328 km² (Fig 2c, blue area). There are no records of *M. capixaba* occurring naturally in the Caparaó region. Although vegetation in this region is classified as Seasonal Semideciduous Forest according to IBGE (2008), a region of Caparaó National Park, especially its eastern portion, has vegetation that is very similar to Mountain Dense Ombrophylous Forest. The proximity of this area to the natural distribution area of *M. capixaba* and its local characteristics suggest that this may be a potential area for species management.

Considering the verified and estimated areas of occurrence, the geographical distribution of species is currently limited to an area of less than 5,000 km². Limited areas of distribution suggest that the classification of this species should be changed from its current status of "Vulnerable" VU-B1ab (iii) to "Endangered" EN-B1ab (iii), according to the criteria adopted in the IUCN Red List of Endangered Species (IUCN Red List Categories, 2001). Taxa at a high risk of extinction in the wild are included in the category "Endangered" - EN-B1ab (iii) if they have a geographical distribution of occurrence that is estimated to be below 5,000 km², severely fragmented, and with an observed continued decline, inferred or projected in the area, extension and/or quality of habitat. These conditions are consistent with the current situation of *M. capixaba*, justifying its reclassification into the category EN "Endangered."

Serra et al. (2012) suggested that the areas of priority for *M. capixaba* conservation are the following conservation units: Mata das Flores State Park, Forno Grande State Park, Pedra Azul State Park, Caparaó National Park, and Cachoeira do Rio Pardo Forest Reserve. Another area that we consider as an area of priority is the Biological Reserve Augusto Ruschi, which is the largest natural fragment within the estimated area of occurrence with an area of approximately 47 km² (4,742 ha). Our sampling could not identify the presence of

M. capixaba in any of these places, and this was probably because of three main reasons. First, the largest regions in these conservation units are at elevations above 1,200 m, with rocky areas and vegetation on top, which are not good conditions for *M. capixaba* colonies. Second, it is possible that there was local extinction because most of the regions are secondary forests and it is common for people to remove natural colonies for beekeeping. Third, in areas with altitudes between 800 and 1,200 m in the dense ombrophylous forest, sampling of bees is quite difficult because of tree height. Thus, absence of records may be due to sampling limitations. Nevertheless, we still consider some areas of priority as the buffer zones for the conservation unit areas, such as the buffer zone of Forno Grande State Park and Pedra Azul State Park (which is the second largest fragment within the area of occurrence of *M. capixaba*) with an area of approximately 43 km² (4,301 ha). This buffer zone is the most promising region in which a program for species management and conservation could be implemented for the maintenance of protected populations (Fig 3B).

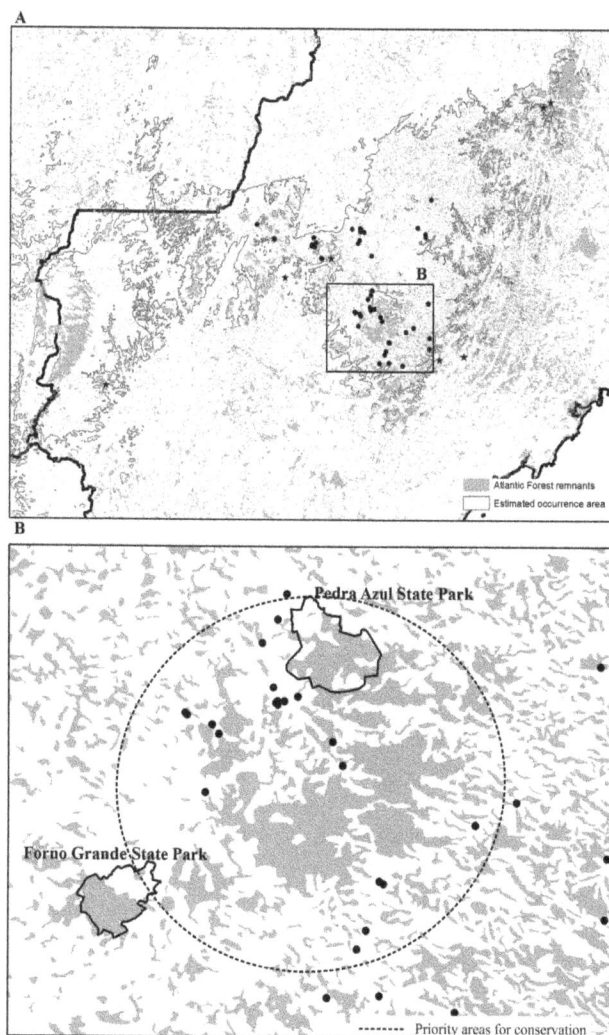

Fig 3. A - Map of Atlantic Forest remnants in the estimated area of *Melipona capixaba* species occurrence. B - Priority areas (hatched line) for the implementation of species conservation and management programs.

*Challenges for the conservation of **M. capixaba***

In the 3,453 km² area calculated for species occurrence, only 1,303 km² (37.37%) corresponds to the remaining fragments of the Atlantic Forest, forming a mosaic of vegetation (Fig 3a). Of these, 93.5% are remaining fragments smaller than 1 km² in size. Habitat fragmentation reduces population sizes and increases isolation of population fragments, resulting in the loss of genetic diversity and consequently increases the risk of species extinction (Frankham et al., 2010). Microsatellite and ISSR markers and mitochondrial haplotypes showed that *M. capixaba* has low genetic variability compared to other insect species (Nogueira et al., 2014).

The remaining forest fragments in the area of *M. capixaba* occurrence are separated by 500 m to more than 5 km in a straight line, which may affect the ability of the species to forage. There is a positive correlation between bee body size and foraging distance (Gathmann & Tscharntke, 2002; Araújo et al., 2004; Greenleaf et al., 2007; Guédot et al., 2009; Zurbuchen et al., 2010; Torné-Noguera et al., 2014). The flight distance of *M. capixaba* has not been determined, but studies of species with similar body sizes suggest that *M. capixaba* can forage in a flight area with a radius of about 2 km. Araújo et al. (2004) estimated that the flight distance of *M. scutellaris* is more than 2 km, and Kuhn-Neto et al. (2009) determined that *M. mandacaia* forages at a distance of 2,100 m.

The main challenge of *M. capixaba* conservation in its natural range is the maintenance of habitats that are capable of providing floral and nesting resources. Considering the size of the remaining fragments and the distance between them, species conservation policies should prioritize not only the maintenance of protected areas but also their connection through ecological corridors that allow adequate foraging in order to support gene flow among isolated colonies. Tewksbury et al. (2002) argued that corridors not only increase the exchange of animals between patches but also facilitate two key animal–plant interactions: pollination and seed dispersal. In addition, these authors suggest that the beneficial effects of the corridors extend beyond the designated area and that increased plant and animal movement through corridors will have positive effects on plant populations and community interactions in fragmented landscapes.

*Prospects for the conservation of **M. capixaba***

One perspective for the conservation of *M. capixaba* is increasing the number of colonies within the area of occurrence through rational rearing of this bee by meliponiculture. Beekeeping of stingless bees in the region of *M. capixaba* occurrence is common among local farmers; however, after inclusion of the species in the List of Threatened Fauna, this practice became prohibited by law. Considering that many colonies of *M. capixaba* lies are maintained by beekeepers, beekeeping can be considered as an alternate, legal way of maintaining *M. capixaba* in its natural habitat for species preservation.

In our view, formation of populations in protected units is the most viable prospect for increasing the current population in natural areas. Reintroduction is a viable option that has been applied for the conservation of several threatened and endangered taxa, and reversal of defaunation is being achieved through the intentional movement of animals to restore populations (Seddon et al., 2014). Winfree (2010) found that the process of bee restoration has been predominantly for agricultural purposes because of the economic importance of maintaining a pollination relationship in agricultural environments. Outside the agricultural context, the primary objective of bee conservation or translocation is improvement of the conservation status of the focal species. From this perspective, it is interesting that species conservation programs take into account the genetic diversity of species and future sustainability of the environment when faced with predicted climate change scenarios.

The forested areas of Pedra Azul and Forno Grande State Park, Biological Reserve Augusto Ruschi, Caparaó National Park, and buffer zones of these conservation units are the major areas for reintroduction of *M. capixaba*, although there are no records of natural colonies in some these sites. In general, choice of the reintroduction area must take into account the known or inferred historical geographical distribution of the focal species or physical evidence of species occurrence. When direct evidence is inadequate to confirm previous occupancy, the existence of a suitable habitat within ecologically appropriate proximity to the proven range may be adequate evidence of previous occupation (IUCN/SSC, 2013). Thomas (2011) argued that there is no need to recreate past ecological communities because of climate change, and, for many species, the only viable option for maintaining the populations of these species in the wild is to translocate them to other locations where the climate is suitable.

Predictions of environmental change in the southeastern portion of the Atlantic Forest indicate a relatively low temperature increase between 0.5°C and 1°C and a rainfall increase of 5–10% by 2040. The temperature will gradually increase by 1.5°C to 2°C and rainfall will increase by 15% to 20% in 2041–2070. By the end of the century (2071–2100), it will become further accentuated with climate patterns between 2.5°C and 3°C increase in temperature and 25–30% increase in rainfall (PBMC, 2014). Considering the current and future climate conditions, the above mentioned places suggested by us for reintroduction should be able to sustain viable populations of *M. capixaba*. They should be able to maintain environmentally and ecologically favorable conditions for the presence of *M. capixaba*, even with the projected temperature increase for the region.

Acknowledgements

We express our thanks to Mr. José Bellon, Mr. Ozenio Zorzal, and Mr. Fábio Caliman for their help in identifying nests and in contacting beekeepers; to INCAPER, IEMA, and FUNARBE for management support; to FAPEMIG, Fundação Grupo Boticário de Proteção à Natureza, and TFCA (Tropical Forest Conservation Act/FUNBIO) for financial support; to CNPq for the doctorate scholarship provided to the first author, to the PPG Genética e Melhoramento of the Universidade Federal de Viçosa, and to CNPq and CAPES for financial grants.

References

Araújo, E.D., Costa, M., Chaud-Netto, J. & Fowler, H.G. (2004). Body size and flight distance in stingless bees (Hymenoptera: Meliponini): inference of flight range and possible ecological implications. Braz. J. Biol., 64(3B): 563-568. doi: 10.1590/S1519-69842004000400003

Brasil (2003). Lista da Fauna Brasileira Ameaçada de Extinção. Instrução Normativa n° 3. 27 May, 2003.

Frankham, R., Ballou J.D. & Briscoe, D.A. (2010). Introduction to conservation genetics, second edition. Cambridge University Press: Cambridge, UK. ISBN 978-0-521-7027-1-3. 1-618 pp.

Freitas, B.M., Imperatriz-Fonseca, V.L., Medina, L.M., Kleinert, A.M.P., Galetto, L., Nates-Parra G. & Quezada-Euan, J.J.G. (2009). Diversity, threats and conservation of native bees in the Neotropics. Apidologie, 40:332–346. doi: 10.1051/apido/2009012.

Gathmann, A. & Tscharntke, T. (2002). Foraging ranges of solitary bees. J. Anim. Ecol., 71:757–764. doi: 10.1046/j.1365-2656.2002.00641.x

Greenleaf, S.S., Williams, N.M., Winfree, R. & Kremen, C. (2007). Bee foraging ranges and their relationship to body size. Oecologia, 153: 589–596. doi: 10.1007/s00442-007-0752-9

Guédot, C., Bosch, J. & Kemp, WP. (2009). Relationship between body size and homing ability in the genus Osmia (Hymenoptera; Megachilidae). Ecol. Entomol., 34:158–161. doi: 10.1111/j.1365-2311.2008.01054.x

IBGE - Instituto Brasileiro de Geografia e Estatística. (1992). Manual técnico da vegetação brasileira. Série Manuais Técnicos em Geociências, n1. Rio de Janeiro, 92p.

IBGE - Instituto Brasileiro de Geografia e Estatística. (2008). Mapa da Área de Aplicação da Lei n° 11.428 de 2006. http://www.ibge.gov.br/home/download/geociencias.shtm. (accessed date: 17 February, 2012).

IUCN – International Union for Conservation of Nature. (2001). IUCN Red List Categories – version 3.1. http://www.iucnredlist.org. (accessed date: 06 October, 2013).

IUCN/SSC - International Union for Conservation of Nature/Species Survival Commission. (2013). Guidelines for Reintroductions and Other Conservation Translocations. Version 1.0. Gland, Switzerland: IUCN Species Survival Commission. 57 pp.

Kuhn-Neto, B., Contrera, F.A.L, Castro, M.S. & Nieh, J.C. (2009). Long distance foraging and recruitment by a stingless bee, Melipona mandacaia. Apidologie, 40:472–480. doi:10.1051/apido/2009007.

Machado, A.B.M., Drummond, G.M. & Paglia, A.P. (Eds.). (2008). Livro Vermelho da Fauna Brasileira Ameaçada de Extinção. Brasília, DF, MMA; Belo Horizonte, MG: Fundação Biodiversitas. v 1. pp 52.

Melo, G.A.R. (1996) Notes on the nesting biology of Melipona capixaba (Hymenoptera, Apidae). J. Kans. Entomol. Soc., 69:207-210.

NASA Land Processes Distributed Active Archive Center (LP DAAC). (2009). ASTER L1B. USGS/Earth Resources Observation and Science (EROS) Center. https://lpdaac.usgs.gov/data_access

Nogueira, J., Ramos J.C., Benevenuto J., Fernandes-Salomão T.M., Resende, H.C., Campos, L.A.O. & Tavares, M.G. (2014). Conservation study of an endangered stingless bee (Melipona capixaba—Hymenoptera: Apidae) with restricted distribution in Brazil. J. Insect Cons., 18:317-326. doi: 10.1007/s10841-014-9639-3

PBMC - Brazilian Panel on Climate Change. (2013) Executive summary: first national assessment report. http://www.pbmc.coppe.ufrj.br/ (accessed date: 10 November, 2014)

Peel, M.C., Finlayson, B.L. & McMahon, T.A. (2007). Updated world map of the Köppen-Geiger climate classification. Hydrol. Earth Syst. Sci., 11:1633–1644. doi:10.5194/hess-11-1633-2007

Potts, S.G., Biesmeijer, J.C., Kremen, C., Neumann, P., Schweiger, O. & Kunin W.E. (2010). Global pollinator declines: trends, impacts and drivers. Trends Ecol. Evol., 25:345-353. doi:10.1016/j.tree.2010.01.007

Serra, B.D.V., de Marco Júnior, P., Nóbrega,C.C. & Campos, L.A.O. (2012) Modeling Potential Geographical Distribution of the Wild Nests of Melipona capixaba Moure & Camargo, 1994 (Hymenoptera, Apidae): Conserving Isolated Populations in Mountain Habitats. Natureza & Conservação, 10:199-206. doi: 10.4322/natcon.2012.027

Seddon, P.J., Griffiths, C.J., Soorae, P.S. & Armstrong D.P. (2014). Reversing defaunation: Restoring species in a changing world. Science, 345:406-412. doi: 10.1126/science.1251818

Silveira, F.A., Melo, G.A.R. & Campos, L.A.O. (2008) Melipona capixaba Moure & Camargo, 1995, In Machado, A.B.M., Drummond, G.M. & Paglia, A.P. (Eds.). Livro Vermelho da Fauna Brasileira Ameaçada de Extinção. Brasília, DF, MMA; Belo Horizonte, MG: Fundação Biodiversitas. v 1. pp 381-382.

Tewksbury, J.J., Levey, D.J., Haddad, N.M., Sargent, S., Orrock, J.L., Weldon, A., Danielson, B.J., Brinkerhoff, J., Damschen, E. & Townsend, P. (2002). Corridors affect plants, animals, and their interactions in fragmented landscapes. PNAS, 99:12923-26. doi: 10.1073/pnas.202242699

Thomas, C.D. (2011).Translocation of species, climate change, and the end of trying to recreate past ecological communities. Trends Ecol. Evol., 26:216-221. doi: 10.1016/j.tree.2011.02.006

Torné-Noguera, A., Rodrigo, A., Arnan, X., Osorio, S., Barril-Graells, H., Rocha-Filho, L.C. & Bosch J. (2014). Determinants of Spatial Distribution in a Bee Community:

Nesting Resources, Flower Resources, and Body Size. PLoS ONE 9(5): e97255. doi:10.1371/journal.pone.0097255

Winfree, R. (2010). The conservation and restoration of wild bees. Ann. N.Y. Acad. Sci., 1195:169–197

Zurbuchen, A., Landert, L., Klaiber, J., Müller, A., Hein, S. & Dorn, S. (2010). Maximum foraging ranges in solitary bees: only few individuals have the capability to cover long foraging distances. Biol. Conserv., 143:669–676. doi:10.1016/j.biocon.2009.12.003

22

Field Evaluations of Broadcast, and Individual Mound Treatments for Red Imported Fire Ant, *Solenopsis invicta* Buren, (Hymenoptera: Formicidae) Control in Virginia, USA

HR ALLEN, DM MILLER

Virginia Tech University, Blacksburg, VA, United States

Keywords
Red Imported Fire Ant, *Solenopsis invicta* Buren, broadcast, control, bait.

Corresponding author
Dini M. Miller
Department of Entomology
Virginia Tech University
216A Price Hall,
Blacksburg, VA 24061, USA
E-Mail: dinim@vt.edu

Abstract
Field evaluations were conducted to determine efficacy, residual activity, and knockdown potential for fire ant control products. Broadcast granular products (Advion, 0.045% indoxacarb; and Top Choice Insecticide, 0.0143% fipronil) were individually evaluated, and compared with a combination of two products applied together, and with individual mound applications of Maxforce Fire Ant Killer Bait (1.0% hydramethylnon). After application, the greatest percent reductions (90 days) were observed in the Advion/Top Choice combination plots (100.0%), followed by Top Choice alone (96.4%). Advion and MaxForce produced significantly lower foraging reductions at 90 days (61.2% and 27.5% respectively). At the conclusion of the test (day 360), significantly fewer ants were collected in the Advion (777.7), Top Choice (972.8), and combination plots (596.2) than in the control plots (1257.8) (*df* 13, *F* = 8.3, *P* < 0.05). The mean number of ants collected from MaxForce treatment plots was not significantly different from controls (*P* > 0.05). Overall, the efficacy and residual studies suggested that the Advion/Top Choice combination produced both the most rapid reduction in ant foraging and the longest lasting control (90%) at 300 days. When evaluating time to knockdown of foraging populations, the Advion/Top Choice combination also provided the most complete and rapid results by day 7, reducing foraging by 100%. While other products also performed well (75.6 - 95.9% reductions), both the MaxForce and Advion plots had significant increases in foraging at 30-90 days. Overall, foraging knockdown was the most complete in the Avion/Top Choice combination plots at 90 days.

Introduction

Prior to 2009, all reported red imported fire ant (RIFA), *Solenopsis invicta* Buren, infestations in Virginia were documented and managed by the Virginia Department of Agriculture and Consumer Services (VDACS). In spite of VDACS' best efforts however, RIFA infestations within the state continued to increase and spread. Therefore, in 2009 the United States Department of Agriculture (USDA) in conjunction with VDACS implemented the Imported Fire Ant Quarantine in the following areas of Virginia: the counties of James City and York, and the cities of Chesapeake, Hampton, Newport News, Norfolk, Poquoson, Portsmouth, Suffolk, Virginia Beach and Williamsburg. Consequently, VDACS is no longer responsible for treating RIFA mounds in the quarantined areas. Fire ant control in the quarantined counties/cities is now the responsibility of homeowners, nurseryman, and pest management

professionals. However, VDACS is still responsible for managing RIFA infestations in cities outside of the designated quarantine areas. As of 2014, the quarantine has not been expanded and VDACS is still responsible for controlling infestations in the large majority of the state.

The standard control method used by VDACS for treating RIFA mounds in Virginia in 2009 was to apply MaxForce® fire ant bait (1.0% hydramethylnon) (Bayer CropScience, Kansas City, MO) around each active mound. Bait applications were followed six weeks later by an acephate mound drench. Although effective, these individual mound treatments (IMT) required that all mounds be located prior to application and then treating them one at a time. While IMTs are labor intensive and time consuming, the direct chemical application to the mound does greatly enhance the amount of insecticide contact with colony members (Barr & Best, 1999).

Baits and liquid insecticides are the typical formulations used for IMTs, but aerosols, granules, and dusts are also frequently used. IMTs are the most useful when 20 or fewer mounds are present in an acre of land (Barr & Best, 2002). IMTs are also beneficial because they are only applied specifically to RIFA mounds, thus preventing native ant mortality. However, because IMTs are only applied to visible mounds, fire ant recolonization can easily occur in treated areas where small mounds are overlooked and not treated. Multiple applications are often necessary to control all the mounds in a particular area.

In contrast to IMTs, broadcast insecticide treatments for RIFA do not require individual mounds to be located. Therefore, broadcast products greatly reduce the time and labor needed to treat a large area. Broadcast fire ant control products are currently formulated as either granules or baits and are applied using either a hand or tractor mounted spreader (Drees et al., 2006). Broadcast products are typically applied in locations where mound densities exceed 30-40 per hectare.

Bait formulations are frequently applied as a broadcast RIFA control method. RIFA baits are usually formulated by combining a slower acting toxicant with soybean oil or some other food matrix that is attractive to foraging fire ants (Williams et al., 2001). Once the ants transport the bait back inside the colony, the ants transfer the active ingredient throughout the colony by trophallaxis. Because the active ingredient must be spread throughout the colony via the worker ants feeding the queens and brood, it may take several weeks to months before significant colony reductions are observable in the field (Drees et al., 2006). Therefore, colony suppression may take significantly longer using broadcast baits when compared with IMTs that provide reductions in a single day. However, a study conducted by Barr and Best (1999) suggested that the benefits of large scale ant suppression that could be achieved with broadcast bait products far out-weighed the delay in short term (but eventual) results (Barr & Best, 1999).

Some broadcast fire ant products have been used to treat fire ants in Virginia, but they have been used infrequently due to VDACS' preference for individual mound treatments. However, now that VDACS is no longer responsible for treating infestations in quarantined counties and cities, residents in these locales have the burden of managing fire ants on their own. With the quarantine implementation, the need for broadcast RIFA product evaluations and other control recommendations are vital, if not for stopping RIFA, at least for slowing the spread of RIFA in Virginia.

Two of the leading broadcast fire ant control products are Advion® fire ant bait (Indoxacarb 0.045%; Syngenta, Research Triangle Park, NC) and Top Choice granular (Fipronil 0.143%; Bayer Environmental Sciences, Cary, North Carolina). Advion is a fast acting bait (Furman & Gold, 2006) that contains the active ingredient, indoxacarb, which belongs to the oxadiazine chemical class. Oxadiazines block sodium channels in the insect nerve axon. Immediately after bait ingestion, ant feeding begins to decrease and target individuals

usually succumb to death within 48 hours (Barr, 2002a). Top Choice contains the active ingredient fipronil which belongs to the phenylpyrazole chemical class. Fipronil is a nerve poison that blocks the passage of chloride ions through GABA receptor and glutamate-gated chloride channels causing nerve hyperexcitation in target insects (Kolaczinski & Curtis 2001).

Previous studies have shown that Advion significantly reduced fire ant foraging 24 hours after treatment (Barr, 2004) and eliminated > 95% of colonies after one week (Hu & Song, 2007) after application. Top Choice has a longer residual activity than Advion but is much slower acting. Barr and Best (2004) reported that Top Choice® reduced the mean number of active fire ant mounds by 80% five weeks after treatment and greater than 90% control was observed 52 weeks later.

The purpose of this study was to evaluate the performance of specific broadcast Red Imported Fire ant treatment in Virginia, and compare their efficacy with that of an IMT. Our goal was to determine which application method might have the longest residual activity, and therefore the greatest potential to prevent fire ant spread. In this study, field applications of the RIFA control products: Advion® fire ant bait (Indoxacarb 0.045%; Syngenta, Research Triangle Park, NC); Top Choice granular (Fipronil 0.143%; Bayer Environmental Sciences, Cary, North Carolina); a combination application of Advion and Top Choice; and an IMT treatment using MaxForce® fire ant bait (1.0% hydramethylnon), were monitored for efficacy for one year. The following year, the same field applications were reapplied to determine the rapidity of initial knock-down.

Materials and Methods

Study Area

Although our initial research plots were established on an infested vacant lot in Hampton Roads, Virginia (2008; prior to the implementation of the Federal Fire Ant Quarantine (FFAQ), the research site came to the attention of a neighboring school facilities manager who demanded that VDACS treat the location. To avoid further conflict within Virginia we moved our research site 161 km due south to North Carolina where the entire state was already under the FFAQ. Our new research plots were established within Fun Junktion Park, a converted landfill located in Elizabeth City, NC. The study was conducted from 5 August 2008 to 26 July 2009. Elizabeth City is located on the northeast coast of North Carolina in Pasquotank County (36°17'44"N; 76°13'30"W). Average monthly temperatures range from a low of 0° C during the winter months to a high of 31.8° C during the summer months. The city receives about 122 centimeters of rainfall annually.

Research Plots

Fourteen 30 x 30 m (900 m²) research plots (Fig 1a-b) were established within three different locations within the park.

Eight plots were located on a driving range that was covered with grass and mowed weekly. Four plots were established in a grass covered field located near an artificial lake. Two plots were located in a weed covered field that was not mowed. An untreated buffer zone (7.6 m) separated each plot to reduce potential ant foraging between research plots. Plots were randomly assigned to different treatments so that each of the four insecticide treatment had three replicates. The two remaining plots served as untreated controls.

Fig. 1a-b. Placement of 1-year RIFA treatment plots located at Fun Junktion Park, Elizabeth City, NC (Google Earth 2010)- Advion (**Adv**), Top Choice (**TC**), MaxForce (**MF**), Advion /Top Choice combination (**Com**) and untreated control (**Con**).

Treatment Products

The broadcast products evaluated in the study were Advion® Fire Ant Bait (0.045% indoxacarb; Syngenta, Research Triangle Park, NC), Top Choice® Insecticide (0.0143% fipronil;

Bayer Environmental Science, Research Triangle Park, NC) and MaxForce® Fire Ant Killer Granular Bait (1.0% hydramethylnon; Bayer Crop Science, Kansas City, MO). Advion and Top Choice were also used in a combination treatment where they were applied together in the same plot. All broadcast products were applied at the label rate (Advion: 1.68 kg/hectare (1.5 lbs./acre), Top Choice: 209 kg/hectare (85 lbs./acre), MaxForce: 14-28g/mound (0.5-1.0 oz./mound), Combination: Advion/Top Choice) using Scott's Handy Green II hand spreaders (Scotts International B.V., Scotts Professional, Geldermalsen, The Netherlands). The MaxForce bait is labelled for application as a broadcast or as an individual mound treatment. However, for the purposes of this study, MaxForce was used as an IMT and was applied directly to individual mounds from the product container. Treatment applications were made on 12 July 2008 between 5:00 p.m. and 7:00 p.m. Each broadcast treatment was applied to three plots. MaxForce bait was applied to seven active fire ant mounds located in three experimental plots.

Sampling Regimen to Quantify Foraging Activity

Prior to treatment applications, slices of uncooked hot dog wieners were used as baits to quantify foraging activity in each of the plots. Pre-treatment bait counts were taken on 11 July 2008 between 5 and 7:00 p.m. Eight beef hot dog (Gwaltney, Smithfield VA) slices (0.5 cm thick) were placed in each plot. The hot dog slices were arranged in two rows of four and each row was spaced 7 m apart. Hot dog slices were left in place for one hour, after which photographic images were taken of each slice with a Sony Cybershot digital camera (Sony Electronics Inc., San Diego, CA). All images were downloaded onto a computer so that the species and number of ants in each hot dog photograph could be counted and recorded. Post-treatment ant sampling with hot dog slices was conducted between 5:00 p.m. and 7:00 p.m. at 3, 7, 14, and 30 days after treatment and every month thereafter for one year. During the initial study, post-treatment data collected on 7, 14, 30, and 60 days were lost after the computer laptop holding that data was stolen. In addition, sampling was not conducted during the winter months (121 and 239 days after treatment) because of low temperatures that eliminated ant foraging.

Product Reapplication to Determine Time to Knockdown

After the one year completion of the study described above, all plots were sampled again (as previously described) to determine ant foraging activity. After determining that the ant pressure had rebounded and was still very high, all products were reapplied. Treatment applications were made on 21 July 2009 between 5:00 p.m. and 7:00 p.m. MaxForce bait was applied to 5 mounds. Control plots (2) were left untreated. Post-treatment sampling was conducted between 5:00 p.m. and 7:00 p.m. Sampling was conducted on days 3, 7, 14, 30, 60, and 90 after treatment to determine the time to knockdown for all treatment products and combinations.

Statistical Analysis

The mean number of foraging fire ants collected per treatment on each sampling date was calculated by adding the total number of ants foraging on all 8 hot dogs in each treatment plot, and dividing that total by number of plots per treatment. To determine if the treatment applications had any effect on the mean number of foraging ants, data were transformed ($\sqrt{(x + \frac{3}{8})}$) (Zar 1984) and subjected to repeated-measures multivariate analysis of variance (MANOVA), with the post treatment date as the repeated measure. Repeated-measures MANOVA was also used to determine if the residual activities of each treatment were significantly ($P \leq 0.05$) different from one another.

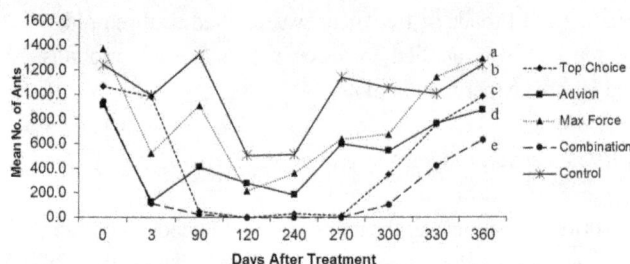

Fig 2. Mean number of foraging RIFA in experimental plots before and after product applications (one-year study). Trend lines followed by the same letter are not significantly different ($\alpha = 0.05$).

Differences in the mean number of RIFA collected in each treatment on each sampling date was determined using two by one way repeated-measures ANCOVA, with the mean number of foraging ants collected on DAT-0 as a covariate. Significant differences among treatment means on each post treatment sampling date were separated by Tukey's HSD test ($P \leq 0.05$). LS Means produced in the ANCOVA were used to calculate percent change in the mean number of RIFA foraging ants after treatment relative to the initial number of foragers on DAT-0 (Vickers 2001). Separate repeated-measures MANOVA and ANCOVA analyses were conducted on both the initial application (year-long test), and re-treatment (knockdown) data.

Results

Product Efficacy Tests

Repeated-measures MANOVA was used to determine whether product applications had any effect on the mean number of foraging ants collected in plots. Results of the repeated measures MANOVA indicated that there was a significant overall treatment effect on the mean number of foraging fire ants ($F = 72.0$; df = 9, $P < 0.0001$) (Figure 2). Contrast comparison tests between the mean number of foraging fire ants collected from treatment plots and control plots indicated that the mean number of foragers collected from

each treatment plot was significantly lower ($P < 0.05$) than that collected in the controls. In addition, contrast comparison tests revealed that the mean number of ants collected in each of the treatment plots were all significantly different from each another.

The ANCOVA was conducted to compare the mean number of foraging ants in each treatment, on each sampling day. LS means calculated by the ANCOVA were used to calculate the percent change in the mean number of active foragers on each post treatment sampling date (Table 1). Three days after treatment the mean number of foraging fire ants in the Advion, MaxForce, and Advion/Top Choice combination plots was significantly lower than that in the untreated controls ($P < 0.05$). The greatest percent reduction in foraging three days after treatment was observed in Advion/Top Choice combination plots (82.7%) followed by Advion alone (79.5%), MaxForce (68.4%), and Top Choice alone (6.6%). Although sampling data was collected between for DAT-7 through DAT-60, these data were lost. When post-treatment sampling resumed on DAT-90 there were significantly fewer ants collected from the Advion (355.5), Top Choice (38.2), and Advion/Top Choice (0.0) plots than in the MaxForce (995.6) and control plots (1369.1). The greatest percent reduction in foraging at DAT-90 was observed in the Advion/Top Choice combination plots (100.0) followed by Top Choice (96.4), Advion (61.2), and MaxForce (27.5). For the remainder of the test (DAT-90 – DAT-360), fewer fire ants were collected in combination and Top Choice treatment plots than in all of the other experimental treatment plots. At the conclusion of the test on DAT-360, there were significantly fewer ants collected in Advion (777.7), Top Choice (972.8), and combination plots (596.2) than in the control plots (1257.8) (df 13, $F = 8.3$, $P < 0.05$). However, the mean number of ants collected from MaxForce treatment plots was not significantly different from controls ($P > 0.05$).

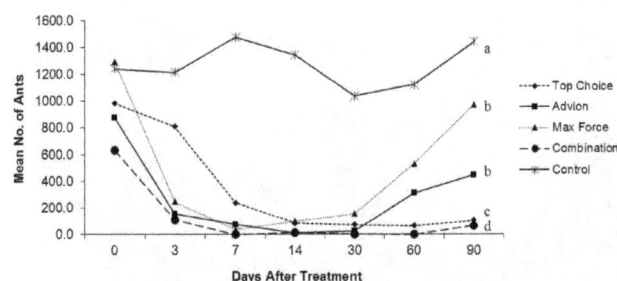

Fig 3. Mean number of foraging RIFA in experimental plots before and after product re-applications (90-day study). Trend lines followed by the same letter are not significantly different ($\alpha = 0.05$).

Overall, the results suggest that the Advion/Top Choice combination, and the Advion treated plots had the greatest reductions in ant foraging by day 3, causing foraging reductions of 82.7 and 79.5 percent respectively. However, Advion, Top Choice, and the Advion/Top Choice combination treatment also provided the longest lasting control with significant reductions in foraging at 360 days.

Table 1 Least square means (± SE) and mean percent change in number of foraging RIFA prior to and days after application.

Treatment	DAT-0	DAT-3	DAT-90	DAT-120	DAT-240	DAT-270	DAT-300	DAT-330	DAT-360
					Day After Treatment (DAT)				
Top Choice									
LS Mean (± SE)	1064.3	994.0^{a} (± 51.1)	38.2^{b} (± 108.7)	-3.3^{b} (± 56.7)	28.5^{c} (± 52.6)	8.7^{c} (± 63.8)	345.0^{bc} (± 101.2)	752.4^{b} (± 70.3)	972.8^{bc} (± 83.9)
Percent Change	-	(6.6)	(96.4)	(100.0)	(97.3)	(99.2)	(67.6)	(29.3)	(8.6)
Advion									
LS Mean (± SE)	916.7	188.1^{bc} (± 52.2)	355.5^{b} (± 111.1)	260.1^{ab} (± 58.0)	186.3^{bc} (± 53.7)	577.3^{b} (± 65.2)	510.2^{ab} (± 103.5)	734.3^{bc} (± 71.8)	777.7^{c} (± 85.8)
Percent Change	-	(79.5)	(61.2)	(71.6)	(79.7)	(37.0)	(44.3)	(19.9)	(15.2)
Max Force									
LS Mean (± SE)	1372.3	433.4^{b} (± 53.7)	995.6^{b} (± 114.3)	239.7^{ab} (± 59.6)	352.7^{ab} (± 55.3)	668.7^{b} (± 67.0)	723.6^{ab} (± 106.4)	1189.4^{a} (± 73.9)	1318.3^{ab} (± 88.2)
Percent Change	-	(68.4)	(27.5)	(82.5)	(74.3)	(51.3)	(47.3)	(13.3)	(3.9)
Advion/Top Choice Combinationo									
LS Mean (± SE)	944.0	63.3^{c} (± 51.9)	-24.4^{b} (± 110.4)	-15.2^{b} (± 57.6)	3.9^{c} (± 53.4)	-18.2^{c} (± 64.8)	78.0^{c} (± 102.8)	396.5^{c} (± 71.4)	596.2^{c} (± 85.2)
Percent Change	-	(82.7)	(100.0)	(100.0)	(99.6)	(100)	(91.7)	(58.0)	(36.8)
Untreated Control									
LS Mean (± SE)	1240.5	943.7^{a} (± 63.1)	1369.1^{a} (± 134.3)	519.6^{a} (± 70.0)	508.9^{a} (± 65.0)	1155.3^{a} (± 78.8)	1076.7^{a} (± 125.0)	1029.6^{ab} (± 86.8)	1257.8^{a} (± 103.7)
Percent Change	-	(23.9)	(10.4)	(58.1)	(59.0)	(6.9)	(13.2)	(17.0)	(1.4)
F	-	58.6	19.1	9.5	11.7	38.1	8.5	12.6	8.3
df	-	13	13	13	13	13	13	13	13
P	-	<0.0001	0.0003	0.003	0.001	<0.0001	0.0046	0.0013	0.005

Means within a column followed by the same letter are not significantly different (Tukeys HSD mean separation test; $\alpha = 0.05$).

Product Reapplication to Determine Time to Knockdown

Repeated-measures MANOVA results indicated that the insecticide products had a significant overall treatment effect on the mean number of foraging fire ants ($F = 76.1$; df=9, $P < 0.0001$) (Table 2). Contrast comparison tests between the mean number of foraging fire ants collected from insecticide treated plots and control plots indicated that the mean number of foragers collected from insecticide treated plots were significantly lower ($P < 0.05$) than that of the controls. Additionally, contrast comparisons also indicated that the greatest reductions in the number of active foragers occurred in the Advion (82.9%), MaxForce (79.6%), and Advion/Top Choice (85.7%) combination plots. These reductions were far greater than reductions observed in Top Choice (17.5%) and control plots (0.9%). The ANCOVA results indicated that throughout the test the mean number of ants collected in all chemical treatments was significantly lower ($P < 0.05$) than the mean number of ants collected in control plots on each sampling date. Additionally, from DAT-3 to DAT-30 the mean numbers of ants collected in chemical treatment plots were significantly lower ($P < 0.05$) than the mean number collected from the control plots. However, the mean number of ants collected from DAT-3 to DAT-30 in each of the treatments plots were not different from each other. However on DAT-60, the mean number of foraging ants increased in all plots except those treated with Top Choice.

While Advion and MaxForce still had significant reductions in foraging at day 60, the reductions were significantly less than those of Top Choice and the Advion/Top Choice combination at 60 days. At the conclusion of test on DAT-90, percent reductions in foraging were greatest in the Advion/Top Choice combination and Top Choice treated plots. At 90 days, the MaxForce bait had the lowest reduction in foraging, but this reduction was not significantly different from that in Advion or Advion/Top Choice combination plots. Overall, the knockdown of foragers was the most rapid and complete in the Advion/Top Choice treated plots on day 7 (100%). However all insecticide treatments produced between 90-100% knockdown in 7-14 days. The Advion/Top Choice combination and Top Choice treatments had the longest lasting effect, suppressing foraging by 89-93% for 90 days.

Discussion

Results obtained from the year-long field efficacy trial indicated that the Advion/Top. Choice combination treatment provided the both the most rapid control of fire ants and the greatest residual activity. While both products were very effective at controlling fire ants in the field, the Advion had most rapid activity although it did not suppress the populations as long as Top Choice. Top Choice did not produce the most rapid knockdown but did have the longest residual activity both alone and in the combination treatment (Table 1).

Table 2. Least square mean ($\pm$ SE) and mean percent change of foraging RIFA prior to and after treatment reapplication.

Treatment	DAT-0	DAT-3	DAT-7	DAT-14	DAT-30	DAT-60	DAT-90
				DAT (Days After Treatment)			
Top Choice							
LS Mean ($\pm$ SE)	972.8	809.0[b] ($\pm$ 64.4)	239.0[b] ($\pm$ 53.4)	83.0[b] ($\pm$ 46.0)	75.0[b] ($\pm$ 49.3)	68.3[c] ($\pm$ 46.7)	107.6[c] ($\pm$ 106.5)
Percent Change		(17.5)	(75.6)	(91.5)	(92.4)	(93.0)	(89.0)
Advion							
LS Mean ($\pm$ SE)	777.7	149.1[c] ($\pm$ 70.0)	70.4[b] ($\pm$ 58.0)	-21.5[b] ($\pm$ 50.0)	39.2[b] ($\pm$ 53.5)	349.7[b] ($\pm$ 50.7)	441.4[bc] ($\pm$ 115.6)
Percent Change		(82.9)	(91.9)	(100)	(95.5)	(59.9)	(49.4)
Max Force							
LS Mean ($\pm$ SE)	1318.3	264.2[c] ($\pm$ 97.7)	53.3[b] ($\pm$ 81.0)	187.0[b] ($\pm$ 69.7)	120.3[b] ($\pm$ 74.7)	429.7[b] ($\pm$ 70.7)	989.6[ab] ($\pm$ 161.4)
Percent Change		(79.6)	(95.9)	(85.5)	(90.7)	(66.8)	(23.5)
Advion/Top Choice Combination							
LS Mean ($\pm$ SE)	596.2	90.2[c] ($\pm$ 106.7)	-18.5[b] ($\pm$ 88.5)	-85.1[b] ($\pm$ 76.1)	44.0[b] ($\pm$ 81.7)	116.5[bc] ($\pm$ 77.3)	44.3[bc] ($\pm$ 176.4)
Percent Change		(85.7)	(100)	(100)	(93.0)	(81.6)	(93.0)
Untreated Control							
LS Mean ($\pm$ SE)	1257.8	1227.9[a] ($\pm$ 99.5)	1489.1[a] ($\pm$ 82.5)	1414.8[a] ($\pm$ 71.0)	1005.8[a] ($\pm$ 76.1)	1039.0[a] ($\pm$ 72.0)	1460.7[a] ($\pm$ 164.4)
Percent Change		(0.9)	20.1	14.1	(18.9)	(16.2)	17.8
F	-	36.7	79.5	91.3	45.5	59.0	20.3
Df	-	13	13	13	13	13	13
P	-	<0.0001	<0.0001	<0.0001	<0.001	<0.0001	0.0002

Means within a column followed by the same letter are not significantly different (Tukeys HSD mean separation test, $\alpha = 0.05$).

Therefore it was no surprise that the two products combined produced results that were superior to either of the broadcast products used alone, and to the IMT using Maxforce bait.

In the 90 day knockdown evaluations, the Advion/Top Choice combination provided the most complete and rapid results by day 7, effectively reducing foraging by 100%. However, the other products also performed very well (75.6 - 95.9% reductions in foraging) and there were no significant differences between the treatments at 7 days (Table 2). Again at Day 14 all of the treatments were equally effective with both Advion and the Advion/Top Choice reducing RIFA foraging by 100%. Interestingly, the trend we observed over 30 to 90 days after application was that there was significant increase in RIFA foraging in both the MaxForce and Advion treatments, yet efficacy for the Top Choice alone and the Advion/Top Choice combination remained relatively high. Foraging suppression was significantly greater in the Avion/Top Choice combination at 90 days than in all other treatments. Overall, the advantage of the combined broadcast treatment application was that one formulation (Advion) was very fast acting, while the slower acting Top Choice provided the persistent residual activity that was limited in the Advion formulation. The data provided in these studies indicated that broadcast fire ant control products can provide longer residual control, and therefore that may slow the spread of RIFA colonies into untreated areas more effectively than IMTs (Williams et al. 2001, Banks et al. 1988). In both the residual study, and 90 day knock-down tests, Advion, Top Choice, and the Advion/Top Choice combination provided faster, longer lasting results than the MaxForce mound treatments.

Studies evaluating Advion conducted by Barr (2002a, 2002b) reported similar rapid knock-down results. In 2002, Barr conducted two tests, one in the summer and one in the fall to evaluate the efficacy of indoxacarb to control RIFA colonies. Both tests were conducted at an airport located in Yoakum, Texas. In both tests, Barr (2002a-b) compared the efficacies of different fire ant products: Amdro® fire ant bait (0.73% hydramethylnon; Ambrands, Atlanta, GA), Extinguish® fire ant bait (0.5% s-methoprene; Wellmark International, Schaumburg, IL), Talstar® 2G (0.2% bifenthrin; FMAC Professional Solutions, Philadelphia, PA) and three different concentrations of indoxacarb (0.025%, 0.05%, and 0.1%). Results from tests conducted in the summer (Barr 2002a) indicated that the three indoxacarb formulations provided faster control of RIFA colonies than the other RIFA products tested. One week after treatment, the mean number of active mounds in all of the indoxacarb treated plots ranged from (0.25 – 1.25) while the mean number of active mounds in plots treated with Amdro was 4.0; Extinguish 16.25, and Talstar was 3.25. However, 6 weeks after treatment the number of active colonies in all the indoxacarb treated plots began to increase. The number of active mounds found in the other treatment plots also began to increase, however fewer active colonies were documented in Extinguish treatment plots. Mound density in plots treated with indoxacarb continued to

increase for the remainder of the test. Barr (2002b) replicated the airport test again in the fall, to determine if colony foraging or reproductive status influenced his summer results. Barr (2002b) found that in the fall the product efficacy results were similar to those of the previous summer. Overall, the three indoxacarb formulations provided more rapid foraging reductions than the other fire ant control products tested.

Similarly, studies evaluating Top Choice also found that the product provided longer residual fire ant control than other fire and products tested. Barr and Best (2004) conducted a test to evaluate the efficacy of two granular formulations of fipronil (0.0143%; 0.00015% fipronil) Amdro Ant Bait (0.73% hydramethylnon, Ambrands, Atlanta, GA), and Talstar 2G (0.2% bifenthrin, FMC, Philadelphia, PA). The study results demonstrated that fipronil provided greater long-term control than the other fire ant control products. Five weeks after treatment, granular applications of 0.0143% fipronil still provided an 83% - 98% reduction in the number of active mounds. By week 52 the number of active mounds began to rebound in all treatment plots except those treated with the fipronil (granular 0.0143%; Barr & Best, 2004).

Because the combination treatment used in our tests consisted of Advion (one of the fastest acting baits on the market) and Top Choice, (which provides long residual control) we expected the combination treatment to outperform the other products and provide long lasting control. Our results indicated that our expectations were correct. However, it should be noted that these products are most efficiently used in large scale situations where many mounds are present, and fire ant spread is a concern. Broadcast products are relatively expensive, costing $7-10 per hectare. Product costs for individual mound treatments are ~25 cents per mound, not including the labor (Drees et al., 2006). Therefore, for small infestations in a residential yard or in some other more contained area, individual mound treatments would still be the most desirable and effective method of fire ant management.

Conclusions

The overall results of this study determined that broadcast fire ant control products tested were faster acting and had a longer residual than The IMT. Presently, VDACS manages fire ant infestations outside of Virginia's fire ant quarantined areas while homeowners and pest control operators are responsible for treating infestations within quarantine borders. VDACS currently uses IMTs to treat all fire ant mounds outside of the quarantine area. Given the evidence provided in this study, it is reasonable to assume that in confined locations where all active fire ant mounds are visible, the IMTs will provide adequate control. However, in large areas that contain many mounds, like a vacant lot, a broadcast application can provide better and longer lasting control. As proven by the implementation of the RIFA Quarantine, the IMT method used alone was not enough to slow the spread of the Red Imported

Fire ant in Virginia. While broadcast RIFA products are more expensive than IMTs they require very little labor to apply. Thus government agencies like, VDACS could possibly save money on the application costs outside the quarantine area by adding broadcast application products to their RIFA arsenal.

Acknowledgements

The authors would like to thank Dr. Clay Scherer of DuPont Professional Products (now Syngenta Professional Products) for providing the funding for this study. We would also like to thank the Virginia Department of Agriculture Office of Plant and Pest Services for their initial oversight of this study prior to our having to move our field site to North Carolina.

References

Banks, W.A., Williams, D. F. & C. S. Lofgren. (1988). Effectiveness of fenoxycarb for control of red imported fire ants (Hymenoptera: Formicidae) Journal of Economic Entomology, 81:83-87.

Barr, C. L. (2002a). The active ingredient indoxacarb as a broadcast bait for the control of red imported fire ants. Texas Coop. Ext., Texas A&M Univ. System. https://insects.tamu.edu/fireant/research/arr/year/99-03/res_dem_9903/pdf/4_ai_indoxacarb.pdf. (accessed date: 12 March 2009)

Barr, C. L. (2002b). Indoxacarb bait effects on mound activity and foraging of red imported fire ants. Texas Coop. Ext., Texas A&M Univ. System. https://insects.tamu.edu/fireant/research/arr/year/99-03/res_dem_9903/pdf/1_indoxacarb_bait_effects.pdf. (accessed date: 12 March 2009)

Barr, C. L. (2004). How fast is fast?: Indoxacarb broadcast bait. P. 46-50. In Proceedings of the Annual Red Imported Fire Ant Conference, Baton Rouge, LA. 201 p.

Barr, C. L., & Best, R. L. (1999). Comparison of Amdro®, Spectracide® fire ant bait and Diazinon using broadcast and individual mound treatment applications. Results and Demonstration Handbook. 1997-1999. Texas Agricultural Extension Service. Bryan, TX. 70 p.

Barr, C.L. & Best R. L. Best. (2002). Product evaluations, field research and new products resulting from applied research. Southwestern Entomologist, Supplement, 25:47-52.

Barr, C. L. & Best, R. L. (2004) Comparison of different formulations of broadcast fipronil for the control of red imported fire ants. Result demonstration handbook 1999–2003. 2004. Texas Agric. Ext. Serv. Bryan, TX. http://fireant.tamu.edu. (accessed date: 18 Feb. 2011).

Drees, B. M., Vinson, S. B., Gold, R. E. Merchant, M. E., Brown, E., Keck, M., Nester, P. Kostrom, D., Flanders, K., Graham, F., Loftin, K. Hopkins, J., Vail, K., Wright, R. Smith, W., Thompson, D. C., Kabashima, J., Layton, B., Koehler, P. G., Oi, D. H. & Callcott, A. M. (2006). Managing imported fire ants in urban areas. Texas Coop. Ext., Texas A&M Univ. System. MP426. 23 p.

Furman, B.D. & R.E. Gold (2006). Determination of most effective chemical form and concentration of indoxacarb, as well as the most appropriate grit size, for use in Advion. Sociobiology. 48: 309-334.

Hu, X, & Song, D. (2007). Field evaluation of label-rate broadcast treatment with baits for controlling the red imported fire ant, *Solenopsis invicta* (Hymenoptera: Formicidae). Sociobiology. 50: 1107-116.

Kolaczninski, J. & Curtis, C. (2001). Laboratory evaluation of fipronil, a phenylpyrazole insecticide, against adult *Anopholes* (Diptera: Culicidae) and investigation of its possible cross-resistance with dieldrin in *Anopheles stephensi*. Pest Managemet Science, 57: 41-45.

Vickers, A. J. (2001). The use of percentage change from baseline as an outcome in a controlled trial is statistically inefficient: a simulation study. BMC Medical Research Methodology, 6: 1-4.

Williams, D. F, Collins, H. L. & Oi, D. H. (2001). The Red Imported Fire Ant (Hymenoptera: Formicidae): A historical perspective of treatment programs and the development of chemical baits for control. American Entomologist, 47:146-149.

Zar, J.H. (1984). Biostatistical Analysis. Second edition. Prentice-Hall Inc., Engelwood Cliffs, New Jersey. 718 pp.

23

Variability of Food Stores of *Tetragonisca fiebrigi* (Schwarz) (Hymenoptera: Apidae: Meliponini) from the Argentine Chaco Based on Pollen Analysis

FG Vossler, GA Fagúndez, DC Blettler

Laboratorio de Actuopalinología, CICyTTP-CONICET/FCyT-UADER, Diamante, Entre Ríos, Argentina

Keywords
honey, "rubiecita", stingless bee, "yateí".

Corresponding author
Favio Gerardo Vossler
Laboratorio de Actuopalinología
CICyTTP-CONICET/FCyT-UADER
Dr. Materi y España, E3105BWA
Diamante, Entre Ríos, Argentina
E-Mail: favossler@yahoo.com.ar

Abstract

Honey and pollen mass samples of *Tetragonisca fiebrigi* (Schwarz) from the same and different nests, seasons, and forest types from the Argentine Chaco region were palynologically analyzed and multivariate techniques were applied. The samples from each forest type (Palosantal and Quebrachal) were grouped separately by Cluster Analysis but the phenological records detected that grouping was determined by the season when samples were taken. Honeys and pollen masses were grouped together or fairly closed for all nests due to similar abundance of the different pollen types. Furthermore, honeys were not clustered together with other honeys but with pollen masses. It can be assumed that both nectar and pollen were gathered from the same plant species, supporting the hypothesis that the Dry Chaco melittophilous vegetation is dominated by plants providers of both pollen and nectar, but not exclusively or predominately of one of them. Results of Principal Component Analysis revealed that the foraging behavior of *T. fiebrigi* was governed by random factors such as local differences of flower availability but not by preferences for some plant families. This idea can also be extended to other species of this genus as they concentrated their foraging on different families according to the local vegetation availability in the studied sites.

Introduction

Honey and pollen stores from *Tetragonisca* species have been reported as a culturally important and appreciated food as well as home medicine since ancient times (Nogueira-Neto, 1997; Arenas, 2003; Cortopassi-Laurino et al., 2006; Zamudio & Hilgert, 2012; Roig-Alsina et al., 2013). Several studies on diet have been done in *Tetragonisca angustula* (Latreille) such as Imperatriz-Fonseca et al. (1984), Carvalho et al. (1999), Novais et al. (2013, 2014) in Brazil, Sosa-Nájera et al. (1994) and Martínez-Hernández et al. (1994) in Mexico, Obregón et al. (2013) in Colombia, Flores & Sánchez (2010) in northwestern Argentina, while only one for the aerial-nesting and aggressive *Tetragonisca weyrauchi* (Schwarz) (Cortopassi-Laurino & Nogueira-Neto 2003). No studies have been performed for the ground-nesting *Tetragonisca buchwaldi* (Friese) neither for *Tetragonisca fiebrigi* (Schwarz).

It is important to study the botanical origin of food stores of *T. fiebrigi* in the Chaco region, a large plain of xerophylous forest of about 1,000,000 km² in southern South America (Prado, 1993). In this region, the "rubiecita/rubiecito" or "rubita/rubito" (*T. fiebrigi*) provides the most reputable honey by local people (Arenas 2003) and it is the most intensively reared stingless bee, together with *Scaptotrigona jujuyensis* (Schrottky) (Roig-Alsina et al., 2013). Furthermore, it is the most frequent Meliponini species in the forest (Vossler, 2012), being their colonies commonly harvested in the field (Roig-Alsina et al., 2013).

Due to the importance of stingless bees in good practices such as meliponiculture and crop pollination, it is desirable to assess the existence of seasonal and environmental foraging tendencies as well as differences in honey and pollen mass composition from the botanical origin of samples.

The aim of this study was to analyze the variability degree in botanical composition of honey and pollen mass samples from the same and different nests, seasons, and forest types in the Chaco region.

Materials and methods

Study area

One sampling site (El Sauzalito, Argentina) was in Palosantal forest while the other three (Miraflores, Juan José Castelli and Villa Río Bermejito, Argentina) were in Quebrachal forest, which are located no more than 250 km away from each other. These sites have similar climate conditions, they are strongly seasonal with very hot summer (December to March) and low temperatures and frost during winter (July to September); there is a manifest yearly variation in rainfall, with a marked dry season in winter-spring and a rainy season from October to April (Prado, 1993). The Palosantal forest is characterized by the dominance of "palo santo" trees (*Bulnesia sarmientoi*, Zygophyllaceae) while the Quebrachal forest by the dominance of "quebracho colorado chaqueño" (*Schinopsis balansae*, Anacardiaceae), "quebracho colorado santiagueño" (*Schinopsis lorentzii*, Anacardiaceae) and "quebracho blanco" trees (*Aspidosperma quebracho-blanco*, Apocynaceae).

Pollen analysis of samples

Honeys and pollen masses from the same nest were analyzed separately (i.e., they were taken into account as independent samples). A total of 11 nests were analyzed from the four seasons and two forest types (Palosantal and Quebrachal). As only honey was found in nest 14 while in nest 18 only the pollen mass, a total of 10 honey and 10 pollen mass samples were studied (Tables 1 and 2). Honey was sampled from different pots of a nest using a disposable plastic syringe and then homogenized. Therefore, one representative sample of honey was kept per nest. Closed pots were preferred for sampling honey. However, when only open pots were available, honey was kept from them. From 8 to 83.8 g of honey corresponding to between 6 and 130 honey pots per nest was studied (Table 1). Honey samples were pure in nests 1, 2, 10, 12, 14 and 15, while they were contaminated with pollen grains from masses during their sampling in nests 5, 6, 7 and 13 (Table 1).

The pollen masses (the content of pollen cerumen pots) of each nest were mixed and analyzed as an only sample. From 12 to 103 g of pollen corresponding to between 5 and 55 pollen pots per nest was studied (Table 2). Honey and pollen mass were weighed on an Ohaus CS200 electronic balance with 0.1 gram of readability. Honey and pollen mass samples were dissolved with a glass rod in 200 ml of distilled water at 80-90 °C and then with a magnetic stirrer for 10- 15 minutes. Five milliliters of this mixture was centrifuged at 472 g (Pendlenton, 2006) and the sediment was dehydrated using acetic acid and acetolyzed (Erdtman, 1960), mounted in slides using a glicerine-gelatin mixture and identified using a Nikon Eclipse E200 light microscope at 400 and 1000 x magnification. Pollen identification was carried out by comparing pollen provision slides with the pollen reference of plants grown in the sites sampled. The identification of the type *Gleditsia amorphoides* was dubious, as it was absent in the sampled area.

For this reason, its family was named as Fabaceae?. The counting of 500 pollen grains per slide was made for honey samples, while a total of 300-500 grains for pollen masses.

The reference pollen collection was made from flower buds of plant species collected in various localities from the Chaco province of Argentina (Juan José Castelli (25°56' S- 60°37' W), Villa Río Bermejito (25°37' S-60°15' W), Miraflores (25°29' S-61°01' W) and El Sauzalito (24°24' S-61°40' W)). These plant specimens were pressed, dried, identified by the author and deposited in the Herbaria of the Museo of La Plata (LP) and the Museo Argentino de Ciencias Naturales "Bernardino Rivadavia" (BA), Buenos Aires, Argentina. Flowering phenology was recorded in these sites during most months except March, May and June (Table 3). Bees were identified by Arturo Roig-Alsina and deposited in the Entomology Collection of the Museo Argentino de Ciencias Naturales "Bernardino Rivadavia", Buenos Aires, Argentina.

Multivariate analysis

The Cluster Analysis and Principal Component Analysis were the two multivariate techniques applied. The PAST statistic package (Hammer et al., 2008) was used. For Cluster Analysis, the algorithm UPGMA (Unweighted pair-group average) and Bray-Curtis distance were applied to the percentage values of the data matrix in Q-mode. The highest similarity level is 1.00 and indicates 100% of similarity among pollen composition of samples. The cophenetic correlation coefficient was taken into account as a distortion measurement of the dendrogram (Sokal & Rohlf, 1962), being values higher than 0.8 indicators of well groupings in the dendrogram compared to the original similarity matrix (Sneath & Sokal, 1973).

For Principal Component Analysis (PCA), a correlation matrix was applied to the data matrix in Q-mode and R-mode. The Scree plot (simple plot of eigenvalues) was used to cut-off the number of significant principal components. After this curve starts to flatten out, the corresponding components may be regarded as insignificant. The eigenvalues expected under a random model (Broken Stick) were also plotted and the ones under this curve represent non-significant components (Jackson, 1993). The first three principal components (PC I, PC II and PC III) that made up the greatest part of the variability were graphed.

Results

Cluster Analysis

The dendrogram of samples of honeys and pollen masses of *T. fiebrigi* showed a high value of cophenetic correlation coefficient (0.879) (Fig 1). Two main groups can be seen within the dendrogram, diverging at 0.1 similarity level (samples of these groups are only 10% similar). Group A (11 samples) is composed of four subgroups diverging at low similarity values (0.2-0.3) while group B of two subgroups diverging at a medium similarity value (0.4) (Fig 1).

Table 1. References of the *Tetragonisca fiebrigi* honey samples studied and abundance of plant families per nest (in percentage). * indicates contaminated honeys; Q = Quebrachal forest; P = Palosantal forest. JJC (J.J.Castelli), (MIR) Miraflores, (ELS) El Sauzalito, (VRB) Villa Río Bermejito.

Honey sample code	Tf1H winter Q	Tf2H winter Q	Tf5H autumn Q*	Tf6H autumn Q*	Tf7H summer P*	Tf10H winter P	Tf12H spring Q	Tf13H spring P*	Tf14H spring P	Tf15H spring P
Amount of honey analyzed (g)	No data	No data	15	8	83.8	30	No data	71	30	20
Number of honey pots analyzed	6	19	10	10	55	60	7	ca. 130	ca. 50	ca. 60
Date of sampling	Aug 2006	Aug 2006	Apr 2008	Apr 2008	Feb 2008	Sep 2008	Oct 2008	Dec 2008	Dec 2008	Oct 2008
Locality of sampling	JJC	JJC	MIR	MIR	ELS	ELS	VRB	ELS	ELS	ELS
Plant family										
Acanthaceae		0.34								
Achatocarpaceae							8.17			6.54
Anacardiaceae	63.46	68.60	52.39	11.30	3.02				30.60	
Apocynaceae		0.34								
Arecaceae			42.82	85.65	0.60		0.48			
Asteraceae		2.05		2.39						1.94
Bignoniaceae					0.60			2.73		6.05
Capparidaceae					4.83	2.28		24.11	3.47	10.41
Celastraceae	30.77	25.94	1.60		18.13	89.35				1.94
Celtidaceae	1.92	0.34		0.65	0.91		12.98	51.15	16.72	2.18
Euphorbiaceae		0.68					2.40			
Fabaceae, Caesalpinioideae					6.34			0.21	0.32	3.87
Fabaceae, Mimosoideae	0.64	0.34			46.83	8.37	2.88	8.60	10.41	13.56
Fabaceae?							2.88			
Malpighiaceae					2.11					1.69
Nyctaginaceae		1.02								
Polygonaceae							0.96		0.63	7.02
Rhamnaceae	3.21	0.34	3.19				59.62		0.32	0.97
Sapotaceae					8.76		3.37		37.54	37.53
Zygophyllaceae					7.85			13.21		6.30
Unidentified 2							3.37			
Unidentified 3							2.88			

Table 2. References of the *Tetragonisca fiebrigi* pollen masses analyzed and abundance of plant families per nest (in percentage). Q = Quebrachal forest; P = Palosantal forest. JJC (J.J.Castelli), (MIR) Miraflores, (ELS) El Sauzalito, (VRB) Villa Río Bermejito.

Code of pollen mass samples	Tf1P Winter Q	Tf2P winter Q	Tf5P autumn Q	Tf6P autumn Q	Tf7P summer P	Tf10P winter P	Tf12P spring Q	Tf13P spring P	Tf15P spring P	Tf18P winter Q
Amount of pollen analyzed (g)	103	No data	71	68	56	12	26	70	60	No data
Number of pollen pots analyzed	40	5	45	55	23	7	20	15	20	No data
Date of sampling	Aug 2006	Aug 2006	Apr 2008	Apr 2008	Feb 2008	Sep 2008	Oct 2008	Dec 2008	Oct 2008	Jul 2009
Locality of sampling	JJC	JJC	MIR	MIR	ELS	ELS	VRB	ELS	ELS	MIR
Plant family										
Acanthaceae	0.07									
Achatocarpaceae	3.07									
Anacardiaceae	73.19	100	40.59	42.67	0.81		2.06		0.03	42.32
Arecaceae	0.13		42.90	56.89	0.95		0.05			49.13
Asteraceae	0.05		6.96	0.15	1.87	0.07				0.12
Bignoniaceae					2.47				9.69	
Bromeliaceae								0.51		
Capparidaceae	0.82		1.53		10.79	12.67	2.20	39.43	2.37	0.05
Celastraceae					30.65	37.60	0.61	14.79	0.77	0.34
Celtidaceae	0.63		5.52	0.29	3.42		3.19	29.43	3.07	0.47
Chenopodiaceae							0.09			
Euphorbiaceae	0.01		1.69				0.14			
Fabaceae, Caesalpinioideae	1.72				3.21	0.47		0.04	13.33	
Fabaceae, Mimosoideae	11.84		0.30		25.29	44.34	30.49	11.85	24.05	7.08
Fabaceae?							24.67	3.81		
Loranthaceae			0.18		0.07			0.14		
Malpighiaceae			0.34		6.77				2.10	
Olacaceae										0.02
Polygonaceae					0.04		0.09		1.54	
Rhamnaceae					6.31		28.33		0.67	0.02
Santalaceae						2.16				0.44
Sapotaceae	8.47				7.34		4.69		42.35	
Simaroubaceae						2.70				
Ulmaceae							3.19		0.03	
Unidentified 1							0.19			

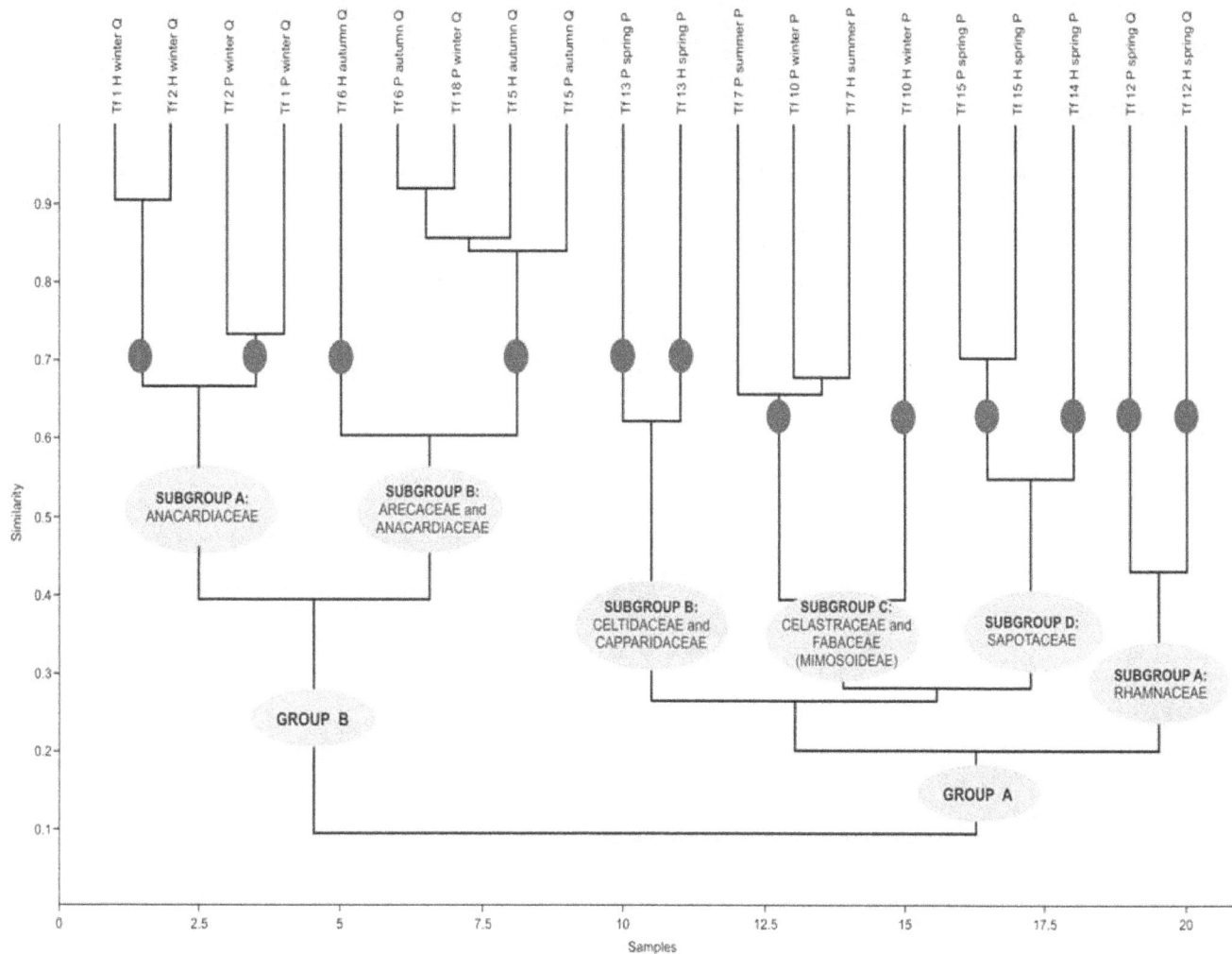

Fig 1. Dendrogram showing the two groups and six subgroups of honey and pollen mass samples of *Tetragonisca fiebrigi*.

Samples from group A are dominated by various plant families, while those from group B are dominated by one or codominated by two families. In each of the four subgroups of group A, different families are important. Subgroup A has important percentages of Rhamnaceae, B of Celtidaceae and Capparidaceae, C of Celastraceae and Fabaceae (Mimosoideae), and D of Sapotaceae. Each subgroup has two sets differing according to the dominance and/or codominance of some of their families. Subgroup A has one set (Tf 12 H spring Q) dominated by Rhamnaceae and another codominated by Fabaceae (Mimosoideae), Rhamnaceae and Fabaceae? (Tf 12 P spring Q); subgroup B has one set (Tf 13 H spring P) dominated by Celtidaceae and in a lesser scale composed of Capparidaceae and Zygophyllaceae, while the other set (Tf 13 P spring P) is codominated by Capparidaceae and Celtidaceae and in a lesser scale composed of Celastraceae and Fabaceae (Mimosoideae); subgroup C has one set (Tf 10 H winter P) dominated by Celastraceae and another codominated and/or dominated either by Celastraceae or Fabaceae (Mimosoideae); and subgroup D is composed of two sets, one of them codominated by Sapotaceae, Fabaceae (Mimosoideae), Capparidaceae (in honey) (Tf 15 H spring P) or Sapotaceae, Fabaceae (Mimosoideae),

Fabaceae (Caesalpinioideae) and Bignoniaceae (in pollen mass) (Tf 15 P spring P) while the other set is codominated by Sapotaceae, Anacardiaceae, Celtidaceae and Fabaceae (Mimosoideae) (Tf 14 H spring P). The subgroup A of group B is dominated by Anacardiaceae, while the subgroup B is codominated by Arecaceae and Anacardiaceae.

Principal Component Analysis

Principal Component Analysis and Cluster Analysis partly agreed in the grouping of samples (Figs 1, 2 and 3). In Q-mode, the Principal Component I (Figs 2 and 3) placed sample Tf 2 H winter Q on the top left corner (dominated by Anacardiaceae and Celastraceae), samples Tf 12 H spring Q and Tf 12 P spring Q (high percentages of Rhamnaceae) on the extreme right and the remaining samples clustered together in the middle (Figs 2 and 3). Families Rhamnaceae, Fabaceae?, Ulmaceae, Unidentified 1 and Chenopodiaceae are the major contributors to the Principal Component I (being 16.35% of the total variability) (Supplementary material 1).

The Principal Component II (Fig 2) separates samples Tf 2 H winter Q and Tf 12 H spring Q to the top of the graph and both samples of nest 15 (the honey sample codominated by Sapotaceae, Fabaceae (Mimosoideae) and Capparidaceae, and the pollen sample by Sapotaceae, Fabaceae (Mimosoideae), Fabaceae (Caesalpinioideae) and Bignoniaceae) to the bottom of the graph and the remaining samples to the middle. Families Bignoniaceae, Fabaceae (Caesalpinioideae), Sapotaceae, Fabaceae (Mimosoideae) and Euphorbiaceae are the major contributors to the Principal Component II (being 15.02% of the total variability) (Supplementary material 1).

The Principal Component III (Fig 3) separates the honey sample of nest 12 (dominated by Rhamnaceae) from the pollen sample

*Variability of nest samples of **Tetragonisca fiebrigi***

The Principal Component I (R-mode) shows that certain families (Anacardiaceae and Arecaceae) are in the extremes of the axes, while Principal Component II shows that Fabaceae (Mimosoideae), Celastraceae and Sapotaceae are the most distant (Fig 4) The association of these families allowed for the differentiation of groups of samples, those from fall and winter (dominated by Anacardiaceae and Arecaceae) (PC I in Supplementary material 1; Group B of Cluster Analysis)

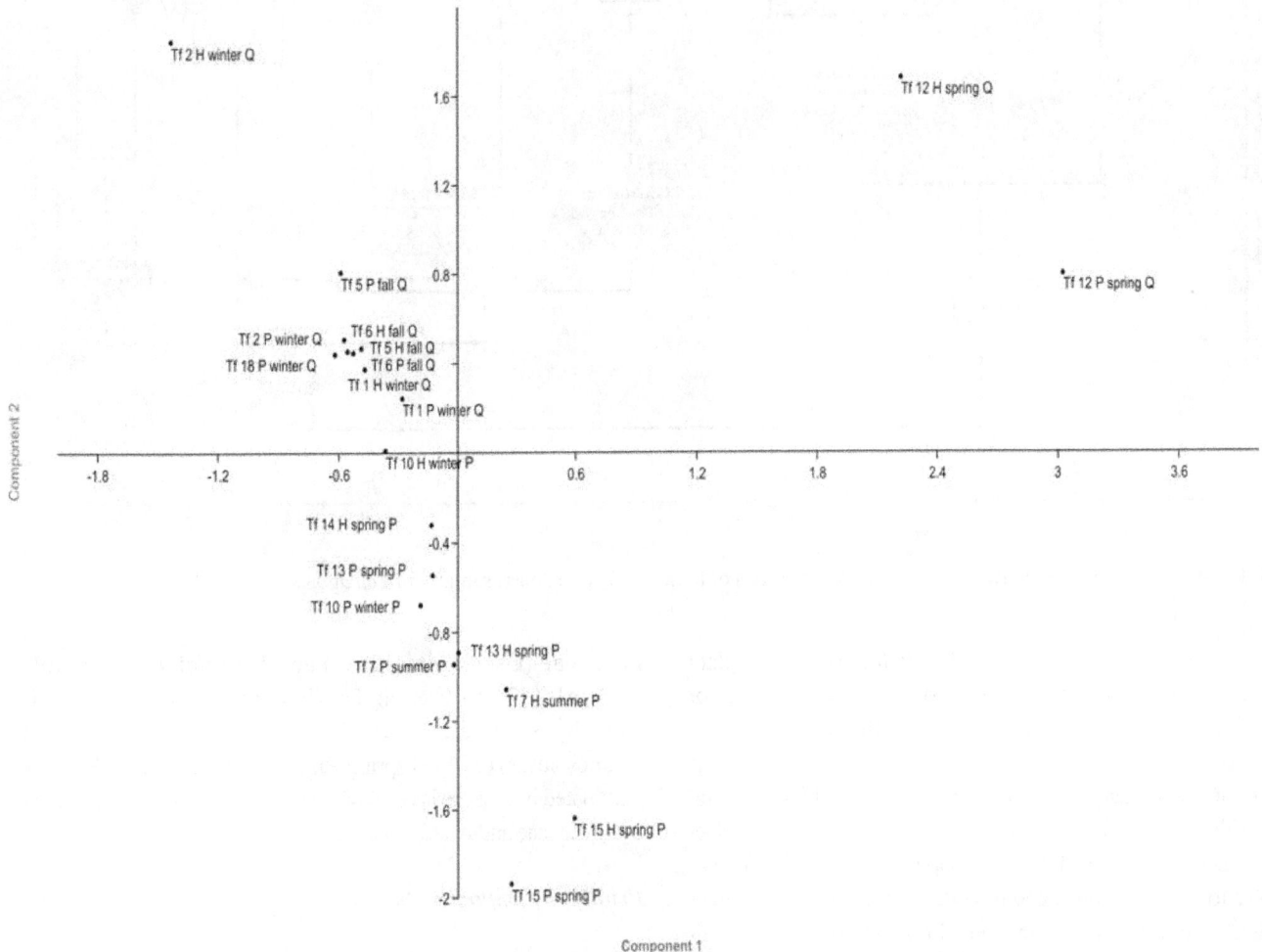

Fig 2. Two-dimensional graph of the Principal Components 1 and 2 showing the distribution of store samples.

of the same nest (codominated by Rhamnaceae, Fabaceae (Mimosoideae) and Fabaceae?) placing them to the bottom and top of the graph, respectively; and the rest in the middle (Fig 3). The three unidentified families (Unidentified 2, 3 and 1), Chenopodiaceae and Ulmaceae are the major contributors to the Principal Component III (being 13.24% of the total variability) (Supplementary material 1).

The first three PC account for only 44.66% of the total variability. However, the greatest part of the variability of samples is made up by the first 8 principal components (83.91%) (Supplementary material 2) as showed by the scree plot (Supplementary material 3).

from those from spring, summer and winter (in which Anacardiaceae and Arecaceae were absent) (PC II in Supplementary material 1; Group A of Cluster Analysis).

The families which were the greater contributors to each Principal Component (Supplementary material 1) were those exclusive in the samples (such as Rhamnaceae, Fabaceae?, Ulmaceae, Unidentified 1 and Chenopodiaceae for Principal Component I). Due to the fact that only a low percentage of the variability (44.66%) was given by the first three principal components (Supplementary material 2), the samples found in the extremes of the axes of each Principal

Table 3. Flowering phenology of the plant taxa whose ascribed pollen types were found in the stores of *Tetragonisca fiebrigi* in the Dry Chaco forest. During March, May and June, flowering was not recorded. Plant life-forms: T = trees, S = shrubs (more than 1 m in height), C = climbers and lianas, H = herbs, semi-shrubs and shrubs less than 1 m in height, E = epiphytes.

Plant family	Plant taxon	J	F	M	A	M	J	J	A	S	O	N	D	Life-form
ACANTHACEAE	Acanthaceae (except *Ruellia*)			-		-	-							H
ACHATOCARPACEAE	*Achatocarpus praecox* Griseb.			-		-	-							S-T
ANACARDIACEAE	*Schinopsis balansae* Engl.			-		-	-							T
	Schinopsis lorentzii (Griseb.) Engl.			-		-	-							T
	Schinus fasciculatus (Griseb.) I.M. Johnst.			-		-	-							S
APOCYNACEAE, APOCYNOIDEAE	*Aspidosperma quebracho-blanco* Schltdl.			-		-	-							T
ARECACEAE	*Trithrinax schizophylla* Drude			-		-	-							S-T
ASTERACEAE, ASTEREAE	*Baccharis breviseta* DC.			-		-	-							H-S
	Baccharis salicifolia (Ruiz & Pav.) Pers.			-		-	-							S
	Baccharis trinervis Pers.			-		-	-							S
	Solidago chilensis Meyen			-		-	-							H
ASTERACEAE, HELIANTHEAE	*Ambrosia* sp.			-		-	-							H
	Bidens spp.			-		-	-							H
	Melanthera latifolia (Gardner) Cabrera			-		-	-							H
	Parthenium hysterophorus L.			-		-	-							H
	Verbesina encelioides A. Gray			-		-	-							H
	Xanthium spinosum L.			-		-	-							H
	Xanthium cavanillesii Schouw			-		-	-							H
ASTERACEAE, INULEAE	*Pterocaulon* spp.			-		-	-							H
	Tessaria dodoneifolia (Hook. & Arn.) Cabrera			-		-	-							S
BIGNONIACEAE	*Fridericia dichotoma* (Jacq.) L.G. Lohmann			-		-	-							C
	Tabebuia impetiginosa (Mart. Ex DC.) Standl.			-		-	-							T
	Tabebuia nodosa (Griseb.) Griseb.			-		-	-							T
BROMELIACEAE	*Aechmea distichantha* Lem.			-		-	-							H
	Bromeliaceae (terrestrial species)			-		-	-							H
CAPPARIDACEAE	*Capparis atamisquea* Kuntze			-		-	-							S
	Capparis retusa Griseb.			-		-	-							S-T
	Capparis salicifolia Griseb.			-		-	-							S
	Capparis speciosa Griseb.			-		-	-							S-T
	Capparis tweediana Eichler			-		-	-							S
CELASTRACEAE	*Maytenus vitis-idaea* Griseb.			-		-	-							S-T
	Moya spinosa Griseb.			-		-	-							S-T
CELTIDACEAE	*Celtis* spp.			-		-	-							T
CHENOPODIACEAE	Chenopodiaceae spp.			-		-	-							H-S
EUPHORBIACEAE	*Cnidoscolus loasoides* (Pax) I.M. Johnst.			-		-	-							H
	Croton argenteus L.			-		-	-							H-S
	Croton bonplandianus Baill.			-		-	-							H-S
	Croton lachnostachyus Baill.			-		-	-							H-S
	Jatropha spp.			-		-	-							S
	Sapium haematospermum Müll. Arg.			-		-	-							T
FABACEAE, CAESALPINIOIDEAE	*Caesalpinia paraguariensis* (D. Parodi) Burkart			-		-	-							T
	Cercidium praecox (Ruiz & Pav. ex Hook.) Harms			-		-	-							T
	Parkinsonia aculeata L.			-		-	-							S-T

Table 3. Flowering phenology of the plant taxa whose ascribed pollen types were found in the stores of *Tetragonisca fiebrigi* in the Dry Chaco forest. During March, May and June, flowering was not recorded. Plant life-forms: T = trees, S = shrubs (more than 1 m in height), C = climbers and lianas, H = herbs, semi-shrubs and shrubs less than 1 m in height, E = epiphytes. (Continuation).

Family	Species	Life-form
	Pterogyne nitens Tul.	T
FABACEAE, MIMOSOIDEAE	*Acacia aroma* Gillies ex Hook. & Arn.	S-T
	Acacia curvifructa Burkart	S
	Albizia inundata (Mart.) Barneby & J.W. Grimes	S-T
	Mimosa detinens Benth.	S-T
	Prosopis alba Griseb.	T
	Prosopis elata (Burkart) Burkart	S-T
	Prosopis kuntzei Harms	T
	Prosopis nigra (Griseb.) Hieron.	T
	Prosopis ruscifolia Griseb.	T
	Prosopis vinalillo Stuck.	T
	Prosopis (hybrids)	S-T
LORANTHACEAE	*Struthanthus uraguensis* (Hook. & Arn.) G. Don	E
MALPIGHIACEAE	*Mascagnia brevifolia* Griseb.	C
NYCTAGINACEAE	*Boerhavia diffusa* L. var. *leiocarpa* (Heimerl) Adams	H
OLACACEAE	*Ximenia americana* L.	S-T
POLYGONACEAE	*Ruprechtia triflora* Griseb.	S-T
RHAMNACEAE	*Ziziphus mistol* Griseb.	T
SANTALACEAE	*Acanthosyris falcata* Griseb.	T
SAPOTACEAE	*Sideroxylon obtusifolium* (Roem. & Schult.) T.D. Penn.	T
SIMAROUBACEAE	*Castela coccinea* Griseb.	S-T
ULMACEAE	*Phyllostylon rhamnoides* (J. Poiss.) Taub.	T
ZYGOPHYLLACEAE	*Bulnesia sarmientoi* Lorentz ex Griseb.	T

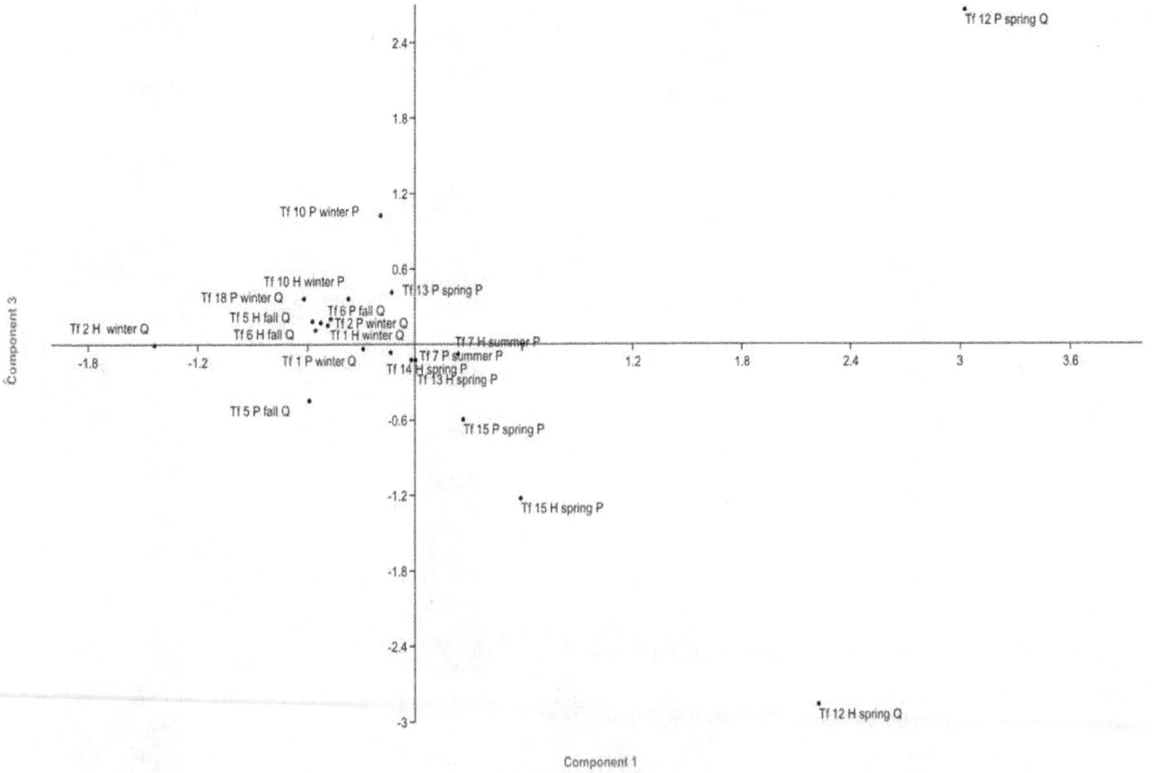

Fig 3. Two-dimensional graph of the Principal Components 1 and 3 showing the distribution of store samples.

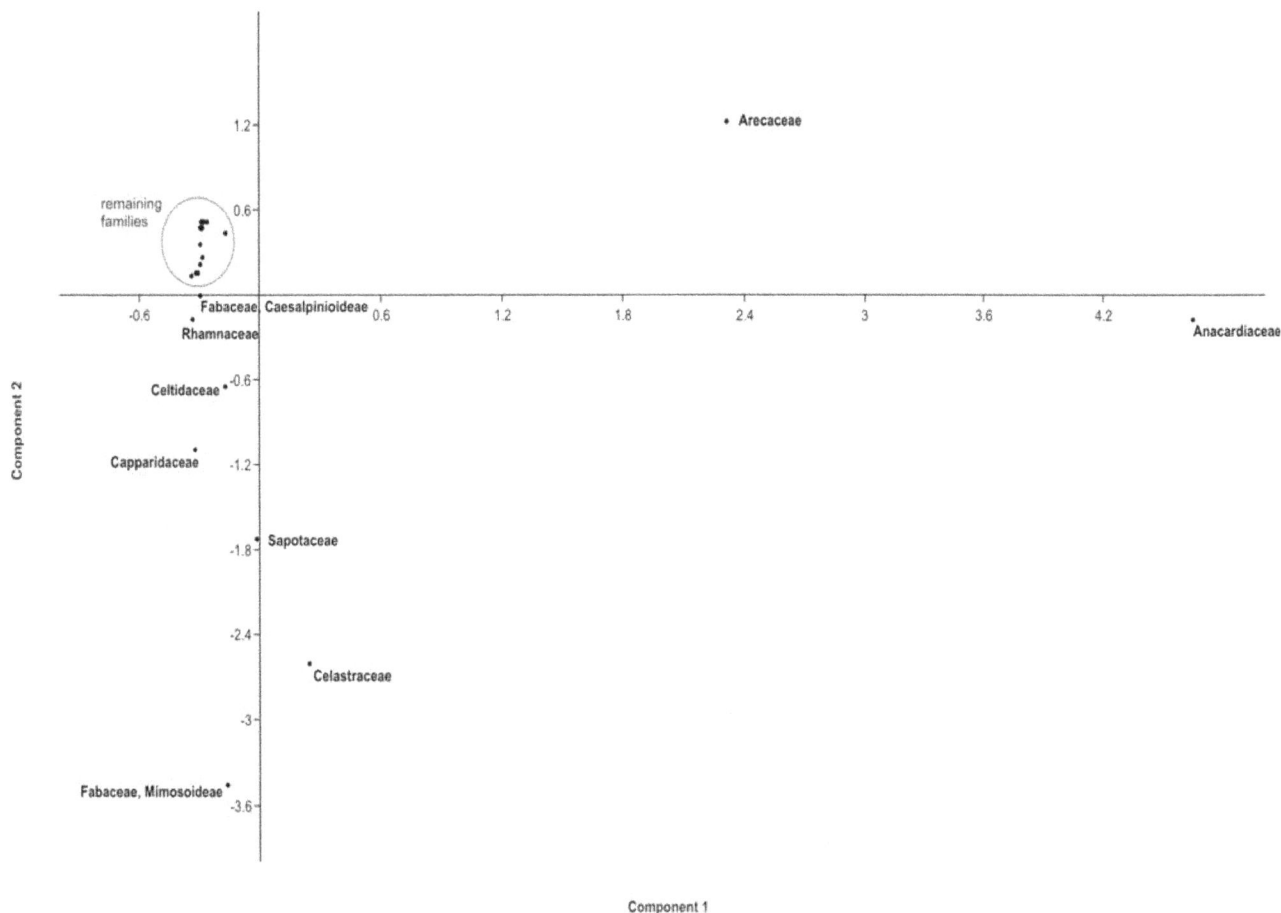

Fig 4. Two-dimensional graph of the Principal Components 1 and 2 showing the distribution of plant families composing the diet of *Tetragonisca fiebrigi*.

Component were different (but not very different) from the remaining samples although they shared many pollen types.

Discussion

Were samples clustered by type of forest or season?

The samples from each forest type were clustered separately by Cluster Analysis with the exception of both provisions of nest 12 from Quebrachal that were grouped together with Palosantal samples. This could suggested that botanical composition of these two groups of samples (Groups A and B) is an indicator of the strong differences existent in the floristic composition of these forest types. For instance, the abundance of *Schinopsis* trees (Anacardiaceae) in the Palosantal is much scarcer than in the Quebrachal (Cabrera, 1971; Prado, 1993). However, samples from group B (all from Quebrachal) were sampled in the fall and winter, while those from group A (all Palosantal plus two Quebrachal samples) mostly in spring suggesting that grouping of samples might not have been achieved by the type of forest. Moreover, the combination of phenological records and pollen analysis of stores detected the cause for these groupings. Fall and winter provisions were composed mainly of floral resources bloomed during summer and fall such as Arecaceae and Anacardiaceae, while spring provisions were composed of spring flowerings alone but not summer-fall floral resources. For instance, the only two winter samples (both stores of nest 10) found in Group A were not clustered in group B together with the remaining samples of winter because they were composed of pollen types from late winter flowerings but not of types foraged during summer-fall. Therefore, it can be said that seasonality strongly influenced the grouping of these store samples. Moreover, Prado (1993) states that no differences in floristic composition exist between Palosantal and Quebrachal plant communities, but only in relative abundance of plant species, supporting the idea that seasonality was the cause of these groupings.

Are there differences in botanical composition among honeys and pollen masses?

Honeys and pollen masses were grouped together or fairly closed for all nests due to similar abundance of the different pollen types. Sampling could be considered as a possible cause of honey contamination as the small cerumen pots of *T. fiebrigi* are densely packed and fragile. However, both pure (six samples) as well as contaminated honeys (four samples) were closely clustered with pollen masses from the same nests. Although

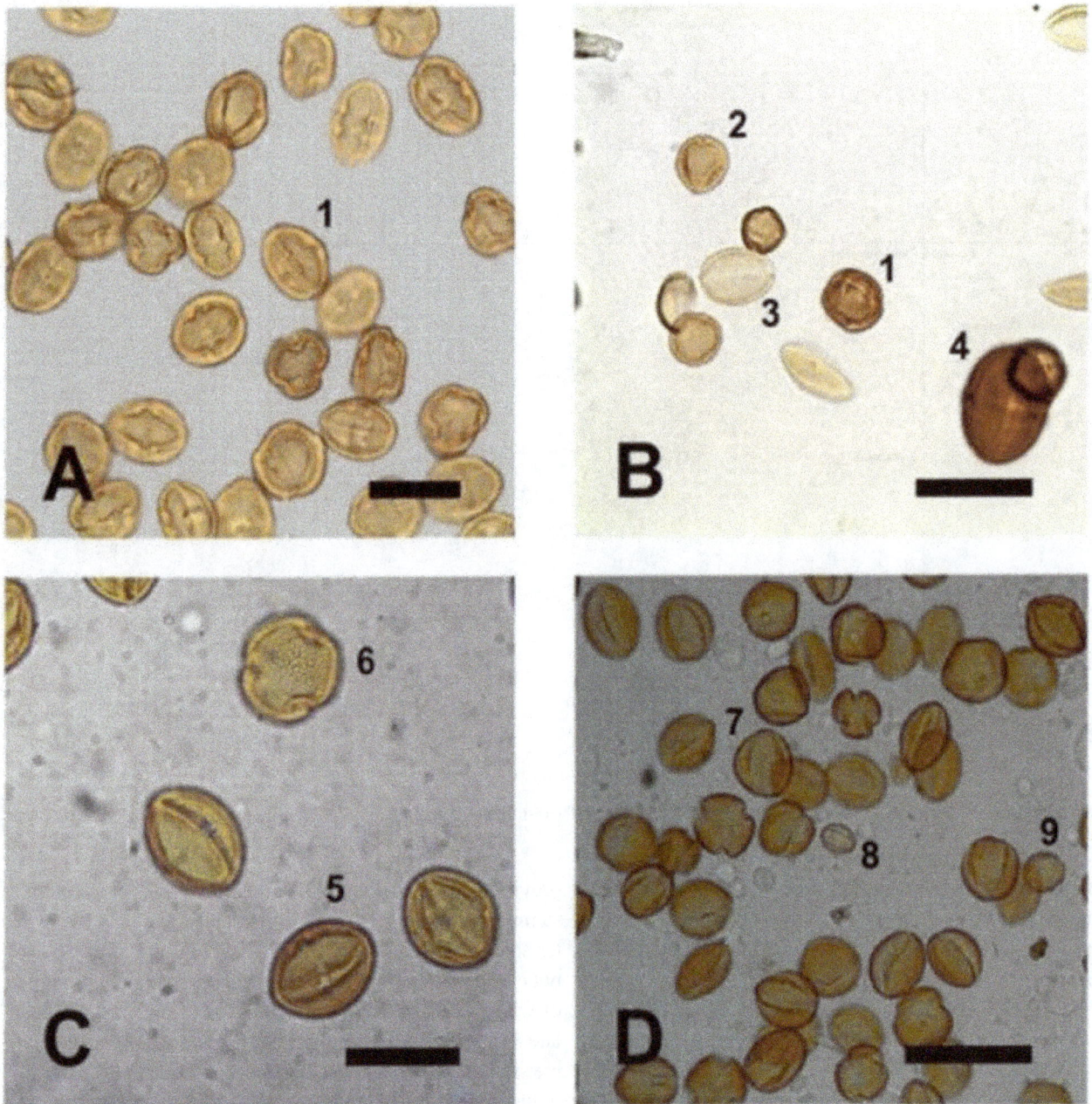

Fig 5. Acetolyzed pollen grains from nest stores, seen in light microscope at 40 x magnification. A–D - Pollen grains of type *Schinopsis* (Anacardiaceae) (1), *Celtis* (Celtidaceae) (2), *Trithrinax schizophylla* (Arecaceae) (3), *Sapium haematospermum* (Euphorbiaceae) (4), *Capparis retusa* (Capparidaceae) (5), *Castela coccinea* (Simaroubaceae) (6), *Prosopis* (Fabaceae, Mimosoideae) (7), type *Maytenus vitis-idaea* (Celastraceae) (8), *Ziziphus mistol* (Rhamnaceae) (9). Bars: 30 μm (A and C); 40 μm (B); 50 μm (D).

contamination can explain the clustering of honey and pollen mass from nests 5, 6, 7 and 13, it does not apply to the remaining six nests (not contaminated honeys). On the other hand, as honeys were not clustered together with other honeys but with pollen masses indicating a similar botanical composition, it can be assumed that both nectar and pollen were gathered from the same plant species. Similar kind of clustering is also observed on the Chaquenian stingless bees *Melipona orbignyi* Guérin and *Geotrigona argentina* Camargo & Moure (Vossler, unpublished data), supporting that the Dry Chaco melittophilous vegetation is dominated by plants providers of both pollen and nectar, but not exclusively or predominately of one of them.

Variability of plant family composition in food stored by Tetragonisca bees

The results of the present study would indicate that the botanical composition of samples of *T. fiebrigi* is governed by random factors such as local differences of flower availability but not by preferences for particular plant families. For this reason, bees of the genus *Tetragonisca* have been associated to different plant families according to the vegetation of the study site where samples were taken. For instance, in two sites of Chiapas (Mexico) *T. angustula* mainly foraged on Fabaceae (Caesalpinioideae), Celtidaceae, Piperaceae,

Celastraceae, Sapindaceae, Amaranthaceae and Clethraceae (Sosa-Nájera et al., 1994) and on Asteraceae, Euphorbiaceae, Rubiaceae, Rutaceae, Celtidaceae, Sapindaceae, Anacardiaceae and Phytolaccaceae (Martínez-Hernández et al., 1994).

Honeys from coffee agroecosystems from Colombia were dominated by Rubiaceae, Rhamnaceae and Malvaceae (Obregón et al., 2013). In Piracicaba (Brazil), this bee species concentrated its foraging on Liliaceae, Myrtaceae, Fabaceae (Mimosoideae), Fabaceae (Papilionoideae), Celtidaceae and Fabaceae (Caesalpinioideae) (Carvalho & Marchini, 1999; Carvalho et al., 1999). In honeys from Paraná (Brazil), the most important families were Apiaceae and Fabaceae (Caesalpinioideae) (Cortopassi-Laurino & Gelli, 1991) while honeys from São Paulo were dominated by Euphorbiaceae, Myrtaceae, Apiaceae and Anacardiaceae (Iwama & Melhem, 1979).

Pollen stores from São Paulo were mainly composed of Euphorbiaceae, Cecropiaceae and Celtidaceae (Imper-atriz-Fonseca et al., 1989). Honeys from different localities of São Paulo state were dominated by either Caricaceae, Asteraceae, Fabaceae (Papilionoideae), Myrtaceae or Faba-ceae (Mimosoideae) (Barth et al., 2013). In semi-arid areas from northeastern Brazil covered by Caatinga vegetation, the pollen types more common in honey were from Malpighiaceae, Asteraceae, Myrtaceae, Fabaceae (Mimosoideae), Solanaceae, Anacardiaceae, Arecaceae and Fabaceae (Caesalpinioideae) (Novais et al., 2013). In this same region, pollen stores were dominated by Fabaceae (Mimosoideae), Solanaceae, Moraceae, Fabaceae (Caesal-pinioideae) and Malvaceae (Novais et al., 2014). In Yungas forest (northwestern Argentina), honeys of *T. angustula* were mainly composed of Fabaceae (Mimosoideae), Asteraceae, Myrtaceae and Rutaceae (Flores & Sánchez, 2010). In the Rio Negro chan-nel (Amazonas, Brazil), pollen stores of *T.* gr. *angustula* were mainly composed of Cecropiaceae and Moraceae (Rech & Absy, 2011). In western Amazonas (Acre, Brazil), honey samples of *T. weyrauchi* were dominated by Myrtaceae (Cortopassi-Laurino & Nogueira-Neto, 2003).

Conclusion

The use of multivariate methods from palynological data allowed the detection of clustering patterns in the food samples of *T. fiebrigi*. Grouping of samples was determined by the season when samples were taken but not by type of forest. Honeys and pollen masses were grouped together or fairly closed for all nests due to similar abundance of the dif-ferent pollen types. Furthermore, honeys were not clustered together with other honeys but with pollen masses. Nectar and pollen were gathered from the same plant species, supporting the hypothesis that the Dry Chaco melittophilous vegetation is dominated by plants providers of both pollen and nectar, but not exclusively or predominately of one of them. The foraging

behavior of *T. fiebrigi* was governed by random factors such as local differences of flower availability but not by preferences for some plant families.

Acknowledgments

We wish to thank Ricardo "Nene" Vossler, Juan Hiperdinger and César Albornoz for their warm hospitality and help during the field studies. We also thank Nora Brea for providing suggestions and comments on the manuscript. This study was supported by CONICET (Consejo Nacional de Investigaciones Científicas y Técnicas).

References

Arenas, P. (2003). Etnografía y alimentación entre los to-ba-ñachilamoleek y wichí lhuku'tas del Chaco Central (Ar-gentina). Buenos Aires: P. Arenas, 562 p.

Barth, O.M., Freitas, A.S., Sousa, G.L., & Almeida-Muradian, L.B. (2013). Pollen and physicochemical analysis of *Apis* and *Tetragonisca* (Apidae) honey. Intercienc., 38: 280-285.

Cabrera, A.L. (1971). Fitogeografía de la República Argentina. Bol. Soc. Arg. Bot., 14: 1-42.

Carvalho, C.A.L. & Marchini, L.C. (1999). Tipos políni-cos coletados por *Nannotrigona testaceicornis* e *Tetragonis-ca angustula* (Hymenoptera, Apidae, Meliponinae). Sci. Agric., 56: 717-722.

Carvalho, C.A.L., Marchini, L.C. & Ros, P.B. (1999). Fontes de pólen utilizadas por *Apis mellifera* L. e algumas espécies de *Trigonini* (Apidae) em Piracicaba (SP). Bragantia, 58: 49-56.

Cortopassi-Laurino, M. & Gelli, D.S. (1991). Analyse pollinique, propriétés physico-chimiques et action antibactérienne des miels d'abeilles africanisées *Apis mellifera* et de Méliponinés du Brésil. Apidologie, 22: 61-73.

Cortopassi-Laurino, M. & Nogueira-Neto, P. (2003). Notas sobre a bionomia de *Tetragonisca weyrauchi* Schwarz, 1943 (Apidae, Meliponini). Acta Amaz., 33: 643-650.

Cortopassi-Laurino, M., Imperatriz-Fonseca, V.L., Roubik, D.W., Dollin, A., Heard, T., Aguilar, I., Venturieri, G.C., Eard-ley, C. & Nogueira-Neto, P. (2006). Global meliponiculture: challenges and opportunities. Apidologie, 37: 275-292.

Erdtman, G. (1960). The acetolysis method, a revised descrip-tion. Sven. Bot. Tidskr., 54: 561-564.

Flores, F.F. & Sánchez, A.C. (2010). Primeros resultados de la car-acterización botánica de mieles producidas por *Tetragonisca angustula* (Apidae, Meliponinae) en Los Naranjos, Salta, Argenti-na. Bol. Soc. Arg. Bot., 45: 81-91.

Hammer, O., Harper, D.A.T. & Ryan, P.D. (2008). PAST-Palaeontological Statistics, ver.1.81.

Imperatriz-Fonseca, V.L., Kleinert-Giovannini, A., Cortopassi-Laurino, M. & Ramalho, M. (1984). Hábitos de coleta de *Tetragonisca angustula angustula* Latreille (Apidae, Meliponinae). Bol. Zool. Univ. São Paulo, 8: 115-131.

Imperatriz-Fonseca, V.L., Kleinert-Giovannini, A. & Ramalho, M. (1989). Pollen harvest by eusocial bees in a non-natural community in Brazil. J.Trop. Ecol., 5: 239-242.

Iwama, S. & Melhem, T.S. (1979). The pollen spectrum of the honey of *Tetragonisca angustula angustula* Latreille (Apidae, Meliponinae). Apidologie, 10: 275-295.

Jackson, D.A. (1993). Stopping rules in principal components analysis: a comparison of heuristical and statistical approaches. Ecology, 74: 2204-2214.

Martínez-Hernández, E., Cuadriello-Aguilar, J.I., Ramírez-Arriaga, E., Medina-Camacho, M., Sosa-Nájera, M.S. & Melchor-Sánchez, J.E. (1994). Foraging of *Nannotrigona testaceicornis, Trigona (Tetragonisca) angustula, Scaptotrigona mexicana* and *Plebeia* sp. in the Tacaná region, Chiapas, Mexico. Grana, 33: 205-217.

Nogueira-Neto, P. (1997). Vida e criação de abelhas indígenas sem ferrão. Edição Nogueirapis: São Paulo, 445 p.

Novais, J.S., Absy, M.L. & Santos, F.A.R. (2013). Pollen grains in honeys produced by *Tetragonisca angustula* (Latreille, 1811) (Hymenoptera: Apidae) in tropical semi-arid areas of north-eastern Brazil. Arthrop. Plant Interact., 7: 619-632.

Novais, J.S., Absy, M.L. & Santos, F.A.R. (2014). Pollen types collected by *Tetragonisca angustula* (Hymenoptera: Apidae) in dry vegetation in Northeastern Brazil. Eur. J. Entomol. (Ceské Budejovice, Print), 111: 25-34.

Obregón, D., Rodríguez-C, Chamorro, F.J. & Nates-Parra, G. (2013). Botanical Origin of Pot-Honey from *Tetragonisca angustula* Latreille in Colombia. In P. Vit, S.R.M. Pedro & D.W. Roubik (Eds.). Pot honey: A legacy of stingless bees (pp. 337-346). New York: Springer.

Pendlenton, M. (2006). Descriptions of melissopalynological methods involving centrifugation should include data for calculating Relative Centrifugal Force (RFC) or should express data in units of RFC or gravities (g). Grana, 45: 71-72. doi:10.1080/00173130500520479.

Prado, D.E. (1993). What is the Gran Chaco vegetation in South America? I. A review. Contribution to the study of flora and vegetation of the Chaco. V. Candollea, 48: 145-172.

Rech, A.R. & Absy, M.L. (2011). Pollen sources used by species of Meliponini (Hymenoptera: Apidae) along the Rio Negro channel in Amazonas, Brazil. Grana, 50: 150-161. doi: 10.1080/00173134.2011.579621

Roig-Alsina, A., Vossler, F.G. & Gennari, G.P. (2013). Stingless bees in Argentina. In P. Vit, S.R.M. Pedro & D.W. Roubik (Eds.). Pot honey: A legacy of stingless bees (pp. 125-134). New York: Springer.

Sneath, P.H.H. & Sokal, R.R. 1973. Numerical taxonomy: the principles of numerical taxonomy. W.H. Freeman, San Francisco, 573 pp.

Sokal, R.R. & Rohlf F.J. 1962. The comparison of dendrograms by objective methods. Taxon, 11: 33-40.

Sosa-Nájera, M.S., Martínez-Hernández, E., Lozano-García, M.S. & Cuadriello-Aguilar, J.I. (1994). Nectaropolliniferous sources used by *Trigona (Tetragonisca) angustula* in Chiapas, southern México. Grana, 33: 225-230.

Vossler, F.G. (2012). Flower visits, nesting and nest defence behaviour of stingless bees (Apidae: Meliponini): suitability of the bee species for Meliponiculture in the Argentinean Chaco region. Apidologie, 43: 139-161. doi: 10.1007/s13592-011-0097-6

Zamudio, F. & Hilgert, N.I. (2012). Descriptive attributes used in the characterization of stingless bees (Apidae: Meliponini) in rural populations of the Atlantic forest (Misiones-Argentina). J. Ethnobiol. Ethnomed., 8: 1-10. doi:10.1186/1746-4269-8-9.

24

Social Information in the Stingless Bee, *Trigona corvina* Cockerell (Hymenoptera: Apidae): The Use of Visual and Olfactory Cues at the Food Site

FMJ Sommerlandt[1], W Huber[2], J Spaethe[1]

1 - University of Würzburg, Würzburg, Germany
2 - University Vienna, Vienna, Austria

Corresponding author
Frank M. J. Sommerlandt
Department of Behavioral Physiology
and Sociobiology, Biozentrum
University of Würzburg, Am Hubland,
97074 Würzburg, Germany
E-Mail: frank.sommerlandt@uni-wuerzburg.de

Keywords
social information, communication, odor marks, visual cues, recruitment.

Abstract

For social insects, colony performance is largely dependent on the quantity and quality of food intake and thus on the efficiency of its foragers. In addition to innate preferences and previous experience, foragers can use social information to decide when and where to forage. In some stingless bee (Meliponini) species, individual foraging decisions are shown to be influenced by the presence of social information at resource sites. In dual choice tests, we studied whether visual and/or olfactory cues affect individual decision-making in *Trigona corvina* Cockerell and if this information is species-specific. We found that *T. corvina* foragers possess local enhancement: they are attracted by olfactory and visual cues released by conspecifics but avoid feeders associated with heterospecific individuals of the species *Tetragona ziegleri* (Friese). Overall, olfactory cues seem to be more important than visual cues, but information by visual cues alone is sufficient for discrimination.

Introduction

When foragers of eusocial insects approach new food sources, their decision to land can be influenced by the presence or absence of visual or chemical information provided by other individuals (Slaa et al., 2003; Danchin et al., 2004; Leadbeater & Chittka, 2007; Yokoi & Fujisaki, 2011). For example, foragers can either be attracted (local enhancement) or repelled (local inhibition) by the presence of conspecifics (von Frisch, 1914; D'Adamo et al., 2000; Slaa et al., 2003).

Social information in general is any information about behavior, physical presence or remnant provided by a sender that reduces uncertainty of a receiver. The information can be produced on purpose or inadvertently and may be of different modalities, as for example visual, olfactory or tactile (Danchin et al., 2004; Dall et al., 2005; Kendal et al., 2005; Gruter & Leadbeater, 2014). Such publicly accessible information provided by conspecific or heterospecific individuals allows the observer to adaptively change its behavior to gain fitness benefits, such as assessing the location and quality of a food source (Chittka & Leadbeater, 2005; Goodale et al., 2010, Yokoi & Fujisaki, 2011). In bees, olfactory information at the food source can be provided by foragers through anal droplets, gland secretions or footprints originating from glandular epithelia of the claw retractor tendon (Nieh et al., 2003; Jarau et al., 2004; Barth et al., 2008; Wilms & Eltz, 2008; Jarau, 2009). In bumblebees and honeybees, for example, scent marks can be either short-lived and repellent, and used to avoid visitation of a recently depleted nectar source, or longer-lasting and attracting, and allows identification of particularly rewarding flowers (Stout & Goulson, 2001). The response to such marks depends also on the bee's previous experience (Hrncir et al., 2009; Saleh & Chittka, 2006) and can be modified through learning (reviewed in Gruter & Leadbeater, 2014). Moreover, chemical information is also provided by the profile of epicuticular hydrocarbons, which is species- and colony-specific (Howard & Blomquist, 1982; Nunes et al., 2011).

Another potential modality depicts the physical presence of conspecifics or heterospecifics that serves as a visual stimulus.

In social wasps, for example, visual information about resident heterospecific wasps at the feeder site can appear attractive or repellent to foraging individuals, depending on characteristics of the involved species (size, aggressiveness, etc.; Raveret Richter & Tisch, 1999). The use of conspecific and heterospecific visual information for decision-making in foragers is also reported for bumblebees (Stout & Goulson, 2001; Dawson & Chittka, 2012), which can learn to use this information to find profitable food sources (Dawson & Chittka, 2012). Under natural conditions, signals and cues from different modalities are used to gain social information about the food source. Findings in the stingless bee *Scaptotrigona mexicana* (Guérin) indicate a redundant function of visual and chemical cues in the bee's recruitment system at the food site (Sánchez et al., 2011). In general, in stingless bees the response to the presence or absence of nestmates or heterospecifics at the food site is usually species specific (Slaa et al., 2003).

The stingless bee *Trigona corvina* Cockerell plays an important role in plant pollination in the neotropics, since pollen from more than 70 different plant species has been found in a single colony (Roubik & Moreno Patiño, 2009). Known to be an aggressive mass-recruiter (Roubik, 1981), the fitness of a *T. corvina* colony relies on an efficient recruitment system, where nestmates are guided to profitable food sources mainly by field-based mechanisms including pheromone trails and deposition of scent marks near the food source (Aguilar et al., 2005; Jarau et al., 2010). *T. corvina* foragers are known to be attracted to the food source by scent marks deposited by themselves (during previous visits), by nestmates or by conspecific non-nestmates, but no evidence was found for the use of heterospecific odor marks (Boogert et al., 2006). Whether *T. corvina* foragers also rely on visual cues provided by the presence of conspecifics and if there is a cross-effect (e.g. additive, redundant or hierarchical) of visual and chemical cues that affect the forager's decision at the food site, needs to be investigated. Here we aim to evaluate the significance of chemical and visual cues on the choice behavior of *T. corvina* foragers. We tested the significance of the presence of conspecific or heterospecific foragers, the effect of epicuticular hydrocarbons and the impact of conspecific forager-deposited odor marks.

In particular, we addressed the following questions:

i) Do foragers of *T. corvina* possess local enhancement or local inhibition when making decisions at food sites?

ii) Which sensory modality (olfaction or vision) is used for decision-making?

iii) Can foragers use visual cues to discriminate between conspecifics and heterospecifics?

Material and Methods

Experiments were performed in March 2013 (experiments 1, 5, 6) and March 2014 (experiments 1, 2, 3, 4, 5) at the Tropical Research Station La Gamba, Golfito, Costa Rica (www.univie.ac.at/lagamba/). Workers of the stingless bee *T. corvina* were trained to a gravity feeder (Spaethe et al., 2014) providing 0.1 M sucrose solution. A small sub-group of bees (up to 10-15 individuals) was then trained from the central feeder to a second feeder that was placed approx. 8-10 m away and that provided 0.3 M sucrose solution. This training procedure allowed us to work with a constant number of foragers during the experiments. The second feeder was made of a piece of foam material (85 mm x 85 mm x 10 mm) with a central hole (15 mm in diameter) covered with filter paper (Fig. 1A). A snap-on cap with sucrose solution (or water during the test) was placed into the hole. This setup allowed a (olfactory) marking of the filter paper by visiting bees (see Fig. 1B; comparable to Sánchez et al., 2011). The experimental procedure consisted of two phases: a training phase and a test phase. During the training phase, the above mentioned second feeder (hereinafter referred to as *training feeder*) provided a sugar reward, and the bees were allowed to forage for 10 minutes to establish the food source. The first training phase of each day was prolonged to 60 minutes to obtain a stable number of foraging bees. Afterwards, during the test phase, a choice test was performed where the bees had to choose between two new, unrewarded feeders (comparable to Boogert et al., 2006) that provided different olfactory and/or visual cues according to the type of experiment (see results for the tested combinations). These feeders, consequently termed *test feeders*, were placed at 30 cm distance between each other. Each bee was individually tested and immediately captured after landing on one of the test feeders to avoid multiple counting. After the decisions of six bees, the test feeders were removed and replaced by a fresh rewarding training feeder to start a new experiment. Following a 10 minute training phase, a new randomly chosen combination of test feeders was presented and the next six bees were tested. During all experiments, the feeder positions were randomized to avoid position effects. We used different types of feeders: To test visual cues, we attached three hexane-washed conspecific individuals (Fig. 1A) or individuals from the related species *Tetragona ziegleri* (Friese). *T. ziegleri* was chosen since it is of similar size as *T. corvina* (about 6 mm body length), but of different color (orange). Thus, both species should be detectable by *T. corvina* foragers at a similar distance, but provide different chromatic information. The hexane-bath was applied for 24 hours and hexane was regularly exchanged to remove all species-specific odors. Before the dummies were used in an experiment, they were sundried for 60 minutes. To assess the impact of epicuticular hydrocarbons, we attached three freshly killed (by freezing) bees. We used olfactory marked feeders with filter papers that were odor-marked by foragers in previous visits to test for the relevance of odor markings of the feeder. The filter papers were marked by 10-15 foragers which completed multiple foraging bouts during the 10 minute training phases (see below). Although we never observed active pheromone deposition by the bees through abdominal droplets, we cannot completely exclude pheromone marking. We chose randomly among a couple of marked filter papers for the ones that were used in the tests. Overall, not more than 30 minutes passed between deposition of odor marks and the use in the experiments. For the control experiment, we used a new and empty control feeder covered with clean filter paper.

Experiments were performed in a randomized order to avoid temporal effects and with inter-test intervals of 10 minutes to obtain stable numbers of recruited foragers at the feeders. This time period corresponded to approximately four foraging flights per individual. We assume all tested bees belonged to a single colony, since no inter-individual aggression behavior was observed at the feeders and *T. corvina* foragers are known to defend food sources against competitors (Roubik, 1981). All captured bees were released at the end of a day. As a consequence, we cannot exclude that some bees were counted multiple times during the following days.

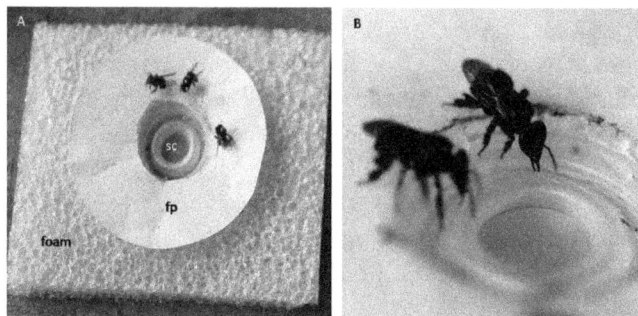

Fig 1. Training and test feeder. A, Feeder with pinned bee dummies. B, Bees feeding on a training feeder which allowed for the deposition of odor marks on the filter paper. fp, filter paper; sc, snap-on cap.

The proportion of choices for all observed landings was calculated for each of the feeder types and shown as bar charts. To calculate the statistics, we used the observed number of landings on either feeder and compared it with the null hypothesis, which is a random choice (1:1 distribution on the two test feeders), using chi-square test. Statistics were performed with IBM® SPSS® Statistics software (Version 20).

Results

Altogether, we counted 450 decisions in 6 different feeder combinations (Fig. 2).

To test if foragers of *T. corvina* possess local enhancement or local inhibition at the feeder site, bees had to choose between a feeder with three freshly killed conspecifics and a clean feeder (experiment 1). The majority of bees (83%) preferred the feeder with the (freshly killed) conspecifics compared to the plain feeder (*p*<0.001, chi²=24.000). To evaluate if odor cues, e.g. epicuticular hydrocarbons play a role in the recognition of nestmates at the feeding site, or if the visual presence of conspecifics is sufficient for the effect of local enhancement, we confronted the foraging bees with one feeder surrounded by three freshly killed conspecifics and a second feeder surrounded by the same number of hexane-washed dead conspecifics (experiment 2). Significantly more foragers landed on the feeder occupied by freshly killed conspecifics (68%) compared to the feeder with odorless dummies (32%; p=0.034, chi²=4.513). In experiment 3, we tested if *T. corvina* foragers can distinguish between conspecifics and heterospecifics (both freshly killed). In this test, conspecifics were more attractive (70%) than heterospecifics of the

species *T. ziegleri* (p=0.021, chi²=5.328). We then eliminated the epicuticular hydrocarbons by washing the dummies of both species with hexane, with the result that the tested foragers could only use visual cues to discriminate between conspecifics and heterospecifics (experiment 4). Both species were similar in body size (approx. 6 mm body length) but differed in their chromatic appearance with *T. corvina* being almost entirely black and *T. ziegleri* possessing a mainly orange-colored body (Jarau et al., 2009). Foragers significantly preferred conspecifics over heterospecifics. 68% of the tested bees decided to land on the feeder occupied by three hexane-washed *T. corvina*, compared to 32% of the foragers that landed in the vicinity of *T. ziegleri* (p=0.034, chi²=4.513).

We then tested the impact of conspecific odor marks (on filter paper) on recruited foragers (experiment 5). Odor marks alone were significantly more attractive (72%) than clean feeders (p=0.002, chi²= 9.649). In experiment 6, we presented two attractive cues in a competition experiment: one feeder hosting hexane-washed conspecifics, whereas the second feeder provided a filter paper containing odor marks released by conspecifics. Foragers possessed a preference towards the odor marks (76%), compared to the visual presence of conspecifics (p=0.032, chi²=4.593).

To test for side-specific preferences of the bees, we also presented two clean feeders and found that bees showed no side-preference but chose randomly between both feeders (p=0.300, chi²=1.086, N=30; data not shown).

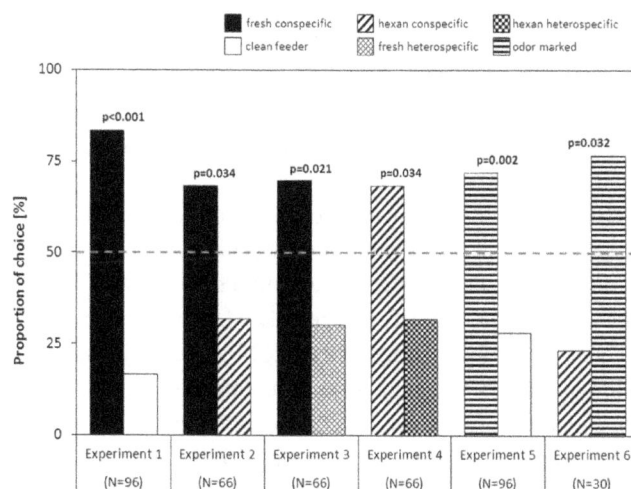

Fig 2. Mean proportion of choices made by *Trigona corvina* foragers towards each of the two feeders offered in each experiment. For each test combination, p-values are shown above the bars. For statistics see text. Numbers in brackets indicate number of choices, dashed line indicates random choice level.

Discussion

Foragers of *T. corvina* use social information when deciding to land on an established food source and make use of olfactory and visual information. In accordance with findings by Boogert et al. (2006), *T. corvina* foragers possess local enhancement, i.e. newly arriving foragers prefer to land close to other conspecifics (a possible distinction between local enhancement and stimulus enhancement and its relevance is reviewed in Heyes, 1994 and was recently discussed in Avarguès-Weber & Chittka, 2014). Our species is described as relatively aggressive (Johnson & Hubbel, 1974; Roubik, 1981); foragers often arrive at resources in groups and defend their foraging territories (Roubik, 1981). Following nestmates to a profitable food source could ensure the monopolization of the food source. *T. corvina* can make use of visual and chemical cues when orienting towards a food source that bears species-specific information, but odor marks seem to be more attractive than the visual presence of a nestmate, as already reported for another stingless bee species, *S. mexicana* (Sánchez et al., 2011).

As previously shown for olfactory cues (Boogert et al., 2006), *T. corvina* foragers can discriminate conspecifics from heterospecifics by their visual appearance alone. In our experiments, feeders surrounded by freshly killed conspecifics – providing the visual information of the presence of the bees combined with the species specific profile of epicuticular hydrocarbons – were preferred by the recruited bees over clean feeders and hexane-washed conspecifics or heterospecifics (experiments 1-3). Thus, the information of cuticular hydrocarbon profiles to recognize nestmates, as shown for several other stingless bee species (Buchwald & Breed, 2005; Nunes et al., 2008; Ferreira-Caliman et al., 2010), plays a role in the recruitment behavior at the food site, as well. Interestingly, conspecifics were also recognized and preferred over heterospecifics when the cuticular hydrocarbon profile and other species-specific odors were absent (due to hexane-washing), indicating that visual information alone is sufficient to identify and discriminate conspecifics from heterospecifics (experiment 4).

Both species used in our experiment were of approximately the same size (about 6 mm body length; Jarau et al., 2009) but differed in color, with *T. corvina* being almost entirely black and *T. ziegleri* possessing a mainly orange-colored body. Thus, aside from flower recognition (Spaethe et al., 2014), color vision may also play an important role in species discrimination at resource sites in stingless bees. Our results concur with findings in other species of the genus *Trigona* (Villa & Weiss, 1990; Spaethe et al., 2014), whereby the relevance of visual cues for decision-making at the food site was reported and the impact of visual cues for individual recognition could have been demonstrated for social wasps (Raveret Richter & Tisch, 1999; Sheehan & Tibbetts, 2011). Whether *T. corvina* foragers can identify heterospecifics that differ from conspecifics in size but not in color, needs to be investigated. When confronted with competing information (experiment 6), olfactory marks (on the filter paper) were preferred over visual cues of hexane-washed bees. We assume the olfactory marks in our experiment to be footprints secreted from the leg tips, as footprints were reported as a type of scent marks at the food source for other stingless bees (*Melipona seminigra* Friese: Hrncir et al., 2004, *Melipona scutellaris* Latreille: Hrncir

et al., 2009). Additionally, we did not observe any deposition of abdominal droplets.

The efficacy of scent marks at the food source (experiment 5) was also shown in *T. corvina* (Boogert et al., 2006) and other stingless bee species (Sánchez et al., 2011) and emphasizes the importance of olfaction as the major sensory modality. This could be caused by either a bias in sensitivity for visual and olfactory stimuli, a difference in the stimuli's action range or a true (innate) preference of the bees for scent marks over visual stimuli. Except for the non-volatile long-chained epicuticular hydrocarbons, odor marks can act over larger distances than visual cues. Estimates of spatial visual resolution in the stingless bee *Tetragonisca angustula* (Latreille) which with 5-6 mm body length is about the size as *T. corvina*, suggest that objects of the size of conspecifics should be visually detectable at relatively short distances of approx. 10 to 20 cm (Zeil & Wittmann, 1993). In contrast, forager-deposited odor marks can be sensed by conspecific stingless bees over long distances up to 10 to 20 m (*Melipona panamica* Cockerell: Nieh, 1998; *Scaptotrigona* aff. *depilis* (Moure): Schmidt et al., 2003).

Here we report that foragers of the stingless bee *T. corvina* possess local enhancement at the food site and that they use visual cues in addition to, or in the absence of, olfactory marks, to identify conspecifics. Whether these bees can also visually discriminate more similar heterospecifics, such as the similar-sized and colored *Trigona fuscipennis* Friese, needs to be investigated.

Acknowledgements

We thank F. Bötzl, J. Kriegbaum, D. Materna and A. Weiglein for help with data collection, S. Jarau for species identification, and the Tropical Research Station La Gamba (www.univie.ac.at/lagamba/), Costa Rica, for making available their laboratory facilities and tropical garden. The Costa Rican Ministerio de Ambiente y Energía kindly granted research permits. We further thank two annonymous reviewers for their thoughtful comments on an earlier draft of this manuscript and J. Plant for linguistic advice. This work was supported by a PhD research scholarship offered by the Free State of Bavaria (Elitenetzwerk Bayern) to FMJS.

References

Aguilar, I., Fonseca, A. & Biesmeijer, J. C. (2005). Recruitment and communication of food source location in three species of stingless bees (Hymenoptera, Apidae, Meliponini). Apidologie, 36: 313-324. DOI: 10.1051/apido:2005005

Avarguès-Weber, A. & Chittka, L. (2014). Local enhancement or stimulus enhancement? Bumblebee social learning results in a specific pattern of flower preference. Anim. Behav., 185-191. DOI: 10.1016/j.anbehav.2014.09.020

Barth, F. G., Hrncir, M. & Jarau, S. (2008). Signals and cues in the recruitment behavior of stingless bees (Meliponini). J Comp Physiol A ,194: 313-327. DOI: 10.1007/s00359-008-0321-7

Boogert, N. J., Hofstede, F. E. & Aguilar Monge, I. (2006). The use of food source scent marks by the stingless bee *Trigona corvina* (Hymenoptera: Apidae): the importance of the depositor's identity. Apidologie, 37: 366-375. DOI: 10.1051/apido:2006001

Buchwald, R. & Breed, M. D. (2005). Nestmate recognition cues in a stingless bee, *Trigona fulviventris*. Anim. Behav., 70: 1331-1337. DOI: 10.1016/j.anbehav.2005.03.017

Chittka, L. & Leadbeater, E. (2005). Social learning: public information in insects. Curr. Biol., 15(21): R869-871. DOI: 10.1016/j.cub.2005.10.018

D'Adamo, P., Corley, J., Sackmann, P. & Lozada, M. (2000). Local enhancement in the wasp *Vespula germanica* - Are visual cues all that matter? Insectes soc., 47: 289-291

Dall, S. R., Giraldeau, L. A., Olsson, O., McNamara, J. M. & Stephens, D. W. (2005). Information and its use by animals in evolutionary ecology. Trends Ecol. Evol., 20: 187-193. DOI: 10.1016/j.tree.2005.01.010

Danchin, E., Giraldeau, L.-A., Valone, T. J. & Wagner, R. H. (2004). Public information: From nosy neighbors to cultural evolution. Science, 305: 487-491

Dawson, E. H. & Chittka, L. (2012). Conspecific and heterospecific information use in bumblebees. PLoS ONE 7(2): e31444 DOI: 10.1371/journal.pone.0031444

Ferreira-Caliman, M. J., Nascimento, F. S., Turatti, I. C., Mateus, S., Lopes, N. P. & Zucchi, R. (2010). The cuticular hydrocarbons profiles in the stingless bee *Melipona marginata* reflect task-related differences. J Insect. Physiol., 56: 800-804 DOI: 10.1016/j.jinsphys.2010.02.004

Goodale, E., Beauchamp, G., Magrath, R. D., Nieh, J. C. & Ruxton, G. D. (2010). Interspecific information transfer influences animal community structure. Trends Ecol. Evol., 25: 354-361 DOI: 10.1016/j.tree.2010.01.002

Gruter, C. & Leadbeater, E. (2014). Insights from insects about adaptive social information use. Trends Ecol. Evol., 29: 177-184 DOI: 10.1016/j.tree.2014.01.004

Heyes, C. M. (1994). Social learning in animals: categories and mechanisms. Biol. Rev., 69: 207-231

Howard, R. W. & Blomquist, G. J. (1982). Chemical ecology and biochemistry of insect hydrocabons. Annu. Rev. Entomol., 27: 149-172

Hrncir, M., Jarau, S., Zucchi, R. & Barth, F. G. (2004). On the origin and properties of scent marks deposited at the food source by a stingless bee, *Melipona seminigra*. Apidologie, 35: 3-13 DOI: 10.1051/apido:2003069

Hrncir, M., Roselino, A. C., Rodrigues, A. V. & Zucchi, R. (2009). Stingless bee footprints at the food sources - repellents, attractants or both? Proceedings of the 46th Annual Meeting of the Animal Behavior Society. Pirenópolis, Brazil: 90-91.

Jarau, S. (2009). Chemical communication during food exploitation in stingless bees. In S. Jarau & M. Hrncir (Eds.). Food Exploitation by Social Insects: Ecological, Behavioral, and Theoretical Approaches (pp. 223-249). Boca Raton (FL), CRC Press.

Jarau, S., Dambacher, J., Twele, R., Aguilar, I., Francke, W. & Ayasse, M. (2010). The trail pheromone of a stingless bee, *Trigona corvina* (Hymenoptera, Apidae, Meliponini), varies between populations. Chem. Senses, 35: 593-601. DOI: 10.1093/chemse/bjq057

Jarau, S., Hrncir, M., Ayasse, M., Schulz, C., Francke, W., Zucchi, R. & Barth, F. G. (2004). A stingless bee (*Melipona seminigra*) marks food sources with a pheromone from its claw retractor tendons. J. Chem. Ecol., 30: 793-804

Jarau, S., Morawetz, L., Reichle, C., Gruber, M. H., Huber, W. & Weissenhofer, A. (2009). Corbiculate Bees of the Golfo Dulce Region, Costa Rica. Wien, Verein zur Förderung der Tropenstation La Gamba, Costa Rica.

Johnson, L. K. & Hubbel, S. P. (1974). Aggression and competition among stingless bees: field studies. Ecology, 55: 120-127

Kendal, R. L., Coolen, I., van Bergen, Y. & Laland, K. N. (2005). Trade-offs in the adaptive use of social and asocial learning. Adv. Study Behav., 35: 333-379. DOI: 10.1016/s0065-3454(05)35008-x

Leadbeater, E. & Chittka, L. (2007). The dynamics of social learning in an insect model, the bumblebee (*Bombus terrestris*). Behav. Ecol. Sociobiol., 61: 1789-1796. DOI: 10.1007/s00265-007-0412-4

Nieh, J. C. (1998). The role of a scent beacon in the communication of food location by the stingless bee, *Melipona panamica*. Behav. Ecol. Sociobiol., 43: 47-58

Nieh, J. C., Ramirez, S. & Nogueira-Neto, P. (2003). Multisource odor-marking of food by a stingless bee, *Melipona mandacaia*. Behav. Ecol. Sociobiol., 54: 578-586 DOI: 10.1007/s00265-003-0658-4

Nunes, T. M., Mateus, S., Turatti, I. C., Morgan, E. D. & Zucchi, R. (2011). Nestmate recognition in the stingless bee *Frieseomelitta varia* (Hymenoptera, Apidae, Meliponini): sources of chemical signals. Anim. Behav. 81: 463-467. DOI: 10.1016/j.anbehav.2010.11.020

Nunes, T. M., Nascimento, F. S., Turatti, I. C., Lopes, N. P. & Zucchi, R. (2008). Nestmate recognition in a stingless bee: does the similarity of chemical cues determine guard acceptance? Anim. Behav., 75: 1165-1171. DOI: 10.1016/j.anbehav.2007.08.028

Raveret Richter, M. & Tisch, V. L. (1999). Resource choice

of social wasps: influence of presence, size and species of resident wasps. Insectes Soc., 46: 131-136

Roubik, D. W. (1981). Comparative foraging behavior of *Apis mellifera* and *Trigona corvina* (Hymenoptera: Apidae) on *Baltimora recta* (Compositae). Rev. Biol. Trop., 29: 177-183

Roubik, D. W. & Moreno Patiño, J. E. (2009). *Trigona corvina*: An ecological study based on unusual nest structure and pollen analysis. Psyche: A Journal of Entomology, 2009: Article ID 268756, 7 pages. DOI: 10.1155/2009/268756

Saleh, N. & Chittka, L. (2006). The importance of experience in the interpretation of conspecific chemical signals. Behav. Ecol. Sociobiol., 61: 215-220. DOI: 10.1007/s00265-006-0252-7

Sánchez, D., Nieh, J. C. & Vandame, R. (2011). Visual and chemical cues provide redundant information in the multimodal recruitment system of the stingless bee *Scaptotrigona mexicana* (Apidae, Meliponini). Insectes Soc., 58: 575-579 DOI: 10.1007/s00040-011-0181-y

Schmidt, V. M., Zucchi, R. & Barth, F. G. (2003). A stingless bee marks the feeding site in addition to the scent path (*Scaptotrigona* aff. *depilis*). Apidologie, 34: 237-248. DOI: 10.1051/apido:2003021

Sheehan, M. J. & Tibbetts, E. A. (2011). Specialized face learning is associated with individual recognition in paper wasps. Science, 334(6060): 1272-1275. DOI: 10.1126/science.1211334

Slaa, E. J., Wassenberg, J. & Biesmeijer, C. (2003). The use of field-based social information in eusocial foragers: local enhancement among nestmates and heterospecifics in stingless bees. Ecol. Entomol., 28: 369-379

Spaethe, J., Streinzer, M., Eckert, J., May, S. & Dyer, A. G. (2014). Behavioural evidence of colour vision in free flying stingless bees. J. Comp. Physiol. A, 200: 485-496. DOI: 10.1007/s00359-014-0886-2

Stout, J. C. & Goulson, D. (2001). The use of conspecific and interspecific scent marks by foraging bumblebees and honeybees. Anim. Behav., 62: 183-189 DOI: 10.1006/anbe.2001.1729

Villa, J. D. & Weiss, M. R. (1990). Observations on the use of visual and olfactory cues by *Trigona spp* foragers. Apidologie, 21: 541-545

von Frisch, K. (1914). Der Farbensinn und der Formensinn der Biene. Jena, Verlag von Gustav Fischer.

Wilms, J. & Eltz, T. (2008). Foraging scent marks of bumblebees: footprint cues rather than pheromone signals. Naturwissenschaften, 95: 149-153. DOI: 10.1007/s00114-007-0298-z

Yokoi, T. & Fujisaki, K. (2011). To forage or not: Responses of bees to the presence of other bees on flowers. Ann. Entomol. Soc. Am., 104: 353-357. DOI: 10.1603/an10053

Zeil, J. & Wittmann, D. (1993). Landmark orientation during the approach to the nest in the stingless bee *Trigona (Tetragonisca) angustula* (Apidae, Meliponinae). Insectes Soc., 40: 381-389

25

Influence of experience on homing ability of foragers of *Melipona mandacaia* Smith (Hymenoptera: Apidae: Meliponini)

F Rodrigues[1,2], MF Ribeiro[3]

1 - Universidade Federal do Vale do São Francisco, Petrolina, PE, Brazil

2 - Universidade Federal Rural do Semiárido (UFERSA), Mossoró, RN, Brazil

3 - Empresa Brasileira de Pesquisa Agropecuária (Embrapa), Petrolina, PE, Brazil

Keywords
Stingless bees, flight range, flight experience.

Corresponding author
Francimária Rodrigues
Universidade Federal Rural do Semiárido, UFERSA
Departamento de Ciências Animais, DCAn
Prédio do DCAn II, Lado Oeste
Av. Francisco Mota, 572
Bairro Costa e Silva, 59.625-900
Mossoró, RN, Brazil
E-Mail: francimaria.rodrigues@ufersa.edu.br

Abstract

The distance a bee can fly to collect food is quite relevant, among other aspects, for successful pollination. However, studies on this aspect concerning stingless bees usually do not take into consideration their homing ability. The objectives of this study were to verify the maximum distance that foragers of *Melipona mandacaia* Smith can fly, and whether experience is relevant for their homing ability in a Caatinga region of Northeast Brazil. Five colonies were used to collect foragers. These were marked and released starting from 100 m from their nests and at every 100 m up to a maximum distance on which there would be no bee returning to the nest. To evaluate the influence of experience, after being marked, another group of bees was put back into colonies, collected again after eight days and released in five distances only (500, 1,000, 1,500, 2,000 and 2,500 m). In both experiments, as the distance increased, the returning success of the bees decreased significantly. In fact, there was a significant negative correlation between their returning success and the distances they were released. The maximum distance a translocate bee returned to its hive was 2,700 m. The percentage of success was very high for bees released at 500 and 1,000 m (100% and 77%, respectively), suggesting this is the common flight range for the species. In most cases, average percentage of success was significantly higher for experienced bees than for other bees reinforcing the idea that experience is quite relevant for homing ability.

Introduction

Stingless bees fly in order to collect the necessary food resources (pollen and nectar) and nest-building materials (resin, mud, etc.), and they need to travel certain distances for that (Roubik, 1989). In fact, bees are faced constantly with the task of navigating back to their nests from remote food sources and this is called 'homing ability'. Honey bees are the most studied bees concerning these aspects. They evolved several methods for doing this, such as compass-direct 'vector' flights, use of learned land marks, and cognitive maps based on spatial memory and on two dimensional snapshots of the surroundings (Anderson, 1977; Cartwright & Collet, 1981, 1983; Gould, 1986; Capaldi & Dyer, 1999; Menzel et al., 2005; Menzel & Giurfa, 2006; Reynolds et al., 2007; Menzel et al., 2012; Cheeseman et al., 2014). Nevertheless, for stingless bees, up to the

moment, no studies concerning homing ability have been performed. Investigations have been restricted to foraging activity, recruitment, flight range and maximum flight distances.

The distances traveled by foragers depend on several factors, as density and seasonality of food source, as well as the bee species (Dornhaus et al., 2006), physiology and body size (Araujo et al., 2004; Greenleaf et al., 2007). Moreover, other aspects, isolated or together, may also affect their flight, as internal colony conditions and climatic factors (Hilário et al., 2000).

Flight range have been the object of some studies in a few stingless bee species. In *Trigona corvina* Cockerell, *Partamona* aff. *cupira* (Smith), *Tetragonisca angustula* (Latreille) and *Nannotrigona testaceicornis* (Lepeletier), the maximum distances reached by bees varied from 623 to 853 m (van Nieuwstadt & Ruano Iraheta, 1996). Among bees of the *Melipona* Illiger genus, the maximum distances were estimated in 2,000 m for *Melipona bicolor* Lepeletier and *Melipona scutellaris* Latreille (Araújo et al., 2004), and 2,100 m for *Melipona mandacaia* Smith (Kuhn-Neto et al., 2009).

The method generally used for such estimations is the training of bees up to a food source. Another method is the capture and recapture of bees, which also allows obtaining information on the maximum flight distance that bees are able to travel to forage (Roubik & Aluja, 1983). There are no studies on the maximum distance a *M. mandacaia* forager can fly using this specific methodology. Moreover, except by the study of Kuhn-Neto et al. (2009), there is no other information concerning the maximum flight distance of this species, specially taking into consideration the experience of bees, which is probably an important factor for homing ability, as demonstrated for honeybees.

M. mandacaia is a very important species for the meliponiculture of Petrolina, Pernambuco state, and Juazeiro, Bahia state, in the Northeastern region of Brazil, being mainly used for honey production (Ribeiro et al., 2012). However, it is relatively little studied, especially concerning its potential for pollination services. In this way, the present study was carried out with the objectives of verifying the maximum distances foragers can fly, their homing ability and the influence of experience on this process.

Material and methods

The experiments were performed at Embrapa Semiárido (09°4'17.53"S 40°19' 10.24" W) in an area of 2,100 ha, at 42 km from the city of Petrolina (Pernambuco state), a semiarid region in Northeast Brazil. The vegetation is typical of hiperxerophile "Caatinga" (Zanella, 2000), a type of savanna. The plants are used to low precipitation, that is restrict to a few months of the year, intense hours of sun and high temperatures, and many of them loose their leaves during the drought period.

M. mandacaia, popularly known as 'mandaçaia', occurs naturally in this Bioma and it is distributed along the São Francisco River, in the states of Bahia, Ceará, Paraíba, Pernambuco e Piauí (Batalha-Filho et al., 2011).

Maintenance of colonies

Five colonies of *M. mandacaia* were used in the experiments. The colonies were installed in hives kept at the Entomology sector of Embrapa Semiárido, in a room maintained at ambient temperature (around 27 °C). Bees had free access to the external environment through a plastic tube in the wall. Supplementary food (*Apis mellifera* Linnaeus honey) was provided in average every eight days, according to a usual beekeeping practice. In the days of the experiments, the colonies did not receive any food.

Two experiments were performed from August 2011 to June 2012, which are described below, totaling 11 consecu-

tive months. Taking into account the possible effects of climate during the experimental period, the release of bees (as explained bellow) occurred in similar conditions of weather and daytime. Thus, bees were released between 8 a.m. to 10 a.m., with average temperature of 27 °C and relative humidity of 65%. In days considered unfavorable (i.e., cloudy, windy, rainy, or with much different conditions concerning temperature and humidity) the release did not happen.

Experiment 1: Homing ability of mixed foragers' group (experienced and inexperienced bees)

In order to be sure only foragers would be used in the experiment, bees were collected at the nest entrance with an insect aspirator when they were arriving from the field. Afterwards, they were placed in acrylic cages (20 x 20 x 20 cm) containing food (*A. mellifera* honey), being one cage for each colony. These bees were then marked on the thorax with plastic nontoxic paint (one color for each investigated distance). Soon after, they were put into wooden boxes and were kept there up to the following day, when they were released.

Initially, because we could not differentiate experienced from inexperienced bees, mixed foragers groups were used. Therefore, mixed foragers groups of 25 bees from each colony were released at distance intervals of 100 m from their nest, up to the distance where no bee returned, i.e. from 100 m, 200 m and so on, up to 2,800 m. The distances from the releasing points were measured with a GPS in a straight line in relation to the nests' entrances (Fig 1). For registering the bees that returned to their nests, small wooden boxes were placed where the original hives were. Those observation boxes had a transparent glass cover, which allowed the observer to check the number and color of bees that returned. The original hives were kept closed throughout the day in order to avoid that other foragers (not marked) returned to the observation boxes. Thus, marked bees were used only once, and for each evaluated distance, new marked bees were used.

The percentage of success was calculated considering the number of bees released, and the number of bees that returned to their nests.

Experiment 2: Homing ability of experienced foragers

During the previous experiment, it was observed that sometimes the bees did not arrive. However, with repetitions of the same distances, other bees were able to return to their nests. Therefore, in order to test the hypothesis that more experienced bees were more capable of recognizing the areas where they were released, and so could find the route back to their nests more easily, another experiment was carried out. The collection and marking of bees was done according to the same methodology already described. Nevertheless, after being marked, the bees were put back into their own colonies. After eight days, these bees were collected again and placed into wooden boxes with

Fig 1. Geographical location of the distances where the *Melipona mandacaia* bees were released. Font: Laboratory of Geoprocessing, Embrapa Semiárido.

food. They were kept there until the next day, when they were released. In this way, these bees had in common eight days of flight experience. However, because we could not collect all marked bees, since some could have died, the number of bees released was defined according to the number of bees that we were able to collect on that day. The registration of returning bees was performed in the same way, but only five distances were tested (500 m; 1,000 m; 1,500 m; 2,000 m; 2,500 m) due to the lower availability of the bees.

At the end of the experiment there were two groups of bees that were compared: a first group, where the experience was not considered (and therefore, could include bees with and/or without experience, i.e., a "mixed group"), and a second group, where only bees with experience were tested ("experienced group").

Statistical analysis

In order to verify if the colonies were statistically different, the Kruskal-Wallis test was applied. To assess the differences found for the returning success between the two groups of bees ("mixed group" and "experienced group"), a Chi-square was applied. The relation between the returning success of the bees and the distance they traveled was tested using a linear regression (Zar, 2010).

Results

Experiment 1: Homing ability of mixed foragers group (experienced and inexperienced bees)

This experiment used 3,225 individuals for releasing, being 25 bees from each colony in each distance. However,

some colonies did not reach the same maximum distance (2,700 m), but smaller ones. Thus, for each evaluated distance, in average 125 bees were released, but from 2,400 m, a smaller number of bees could be released (75-100 bees).

Nevertheless, colonies did not show differences among themselves (Kruskal-Wallis, P= 0.463; n= 27 distances) and for this reason the data related to the returning success of the bees were analyzed together.

The number of bees that returned to their nests decreased gradually with increasing distance. As mentioned above, they reached a maximum distance of 2,700 m (Fig 2).

In general, bees had low success in coming back to their nests when released up to the distance of 2,300 m, 2,500 m and 2,700 m (4% of success, for these distances). It was clear that with increasing distance the percentage of success decreased, and there was a strong correlation between these two factors (Fig 2). Indeed, there was a highly significant negative correlation between the returning success of the bees and the distance from where they were released (Spearman ranking correlation, rho= -0.937, P= 0.000, N= 27 distances; Fig 2).

Experiment 2: Homing ability of experienced foragers

In this experiment, 113 individuals were released, being 12-35 bees in each distance, according to the availability of bees, as mentioned above.

The returning success of the bees reached 100% when they were released up to 500 m (Fig 2). Even at the distance of 1,000 m the percentage of success was quite high (77%), reinforcing the idea that experience must be important for their homing ability. This percentage was twice as big as the one found for bees of the previous experiment (Fig 2).

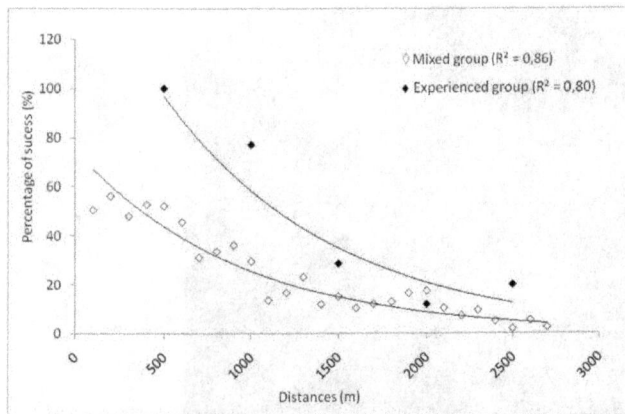

Fig 2. Percentages of success of *Melipona mandacaia* bees returning to their nests in relation to the distance from where they were released for the "Experienced group" and "Mixed group".

When the two groups of bees ("mixed group" and "experienced group") were compared, most the results were higher when experienced bees were used (Table 1). The only exception was for 2,000 m. Thus, both groups of bees were very different in returning success (P values were highly significant at the level of 0.001, Table 1). In the same way, as the distance increased, the returning success decreased significantly, showing a negative correlation (rho= -0.949, P= 0.014, N= 5 distances; Fig 2).

Table 1. Comparison between the number of bees released and returning success of the bees (as well the percentage of success) for the two groups of *Melipona mandacaia* bees ("mixed group" and "experienced group"), and statistical analysis ($\chi c2$ and P values, Chi-square test), for the different distances.

Distances (m)	Number of bees released		Number of returned bees		Percentage of success (%)		Xc^2 (P values)
	Mg	Eg	Mg	Eg	Mg	Eg	
500	125	12	65	12	52.0	100.0	15.2 (P<<0.001)
1,000	125	13	37	10	29.6	77.0	21.1 (P<<0.001)
1,500	125	35	19	10	15.2	28.6	4.1 (P<<0.001)
2,000	125	33	22	4	17.7	12.1	1.0 (P<<0.001)
2,500	100	20	2	4	2.0	20.0	14.7 (P<<0.001)

Mg: "Mixed group"; Eg: "Experienced group".

Discussion

Roubik and Aluja (1983) performed studies using a similar method applied in the present study and found that *Melipona fasciata* Latreille bees returned to their nests when released at the distance of 2,100 m, and *Cephalotrigona capitata* (Smith) returned from 1,500 m. These authors observed that there was a relation between the head size and the distance the bees could fly. Moreover, they also made an estimative through regression tests, and verified that the maximum distances would be 2,400 m and 1,700 m respectively.

Kuhn-Neto et al. (2009) studied the flight of *M. mandacaia* through the training of foragers to the food source. They verified that the maximum distance reached was 2,100 m for larger bees, and 1,560 m for the smaller ones. This confirms what was found by Nieuwstadt and Iraheta (1996) when studying the relationship between the size of some bees (*Trigona corvina*, *Partamona* aff. *cupira*, *Tetragonisca angustula* and *Nannotrigona testaceicornis*) and their foraging range. The authors emphasized that the maximum distances obtained in the experiment of capture and recapture increased ca. 300 m in comparison to the experiment that used the artificial feeder.

In the present study, with another methodology, *M. mandacaia* bees traveled larger distances (2,700 m) than in the study of Khun-Neto et al. (2009), i.e., from 2,100 to 1,560 m. We did not analyzed the size of the bees. However, it is possible that the discrepancy between the results found by us and by Khun-Neto et al. (2009) were due to the different methods used. On the other hand, both methods present limitations. The feeder method could underestimate the flight range since a feeder would not be as attractive for the bee as a flower. In addition, the method of releasing bees could fail when bees do not know the location where they are released (Niewstadt & Ihareta, 1996).

Another fact to be considered is that *M. mandacaia* is a stingless bee endemic from the "Caatinga" region (Zanella, 2000; Batalha-Filho et al., 2011) and this Bioma is characterized by a low density of natural sources and a prolonged drought (Drumond et al., 2000). Thus, plant physiognomy and biology could force the bees to fly longer distances to feed. Although both localities of the experiments (from Khun-Neto et al, 2009, and our study) are in "Caatinga" areas, it is possible that in our case the region presented harder conditions (as the extreme drought of the last years).

Capaldi and Dyer (1999), studying the homing ability of *A. mellifera*, concluded that several factors influence the performance of bees, as for example their learning in relation to nest location and place where they were released, as well as whether they are experienced in foraging. In fact, the data found in the present study with *M. mandacaia* suggest that flight experience is indeed important to bees for homing ability.

Sánchez et al. (2007) studied foraging experience in *Scaptotrigona mexicana* (Guérin) and found that more experienced bees tend to change to other food sources more easily than less experienced bees. When analyzing the effect of experience on the distance reached by *S. mexicana*, they observed that indeed the experience and not the distance of the feeders was the most relevant factor that affected the choice for the food source. These authors concluded that foraging experience could be an advantage for the colony since it allows the exploitation of new food sources, as contributes for diminishing the competition among foragers.

The experience is usually neglected in flight distances studies of stingless bees. The homing ability of an experienced bee in relation to the ability of a naive (or less experienced) one is remarkable in our results. In fact, the present study is the first one to investigate this factor. In addition, as it was observed, the returning success increased about 30% when more experienced bees were released at 500 m, according to Table 1.

In our experiments, we found that the returning success of the bees was higher for "experienced bees" than for the "mixed group" in all distances, except for 2,000 m (Table 1). For this distance, the result was opposite, and the reason could be that, by chance, the bees released in that distance had a longer experience in the "mixed group" than in the other group. This question remains to be clarified. However, it was remarkable that in all other tested distances the number of succeeded bees was significantly higher for the "experienced group".

The mechanisms by which bees find their way back to their nests were not investigated in this study. However, it is possible that stingless bees use the same learning tools as honeybees, such as landmarks and spatial memory, as previously mentioned. In bumblebees, it was observed that bees that were released presented a 'circling' behavior: flying on circles over the release site (Goulson & Stout, 2001). The same was registered in our experiments and probably was used for initial orientation of the bees.

Conclusions

Although the maximum flight distance reached for *M. mandacaia* was 2,700 m, it was outstanding that all bees released at 500 m were able to return to their nests, and even at 1,000 m the large majority returned, suggesting that these distances are part of the common flight range of foragers. As mentioned by Nogueira-Neto (1997), studies on flight distances of stingless bees are relevant since they provide information on flight range of bees in relation to availability of food sources and possibilities for productivity. Thus, the results presented in this study may be useful in crop pollination programs since the distance of hives in relation to cultivated areas may influence pollination efficiency, and consequently, productivity of crops. Moreover, adequate bee pasture should be included in the common flight range of bees in order to guarantee their production.

Finally, our results demonstrated for the first time that experience can limit or improve the homing ability of stingless bees, and this aspect should be considered in future investigations.

Acknowledgments

To Juliara Reis Braga, Cândida Beatriz da Silva Lima and Emison Marcelino Borges, for helping during data collection; to the beekeeper Francisco Camilo de Sousa, for the donation of some colonies used in the experiments and honey and pollen used for feeding the bees; to Ms.C. Tatiana Ayako Taura, for helping with the satellite images; to Dr. Airton Torres Carvalho, for critical reading of the manuscript; to Dr. Marlon da Silva Garrido, Dr. Paulo Gustavo Serafim de Carvalho, Dr. Sérgio Dias Hilário and one anonymous referee, for the valuable suggestions, Erica M. T de Alencar, for language advices; to Coordenação de Aperfeiçoamento de Pessoal de Nível Superior (CAPES), for the post-graduation scholarship to F. Rodrigues; to Empresa Brasileira de Pesquisa Agropecuária (EMBRAPA SEMIÁRIDO) (Cod. SEG 02.11.01.029.00.00) for the facilities.

References

Anderson, A.M. (1977) A Model for Landmark Learning in the Honey-Bee. J. Comp. Physiol. 114: 335-355. doi: 10.1007/BF00657328

Araujo, E.D., Costa, M., Chaud-Netto. J. & Fowler, H.G. (2004). Body size and flight distance in stingless bees (Hymenoptera: Meliponini): Inference of flight range and possible ecological implications. Braz. J. Biol. 64: 563-568. doi: 10.1590/S1519-69842004000400003

Batalha-Filho, H., Waldschmidt, A.M. & Alves, R.M.O. (2011). Distribuição potencial da abelha sem ferrão endêmica da caatinga, (Hymenoptera, Apidae) Melipona mandacaia. Magistra 3: 129-133.

Capaldi, E.A. & Dyer, F.C. (1999). The role of orientation flights on homing performance in Honeybees. J. Exp. Biol. 202: 1655-1666. Retrieved from http://jeb.biologists.org/content/202/12/1655.full.pdf

Cartwright, B.A. & Collett, T.S. (1982). How honey bees use landmarks to guide their return to a food source. Nature 295: 560-564. doi:10.1038/295560a0

Cartwright, A. & Collett, T.S. (1983). Landmark Learning in Bees. J. Comp. Physiol. 151: 521-543. doi 10.1007/BF00605469

Cheeseman, J.F., Millar, C.D., Greggers, U., Lehmann, K., Pawley, M.D.M., Gallistel, C. R., Warman, G.R. & Menzel, R. (2014). Reply to Cheung et al.: The cognitive map hypothesis remains the best interpretation of the data in honeybee navigation. Proc. Natl. Acad. Sci. USA 11(42): 4398. doi: 10.1073/pnas.1415738111

Dornhaus, A., Klügl, F., Oechslein, C., Puppe, F. & Chittka, L. (2006). Benefits of recruitment in honey bees: effects of ecology and colony size in an individual based model, Behav. Ecol. 17: 336-344. doi:10.1093/beheco/arj036

Drumond, M. A., KillL, L. H. P., Lima, P. C. F., Oliveira, M. C., Oliveira, V. R., Albuquerque, S. G., Nascimento, C. E.S. & Cavalcanti, J. (2000). In Avaliação e identificações de ações prioritárias para a conservação, utilização sustentável e repartição dos benefícios da biodiversidade do bioma caatinga. Seminário "Biodiversidade da Caatinga", Petrolina: Embrapa Semiárido, 23p. Retrieved from http://ainfo.cnptia.embrapa.br/digital/bitstream/item/33873/1/usosustentavel.pdf

Greenleaf, S. S.; Williams, N.M.; Winfree, R. & Kremen, C. (2007). Bee foraging ranges and their relationship to body size. Oecologia 153: 589-596. doi: 10.1007/s00442-007-0752-9

Gould, J.L. (1986). The Locale Map of Honey Bees: Do Insects Have Cognitive Maps? Science 232 (4752): 861-863. doi: 10.1126/science.232.4752.861

Goulson, D. & Stout, J. (2001). Homing ability of the bumblebee *Bombus terrestris* (Hymenoptera: Apidae). Apidologie 32: 105-111. doi: 10.1051/apido:2001115

Hilário, S.D., Imperatriz-Fonseca, V.L. & Kleinert, A.M.P. (2000). Flight activity and colony strength in the stingless *Melipona bicolor bicolor* (Apidae, Meliponinae), Rev. Bras. Biol. 60: 299-306. doi: 10.1590/S0034-71082000000200014

Kuhn-Neto, B., Contrera, F.A.L., Castro, M.S. & Nieh, J.C. (2009). Long distance foraging and recruitment by a stingless bee, *Melipona mandacaia*. Apidologie 40: 472-480. doi: http://dx.doi.org/10.1051/apido/2009007

Menzel, R. & Giurfa, M. (2006). Dimensions of Cognition in an Insect, the Honeybee. Behav. Cogn. Neurosci. Rev. 5: 24-40. doi: 10.1177/1534582306289522

Menzel, R., Greggers, U., Smith, A. Berger, S., Brandt, R., Brunke, S., Bundrock, G., Hulse, S., Plumpe, T., Schaupp, F., Schuttler, E., Stach, S., Stindt, J., Stollhoff, N. & Watzl, S. (2005). Honey bees navigate according to a map-like spatial memory. Proc. Natl. Acad. Sci. USA 102: 3040-3045. doi: 10.1073/pnas.0408550102

Menzel, R., Lehmann, K., Manz, G., Fuchs, J., Koblofsky, M. & Greggers, U. (2012). Vector integration and novel shortcutting in honeybee navigation. Apidologie,43: 229-243. doi: 10.1007/s13592-012-0127-z

Nogueira-Neto, P. (1997). Vida e criação de abelhas indígenas sem ferrão. São Paulo: Nogueirapis, 446 p.

Reynolds, A.M., Smith, A.D., Menzel, R., Greggers, U., Reynolds, D.R. & Riley, J.R. (2007). Displaced honey bees perform optimal scale-free search flights. Ecology 88(8): 1955-1961. doi: 10.1242/jeb.009563

Ribeiro, M.F., Rodrigues, F. & Fernandes, N. S. (2012). A mandaçaia (*Melipona mandacaia*) e seus hábitos de nidificação na região do polo Petrolina (PE)- Juazeiro (BA). Mens. Doce 115: 6-10.

Roubik, D.W. (1989). Ecology and natural history of tropical bees. New York: Cambridge University Press, 514p.

Roubik, D.W. & Aluja, M. (1983). Flight ranges of *Melipona* and *Trigona* in tropical forest. J. Kans. Entomol. Soc. 56: 217-222.

Sánchez, D., Bernhard Kraus, F., Hernández, M.J. & Vandame, R. (2007). Experience, but not distance, influences the recruitment precision in the stingless bee *Scaptotrigona mexicana*. Naturwissenschaften 94: 567-573. doi:10.1007/s00114-007-0229-z

Van Nieuwstadt, M.G.L. & Iraheta, C.E.R. (1996). Relation between size and foraging range in stingless bees (Apidae, Meliponinae). Apidologie 27: 219-228. doi: 10.1051/apido:19960404

Zanella, F.C.V. (2000). The bees of the Caatinga (Hymenoptera, Apoidea, Apiformes): a species list and comparative notes regarding their distribution. Apidologie 31: 579-592. doi: 10.1051/apido:2000148.

Zar, J. H. (2010). Biostatistical Analysis, 5th ed., New Jersey: Prentice Hall, 943 p.

26

Rescue of Stingless bee (Hymenoptera: Apidae: Meliponini) nests: an important form of mitigating impacts caused by deforestation

L Costa[1], RM Franco[1], LF Guimarães[1], A Vollet-Neto[1,2], FR Silva[1,2], GD Cordeiro[1,2]

1 - ARCADIS Logos, Divisão de Meio Ambiente, São Paulo, Brazil
2 - Universidade de São Paulo, São Paulo, Brazil

Keywords
Fauna rescue, Pollinators, Dam, Amazon Basin, Meliponiculture.

Corresponding author
Guaraci Duran Cordeiro
Universidade de São Paulo
Departamento de Biologia, FFCLRP
Av. Bandeirantes, 3900, Ribeirão Preto
14040-901, São Paulo, Brazil
E-Mail: guaradc@usp.br

Abstract
As stingless bees are important pollinators of wild and cultivated plants, their preservation is of vital importance to sustain the global ecosystem and to safeguard human food resources. The construction of large dams for the production of energy involves the removal of wide extents of riparian vegetation, where many species of bees, especially Meliponini, build their nests. The rescue of bee colonies is essential, not only in the conservation of pollinators, but also in the use of these colonies in meliponiculture and biological research. The aim of this work was to describe the procedures used in the rescue of stingless bee colonies at the time of deforestation, prior to initiating construction of a large dam in the Madeira River (Amazon Basin, Brazil). With simple equipment and widely known methods of meliponiculture 287 stingless bee nests were rescued, of which 15.7% were reallocated and 26.5% perished. The remaining 57.8% recovered well and were donated to local stingless beekeepers. The rescue of Meliponini nests during deforestation, besides resulting in the conservation of numerous colonies of various species, also contributes to the generation of environmental and social benefits.

Introduction

Pollinators, and especially bees, are responsible for the production of fruits and seed crops that are essential to guarantee human food resources, as well as in the maintenance of worldwide economy (Tepedino, 1979; Slaa et al., 2006; Klein et al., 2007). Unfortunately, pollinator diversity and abundance has decreased worldwide, due to deforestation, habitat fragmentation and the use of pesticides in agriculture (Garibaldi et al., 2011; González-Varo et al., 2013). In natural habitats, the lack of pollinators detrimentally affects wild plant reproducibility, thereby causing local extinction, and adversely affecting other dependent species (Allen-Wandell et al., 1998; Biesmeijer et al., 2006; Steffan-Dewenter et al., 2006; Ramírez et al., 2011).

Stingless bees (Meliponini) have been identified as important pollinators in both natural environments and crops (Imperatriz-Fonseca et al., 2006; Slaa et al., 2006). Hence their preservation is of the utmost importance for the sustenance of ecosystems and food resources worldwide. Many species of Meliponini are especially vulnerable to environmental degradation. In *Melipona* Illiger bees, for example, inbreeding can lead to decline or even the extinction of native bee populations through the presence of diploid males (Kerr, 1987; Carvalho, 2001; Alves et al., 2011; Francini et al., 2012).

Meliponini bees inhabit tropical and subtropical regions of the world, their diversity and abundance reaching the highest expression in the Amazon Basin (Michener, 2007; Camargo & Pedro, 2012). Thus, this region has become an essential area for research and conservation.

In Brazil, as in other tropical countries worldwide, economic growth has given rise to the construction of large infrastructure projects, especially power plants. The construction of large dams for the production of energy involves the removal of wide extents of riparian vegetation prior to the formation of reservoirs, thereby causing damage to the entire aquatic and riparian environment (Junk & Mello, 1990). Riparian environments are especially important ecosystems for wildlife, where many bee species, especially Meliponini, find the appropriate conditions for nesting (Roubik, 1989, 2006; Camargo, 1994).

In this context, the inclusion of stingless bee rescue programs in infrastructure projects, such as power plant construction, is an effective way of mitigating environmental damage caused by deforestation. Furthermore, the rescue of stingless bees nests, besides providing an unusual opportunity of data sampling for research in various aspects of Meliponini biology, can be a source of colonies appropriate for meliponiculture, an important activity in sustainable land use and environmental education (Kerr et al., 1996; Souza et al., 2012).

The aim of this study was to describe plausible procedures for rescuing stingless bee nests, and to discuss the possibilities for improvements in deforestation activities. Data was also presented on the various bee-species found during deforestation, prior to building the Santo Antonio hydroelectric power plant on the margin of the Madeira River, Rondonia State, Brazil.

Material and Methods

Study area and equipment used

The rescue of stingless bee nests was undertaken within the south-western Amazon Basin in the state of Rondonia, northern Brazil, on the left margin of the Madeira River (8° 47' 29.49" S and 63° 58' 58.5" W). This area was undergoing intense deforestation prior to construction of the Santo Antonio hydroelectric power plant. The area consists of 1,620ha of preserved riparian forest. The regional climate is equatorial Af (Köppen, 1948), with an average annual rainfall of 2,300mm, and an average annual temperature of 26°C. The intense dry season lasts from May to August.

Rescue was carried out between August 2010 and October 2011. The rescue teams consisted of a specialist in stingless beekeeping (meliponiculture) and two field assistants with experience of working in forest environments. One of the field assistants latter was a chainsaw operator. All the members used personal protective equipment for fauna rescue. The tools used were those traditionally employed in meliponiculture for handling colonies (Nogueira-Neto, 1997) (Table 1). Beekeeper suits were also used for protection against defensive species. A four-wheel-drive pickup was the means of access to areas of deforestation and the transportation of rescued colonies.

Search and rescue of nests

The search lasted eight hours a day in four distinct situations: coincident with deforestation (forefront) during cutting; lugging to storage; log stacking; and the final stage, loading for transportation out. In all of the situations, active search went ahead.

Binoculars were used in the search for nests, while simultaneously everybody was on the look out for bees among

Table 1. List of basic material for one stingless bees rescue team (three people).

Basic tool for the rescue of Meliponini bees	
Description	Quantity
Chainsaw	1
Bucksaw	1
Hatchet	1
Metal wedges	2
Crowbar	1
sledgehammer	1
Stone chisel	1
Machete	1
Knife	1
Painting spatula	1
3" Paintbrush	1
Syringe >20ml	Material of continuous use
Plastic bottles 500ml	Material of continuous use
Plastic bag >1l	Material of continuous use
Plastic tray	2
Insect aspirator	1
Samples containers	Material of continuous use
Alcohol 96%	Material of continuous use
5L bottle with water	2
Beekeeper suit	3
Wood stapler	1
Metal net of fine mesh	Material of continuous use
Strong scissors	1
Stingless bees' hives	Material of continuous use
Striped warning tape	Material of continuous use
Adhesive tape	Material of continuous use
Rope	>10 m
Binoculars	1
Camera	1
GPS	1

logs and branches. On coming across a nest, the surrounding area was taped off. GPS data of the located colonies, photos and samples of worker bees (n~10) in alcohol 96%, were collected for species identification.

After localizing and signalizing the colonies, in accordance with nesting biology, a decision was taken as to whether to transfer to a beehive or to leave wherever the colony had been originally found, in the original tree trunk or branch, in a termite nest, or an external nest.

Vertical modular hives, adapted from Venturieri (2004) and Carvalho-Zilse et al. (2005) were used. These came in two sizes: inner space 12x12x7 cm, for small colonies, and inner space 20 x 20 x 7cm, for larger ones. Cube-shaped boxes, inner space 40 x 40 x 40cm, were used for species in which nest architecture was not adjustable to vertical boxes.

If the decision was to do the transference to a beehive, the trunk or branch was opened with a chain saw, according to instructions by Nogueira-Neto (1997) and Coletto-Silva (2005), but with adaptations. The transference of colonies associated with termitaria and external nests was with stone chisels, hammers, matches and knives.

Pollen and honey pots were not transferred to beehives, so as to avoid parasites. Only the light colored, mature brood combs (pupal stage and the last larval stage) and cerumen were transferred. Food storage pots were placed in plastic bags or boxes and stored in a refrigerator. Pollen and honey were used for feeding the colonies later on. Brood combs containing larval food were discarded.

After an interval of approximately 24 hours, usually on the following day, the colonies that had been transferred to beehives were sealed in, by closing the entrance, and then transported to our field base, which was 3 to 32km from the rescue areas, for a period of observation and care. This was done at the end of the day, in order to capture the maximum number of forager workers in transit. During the observation and care period (c. one month), the colonies were nourished with sugar syrup (or their own honey) and their own pollen. Internal and external feeders were used for sugar syrup (Kerr et al., 1996; Nogueira-Neto, 1997), whereas only internal ones were for pollen. Cerumen, i.e., a mixture of bee wax and resin used for building nest structures, was collected, washed and returned to the nests for reutilization by the workers. Vinegar traps were employed against Phoridae fly infestation (see Nogueira-Neto (1997) for details of parasite control methods).

In the case of colonies to be left within their original log, the nest size was first estimated according to the species in question, whereupon the log was trimmed at both ends with a chainsaw. Afterwards, these were sealed by way of a thin metal mesh, to so avoid the entrance of Phoridae flies and the exit of bees. In cases of species associated with termites or those with external nests, and depending on the possibility, the entire structure was detached from the tree.

These colonies were transported to our field base or re-allocated in the nearby forest at a level higher than the final water level of the reservoir. The decision for the right procedure depended on the biology of the species found, or in another words, nesting habits, behavior, chances of survival in a bee hive, and the size of the trunk or external nest.

The colonies transported to the field base after the recuperation period, were donated to those local, experienced stingless beekeepers who presented the necessary conditions for giving special care. Bee samples were later identified by Dr. Silvia R. M. Pedro (FFCLRP, Universidade de São Paulo, Brazil).

Results and Discussion

Throughout the deforestation process, 416 colonies of stingless bees were found. From these, bee-specimens were collected from 118 nests (28% of the total found), comprising 36 species of 15 of the 33 known Neotropical genera of Meliponini (Table 2). It shows that the number of species in the area is certainly higher and the rescue of Meliponini is essential during deforestation activities, not only in the Amazon region, but also in other tropical forests.

Being present at the forefront of deforestation was an important step in the rescue of stingless bees, since the impact of large trees falling to the ground often resulted in cracks in the hollow branches or trunks where nests were located. Most of the Meliponini nests found were in branches (up to 20m high). Since brood combs, as well as honey and pollen pots, were seriously damaged after the fall, colonies had to be quickly transferred to beehives. As external nests (*Trigona* Jurine) and those in termitaria (mainly *Partamona* Schwarz) were also often severely damaged with the fall, there was no other choice, but to attempt transferring them to a beehive, as well.

A special problem in the forefront of deforestation, especially in the case of *Melipona* bees, was the destruction of colonies by deforestation workers to collect honey. The workers knew from tradition that *Melipona* bees are harmless and produce good honey. In this case, the presence and orientation given by the rescue team were fundamental in preserving colonies of a harmless species. There was also evidence of colonies of *Melipona* in cracked branches, some days after felling, having fallen prey to *Eira barbara* Linnaeus (Mustelidae) and *Potos flavus* Schreber (Procyonidae), mammal species common in the area, but of minor problem. The attacks of these animals were also recorded by Nogueira-Neto (1997).

At the second stage in the deforestation process, during lugging to storage, colonies, which were not evident at the time of cutting, were discovered, whereupon machine operators usually gave aid in signalizing. However, upon removal from one place to another, a further mishap appeared. Several colonies presented problems with strong insolation, honey fermentation and infestation by Phoridae and *Hermetia* Latreille. In this case, transfer to a beehive was considered to be the best, immediate, option.

During the process of stacking, the rescue team was able to avoid loosing species with cryptic behavior, such as *Plebeia* Schwarz. In the final stage, when the logs were being loaded for transportation, colonies, such as of *Tetragonisca* Moure, *Nannotrigona* Cockerell, *Scaptotrigona* Moure, *Frieseomelitta* Ihering, and *Trigona* Jurine, were still found.

Considering all the stages in the deforestation process, 416 stingless bee nests were found. However, 31% of these had been destroyed (n = 129) either during felling, in attempts to abstract honey, or by parasites. However, the remainder (69%, n = 287) presented mature brood combs and queens and could consequently be recuperated.

As regards species of the genera *Melipona, Scaptotrigona, Tetragona* Lepeletier & Serville, *Frieseomelitta, Nannotrigona*, and *Tetragonisca*, there were no relevant problems for most following rescue with the methods used for transference to beehives. The importance of the non-transference of brood

Table 2. Species identified during the Meliponini rescue at Santo Antônio Dam, Madeira River, Brazil.

Species	N. nests	Substratum
Cephalotrigona femorata (Smith, 1854)	9	trunk/branch
Frieseomelitta silvestrii (Friese, 1902)	1	trunk/branch
Frieseomelitta trichocerata Moure, 1990	7	trunk/branch
Melipona (*Michmelia*) *brachychaeta* Moure, 1950	2	trunk/branch
Melipona (*Michmelia*) *seminigra abunensis* Cockerell, 1912	5	trunk/branch
Melipona (*Michmelia*) sp. 1 (gr. *rufiventris*)	1	trunk/branch
Melipona (*Michmelia*) sp. 2 (gr. *melanoventer*)	1	trunk/branch
Nannotrigona melanocera (Schwarz, 1938)	3	trunk/branch
Oxytrigona cf. *flaveola* (Friese, 1900)	3	trunk/branch
Oxytrigona obscura (Friese, 1900)	1	trunk/branch
Partamona ailyae Camargo, 1980	7	termitaria/trunk
Partamona batesi Pedro & Camargo, 2003	5	termitaria
Partamona sp.	1	termitaria
Partamona testacea (Klug, 1807)	4	termitaria
Partamona vicina Camargo, 1980	1	termitaria
Plebeia alvarengai Moure, 1994	1	trunk/branch
Ptilotrigona lurida (Smith, 1854)	8	trunk/branch
Scaptotrigona polysticta Moure, 1950	1	trunk/branch
Scaptotrigona sp. 1	1	trunk/branch
Scaptotrigona sp. 2	1	trunk/branch
Scaptotrigona sp. 3	3	trunk/branch
Scaptotrigona sp. 4 (gr. *bipunctata*)	1	trunk/branch
Tetragona clavipes (Fabricius, 1804)	9	trunk/branch
Tetragona essequiboensis (Schwarz, 1940)	1	trunk/branch
Tetragona goettei (Friese, 1900)	4	trunk/branch
Tetragona truncata Moure, 1971	1	trunk/branch
Tetragonisca angustula (Latreille, 1811)	14	trunk/branch
Trichotrigona sp. n	1	*
Trigona branneri Cockerell, 1912	2	external
Trigona chanchamayoensis Schwarz, 1948	1	external
Trigona crassipes (Fabricius, 1793)	5	trunk/branch
Trigona dallatorreana Friese, 1900	1	external
Trigona guianae Cockerell, 1910	5	trunk/branch
Trigona pallens (Fabricius, 1798)	1	trunk/branch
Trigona truculenta Almeida, 1984	1	trunk
Trigona williana Friese, 1900	5	trunk/branch

combs containing larval food and storage pots to the beehives becomes evident, since the problem of parasite infestation, especially by Phoridae flies, was thus avoided. Most species recovered well, when transferred to beehives and brought to our field base for the period of observation and care. In the case of the presence of Phoridae or *Hermetia* flies, the problem was solved by using vinegar traps. This technique is widely known in meliponiculture (Nogueira-Neto, 1997). As 166 colonies, 57.8%

of all those rescued, were considered to be in good condition, these were donated to local stingless beekeepers.

In the case of some species, transfer to beehives compensated little, due to hardy defense and sensitivity to rescue, leading to perishing after disturbance. This occurred in the case of *Ptilotrigona lurida* Smith, *Trigona truculenta* Almeida and other species of *Trigona*, mainly those with external nests. In the case of these species, *Oxytrigona* spp., due to their defensiveness, and colonies attached to or inside very large logs, they were considered preferable to reallocate the entire colony, as it was, to the nearby forest with the aid of deforestation machinery, up to the very margin of the future reservoir. We reallocated 15.7% of the rescued colonies (n = 45), including mainly *P. lurida*, *Trigona* spp., *Partamona* spp. and *Oxytrigona* spp.

However, attention is called to the difficulty in finding a safe place to leave the colonies, since besides the possibility of exceptional flooding, there is the lack of adequate and appropriate programs to monitor colony survival. Furthermore, as already pointed out, most colonies were severely damaged during felling, with frequent parasite infestation. As regards colonies brought to the field base for care and observation, 26.5% (n = 76), among species with external nests, colonies left inside logs and termitaria, perished, most after one to two months due to Phoridae fly infestation. These factors led to suppose that the probability of reallocated colony survival would be low.

Thus, apart from the destiny chosen for rescued colonies (re-allocation or donation to stingless beekeepers or research centers), a monitoring program is called for as a way of evaluating colony survival and guiding future rescue action. As part of this monitoring process, and in the case of donations to beekeepers, it is also important to promote training courses for updating knowledge (Fig 1). Furthermore, it is necessary to have a larger number of available rescue teams, at least one per forefront of deforestation, for improving results. Stingless-bee rescue, concomitant with deforestation, is imperative as a means of aiding in the conservation of pollinators, for providing singular opportunities for data sampling on stingless bee biology, and in support of meliponiculture. By using the simple methods employed in meliponiculture, the rescue of Meliponini can mitigate the various environmental impacts caused by deforestation, besides generating social and cultural benefits.

Acknowledgments

We are very grateful to Dr. Silvia R. M. Pedro for species identification, Felipe A.s L. Contrera and Giorgio C. Venturieri for suggestions on the manuscript, Arcadis Logos (Sandra Favorito, Laerte Viola, Beatris Beça, and Alex Aurani) for all the support during rescue, and to all field colleagues.

References

Allen-Wardell, G., Bernhardt, P., Bitner, R., Burquez, A., Buchmann, S., Cane, J.H., Cox, P. A., Dalton, V., Feinsinger, P., Inouye, D., Ingram, M., Jones, C. E., Kennedy, K., Kevan, P., Koopowitz, H., Medellin, R., Medellin-Morales, S., Nabhan, G.P., Pavlik, B., Tepedino, V., Torchio, P. & Walker, S. (1998). The Potential consequences of pollinator declines on the conservation of biodiversity and stability of food crop yelds. Conserv. Biol. 12: 8-17.

Alves, D.A., Imperatriz-Fonseca, V.L., Francoy, T.M., Santos-Filho, P.S., Billen, J. & Wenseleers, T. (2011). Successful maintenance of a stingless bee population despite a severe genetic bottleneck. Conserv. Genet. 12: 647-658. doi: 10.1007/s10592-010-0171-z

Biesmeijer, J.C., Roberts, S.P.M., Reemer, M., Ohlemuller, R., Edwards, M., Peeters, T., Schaffers, A.P., Potts, S.G., Kleukers, R., Thomas, C.D., Settele, J. & Kunin, W.E. (2006). Parallel declines in pollinators and insect-pollinated plants in Britain and the Netherlands. Science 313: 351-354.

Camargo, J.M.F. (1994). Biogeografia de Meliponini (Hymenoptera, Apidae, Apinae): a fauna Amazônica. *In* Anais do Encontro Sobre Abelhas, 46-59.

Camargo, J.M.F. & Pedro, S.R.M. (2012). Meliponini Lepeletier, 1836. *In* J.S. Moure, D. Urban & G.A.R. Melo (Eds.), Catalogue of Bees (Hymenoptera, Apoidea) in the Neotropical Region - online version. http://www.moure.cria.org.br/catalogue. (accessed date: 29 September, 2014).

Carvalho-Zilse, G.A. (2001). The number of sex alleles (CSD) in a bee population and its practical importance (Hymenoptera: Apidae). J. Hymenop. Res. 10: 10-15.

Carvalho-Zilse, G.A., Silva, C.G.N., Zilse, N., Vilas-Boas, H.C., Silva, A.C., Laray, J.P., Freire, D.C.B. & Kerr, W.E. (2005). Criação de abelhas sem ferrão. Brasília: Instituto Brasileiro de Meio Ambiente e Recursos Naturais Renováveis, Projeto Manejo dos Recursos Natuais da Várzea.

Coletto-Silva, A. (2005). Captura de enxames de abelhas sem ferrão (Hymenoptera, Apidae, Meliponinae) sem destruição de árvores. Acta Amaz. 35: 383-388.

Francini, I.B., Nunes-Silva, C.G. & Carvalho-Zilse, G.A. (2012). Diploid male production of two Amazonian Melipona Bees

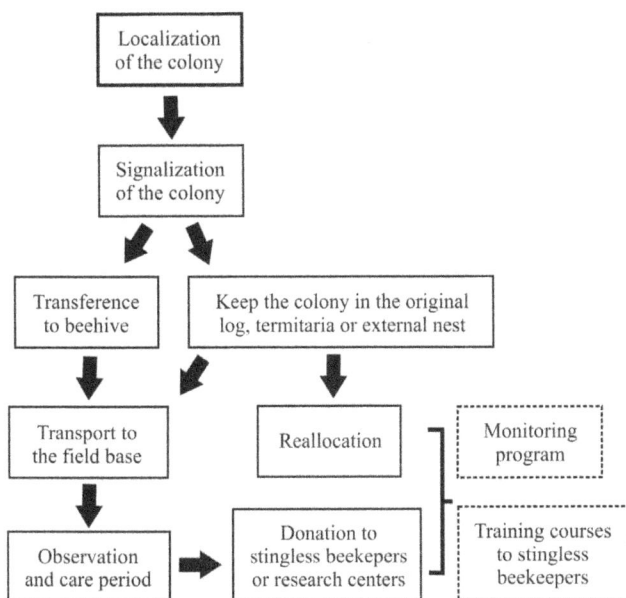

Fig 1. Diagram of steps adopted for the rescue of stingless bees during the deforestation to build the Santo Antonio hydroelectric power plant on the margin of the Madeira River and suggested post rescue actions (squares surrounded by dotted line).

(Hymenoptera: Apidae). Psyche 2012: 1-7. Available at: http://www.hindawi.com/journals/psyche/2012/484618/.

Garibaldi, L.A., Aizen, M.A., Klein, A.M., Cunningham, S.A. & Harder, L.D. (2011). Global growth and stability of agricultural yield decrease with pollinator dependence. Proc. Natl. Acad. Sci. 108: 5909-5914.

González-Varo, J.P., Biesmeijer, J.C., Bommarco, R., Potts, S.G. & Schweiger, O. (2013). Combined effects of global change pressures on animal-mediated pollination. Trends Ecol. Evol. 28: 524-530.

Imperatriz-Fonseca, V.L., Saraiva, A.M. & De Jong, D. (2006). Bees as pollinators in Brazil: assessing the status and suggesting best practices. Ribeirão Preto: Holos Editora, 112 pp.

Junk, W.J. & Mello, J.A.S. (1990). Impactos ecológicos das represas hidrelétricas na bacia amazônica brasileira. Rev. Estud. Avançados 4: 126-143.

Kerr, W.E. (1987). Sex determination in bees XXI. Number of xo-heteroalleles in a natural population of *Melipona compressipes fasciculata* Apidae. Insectes Soc. 34: 274-279

Klein, A.M., Vaissière, B.E., Cane, J.H., Steffan-Dewenter, I., Cunningham, S.A., Kremen, C. & Tscharntke, T. (2007). Importance of pollinators in changing landscapes for world crops. Proc. R. Soc. B. 274: 303-313.

Köppen, W. (1948). Climatologia: con un estudio de los climas de la tierra. México: Fondo de Cultura Econômica, 479 p.

Michener, C.D. (2007). The Bees of the World. 2nd. Ed. Baltimore: Johns Hopkins University Press. 913 p.

Nogueira-Neto, P. (1997). Vida e Criação de abelhas indígenas sem ferrão. São Paulo: Editora Nogueirapis, 445 p.

Ramírez, S.R., Eltz, T., Fujiwara, M.K., Gerlach, G., Goldman-Huertas, B., Tsutsui, N.D. & Pierce, N.E. (2011). Asynchronous diversification in a specialized plant-pollinator mutualism. Science 333: 1742-1746.

Roubik, D.W. (1989) Ecology and natural history of tropical bees. New York: Cambridge University Press, 526 p.

Roubik, D.W. (2006) Stingless bee nesting biology. Apidologie 37: 124-143. doi: 10.1051/apido:2006026

Slaa, E.J., Sánchez-Chaves, L.A., Malagodi-Braga, K.S. & Hofstede, F.E. (2006) Stingless bees in applied pollination: practice and perspectives. Apidologie 37: 293-315.

Souza, B.A., Lopes, M.T. & Pereira, F.M. (2012). Cultural aspects of meliponiculture. In P. Vit & D.W. Roubik (Eds.), Stingless bees process honey and pollen in cerumen pots, (pp. 1-6). Mérida: SABER-ULA, Universidad de Los Andes.

Steffan-Dewenter, I., Klein, A.M., Gaebele, V., Alfert, T. & Tscharntke, T. (2006). Bee diversity and plant-pollinator interactions in fragmented landscapes. In N.M. Wasser & J. Ollerton (Eds.), Plant-pollinator interaction from specialization to generalization (pp. 387-407). Chicago: The University of Chicago Press.

Tepedino, V.J. (1979) The importance of Bees and other insect pollinators in maintaining floral species composition. Great Bas. Nat. 3: 139-150.

Venturieri, G. (2004) Meliponicultura I: Caixa Racional de Criação. Comunicado Técnico - Embrapa 123: 1-3.

27

Population Structure of *Melipona subnitida* Ducke (Hymenoptera: Apidae: Meliponini) at the Southern Limit of its Distribution Based on Geometric Morphometrics of Forewings

CBS Lima[1], LA Nunes[2], MF Ribeiro[3] CAL de Carvalho[1]

1 - Universidade Federal do Recôncavo da Bahia, Cruz das Almas, BA, Brazil

2 - Universidade Estadual do Sudoeste da Bahia, Jequié, BA, Brazil

3 - Empresa Brasileira de Pesquisa Agropecuária, Petrolina, PE, Brazil

Keywords

Meliponiculture, jandaira bee, MANOVA, Procrustes superposition, shape.

Corresponding author

Cândida Beatriz da Silva Lima
Universidade Federal do Recôncavo da Bahia, Rua Ruy Barbosa, 710, Centro, Cruz das Almas/BA, Brazil
E-Mail: candidab.lima26@gmail.com

Abstract

Bees provide fundamental services to humanity, and many researchers have been concerned about the rapid loss of genetic diversity that these organisms have been suffering. The stingless bee *Melipona subnitida* Ducke is endemic to Northeastern Brazil and has high potential for the production of honey and wax; it is also an important pollinator in the Caatinga biome. Populations of *M. subnitida* have increasingly declined due to predatory extractivism and habitat destruction. However, knowledge about its population structure could give insights on strategies for monitoring and conservation of this species. Here we collected workers from nine sites located at the southern limit of the species distribution and employed geometric morphometric techniques on their forewings in search of covariance between sampling site and wing morphology. A very significant correlation between both variables was observed, indicating that the divergence among the sampled populations of *M. subnitida* was due to geographical distance among the sampling sites and, hence, suggesting the formation of different groups of populations along the studied geographical zone, each one with specific characteristics. Since *M. subnitida* habitat has been increasingly fragmented thus hindering the genetic flow among populations, our findings will contribute to the formulation of management and conservation plans for this species in order to preserve its genetic diversity and, therefore, to collaborate to the generation of income for beekeepers in meliponiculture programs.

Introduction

Stingless bees form an important group of pollinators in the Neotropics, playing a significant role on the maintenance of ecosystems and agricultural production (Heard, 1999; Slaa et al., 2006). The stingless bee *Melipona subnitida* Ducke, locally known as "jandaira", is endemic to the Caatinga (Zanella, 2000), one of the biomes found in the Northeastern Brazil, which is characterized by a semiarid climate and xerophylous vegetation (Andrade-Lima, 1981). *M. subnitida* is found mainly in the states of Bahia, Ceará, Rio Grande do Norte and Paraiba (Martins, 2002; Camargo & Pedro, 2007; Webbee, 2014), playing an important role on pollination of cultivated plants, as well as on honey and wax production (Cruz et al., 2005).

The predatory exploitation of nests of native bees by humans has become a major concern for scientists (Brown & Paxton, 2009), but deforestation, agriculture intensification and introduction/spread of exotic competing bee species are considered the main threats to most indigenous species (Freitas et al., 2009). On the other hand, knowledge about the population structure of these organisms has been considered critical for management and conservation plans of endangered species (Nunes et al., 2008; Bonatti et al., 2014).

The employment of geometric morphometrics on identification and assessment of population diversity/structure has been proven efficient (Breuker et al., 2006; Wappler et al., 2012). Geometric morphometrics allows detecting relationships between animal populations and their geographic locations through a rigorous analysis of variations in the shape of a model structure. The most common application of the geometric morphometrics is the identification of landmark configurations in several morphological characters (Klingenberg, 2002). Cartesian coordinates provide the relative positions of each point, and, therefore, they make possible the reconstruction of shape and the identification of shape variations (Rohlf & Marcus, 1993; Rohlf, 1998). This method is very sensitive to subtle variations and, hence being suitable for detection of groups and subgroups (Francoy et al., 2011).

Studies on population structure and geographic variation of bees have been carried out based on morphometric data comparing races or populations (Ferreira et al., 2011; Lima Junior et al., 2012; Nunes et al., 2013). Those analyses are usually based on wing characters due to their high heritability and because they are strongly affected by the environment (Diniz-Filho & Bini, 1994). A possible disadvantage of this technique is that the biological structures that will be studied need to be completely undamaged. However, such disadvantage is rarely a critical factor (Lyra et al., 2010).

Considering the ecological and economical importance of stingless bees for the Neotropics and the necessity of studies on the population structure of endemic species of the Caatinga biome, the aim of this study was to determine the population diversity of M. subnitida from sites located at the southern limit of its distribution based on geometric morphometrics of forewings.

Material and methods

We collected 630 workers from 63 colonies of M. subnitida in nine localities of Northeastern Brazil, from August to December 2012 and in July 2013 (Table 1).

We removed the bees' forewings, placed them between microscope slides, and photographed them with a digital camera coupled to a stereomicroscope for the analyses of venation. We transformed the photographs in the software tpsUtil 1.40 (Rohlf, 2008a). We inserted ten landmarks (Fig. 1) at the vein

intersections of each forewing using tpsDig version 2.17 (Rohlf, 2008b). We used the data obtained as variables for multivariate analyses, such as principal component analysis (PCA), canonical correlation analysis, Mahalanobis distance (D^2), and Procrustes distance in the software MorphoJ 1.03 (Klingenberg, 2011). We also made cluster analyses based on UPGMA (Unweighted Pair Group Method with Arithmetic Mean) and a Mantel test based on geographic distance, and size and shape of bee wings in the software Past 2.17.

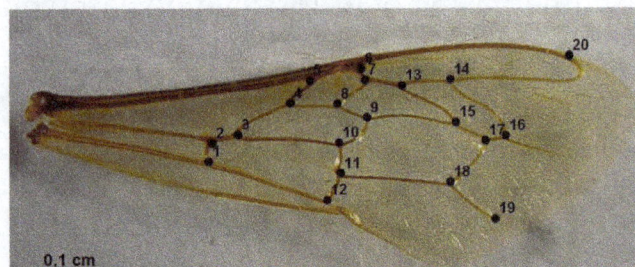

Fig 1 Right forewing of *Melipona subnitida* showing the 10 landmarks located at wing vein intersections, which were used in the morphometric analysis.

Results

In the PCA applied to the populations of *M. subnitida*, the first four principal components explained 62.96% of the total variation among individuals of different communities: PC1 = 20.61%, PC2 = 16.26%, PC3 = 14.20%, and PC4 = 11.88%. We represented the group distribution in a bidimensional space, using the scores of the first two components, to test for dispersal among groups.

Based on the Mahalanobis and Procrustes distances, the largest difference was observed between the populations from Exu and Água Branca (305 km apart) whereas the highest morphological proximity occurred between the populations from Água Branca and Mata Grande (30 km apart) (Table 2). Based on a dendrogram (UPGMA; Fig. 2), we observed morphological differences between the populations from Exu, Passira and Taquaritinga do Norte, which were grouped in a

Table 1. Origin and geographic location of the *Melipona subnitida* samples at the Southern Limit of its Distribution natural in Brazil. N- number of samples.

Sites/ State	N	Latitude (S)	Longitude (W)	Altitude (m)	Climate / Relief
Água Branca/Alagoas	9	9°10'24.7"	37°51'41.9"	380	Semiarid/ massive
Cumaru/Pernambuco	6	8°1'58.55"	35°45'3.11"	348	Semiarid/ low hills
Exu/Pernambuco	3	7°20'22.66"	39°54'58.54"	887	Semiarid/low mountains
Joá/Paulo Afonso/ Bahia	6	09°31'08.8"	38°25'36.7"	243	Semiarid/ upland
Mata Grande/Alagoas	12	09°11'09.3"	37°50'09.8"	424	Semiarid/ massive
Passira/Pernambuco	5	7°55'37.85"	35°30'14.01"	160	Semiarid/ high massive
Riacho das Almas/Pernambuco	6	8°3'40.79"	35°49'9.62"	413	Semiarid/ low hills
São José/ Paulo Afonso /Bahia	8	09°39'04.8"	38°22'43.2"	243	Semiarid/upland
Taquaritinga do Norte/Pernambuco	8	7°56'14.15"	36°7'05.7"	785	Tropical/massive

more isolated branch, standing out from populations of other regions, with a correlation index (P< 0.0001), which shows that the divergence between population is highly significant.

The correlation between shape, size, altitude, and geographic distance among 63 *M. subnitida* colonies compared with a Mantel test (Table 3) indicated a positive correlation between wing shape and geographic distance among all populations (P < 0.01). The other correlations had high P-values, which points to no relationship among variables.

There was variation in the shape of the forewings among *M. subnitida* populations, which is reflected in the formation of groups.

Table 2. Mahalanobis distances (lower half of the matrix) and Procrustes distance (upper half of the matrix) among populations calculated from the canonical correlation analysis. Sites: AB: Água Branca; CU: Cumaru; EX: Exu; JO: Joá; MT: Mata Grande; PA: Passira; RA: Riacho das Almas; SJ: São José; TN: Taquaritinga do Norte.

	AB	CU	EX	JO	MT	PA	RA	SJ	TN
AB		0.017	0.020	0.015	0.008	0.019	0.017	0.015	0.014
CU	2.93		0.015	0.012	0.014	0.015	0.010	0.013	0.016
EX	3.68	2.54		0.019	0.017	0.014	0.017	0.014	0.015
JO	2.96	1.78	3.08		0.012	0.019	0.013	0.011	0.016
MT	1.33	2.32	3.27	2.37		0.017	0.018	0.010	0.017
PA	3.37	2.17	2.84	2.52	2.77		0.017	0.012	0.017
RA	3.21	1.68	2.85	2.11	3.10	2.66		0.015	0.013
SJ	2.87	1.94	2.63	1.83	2.13	1.71	2.31		0.016
TN	2.47	2.65	2.48	2.87	2.72	3.10	2.41	2.70	

Discussion

The geometrical morphometrics of forewings was very efficient in detecting variability among the populations and, hence, in unveiling the population structure of *M. subnitida*. The results of the morphometric analysis pointed to morphological variability in wing shape among populations along their geographic distribution. This variation was associated with geographic distance among the sampling sites and is also probably related to the environmental variability among sampling localities.

The populations from Exu and Água Branca showed the highest index of morphometric divergence, with a dissimilarity

Table 3. Mantel test used to compare matrices of shape, size, altitude, and geographic distance, based on wing measurements of Melipona subnitida with 5,000 permutations. NS not significant; * significant

COMPARED MATRICES	R^2	P
Shape x Altitude	0.009	0.473^{NS}
Shape x Geographic distance	0.506	0.008^{*}
Size x Shape	0.282	0.972^{NS}
Size x Altitude	0.072	0.537^{NS}
Size x Geographic distance	0.045	0.512^{NS}

value of 3.68, which may be explained by an interaction between geographic distance and climatic/relief differences between regions. Exu is characterized by a semi-arid climate: warm in the summer and cold in the winter; it is located between massifs and low mountain ranges (300 – 800 m height); on the other hand, Água Branca is characterized by a tropical semi-arid climate with summer rains, and it is located between massifs and high mountain ranges (650 – 1,000 m height) (CPRM, 2014). The less divergent populations were those of Água Branca and Mata Grande, both located in the state of Alagoas (dissimilarity value = 1.33). These areas are geographically close to each other (3 km) and have similar climate, vegetation, and relief.

In the cluster analysis, the formation of groups reflected morphological differences between the populations and, in general, we found a correlation between forewing morphology and geographical distance. However, the populations from Exu and Passira, which are more than 300 km apart, curiously were found to form a distinct group, indicating morphological similarities between the two far apart populations. These similarities might be a result of the similar environmental conditions found in Exu and Passira, which, despite the distance, present suchlike vegetation and relief (CPMR, 2014).

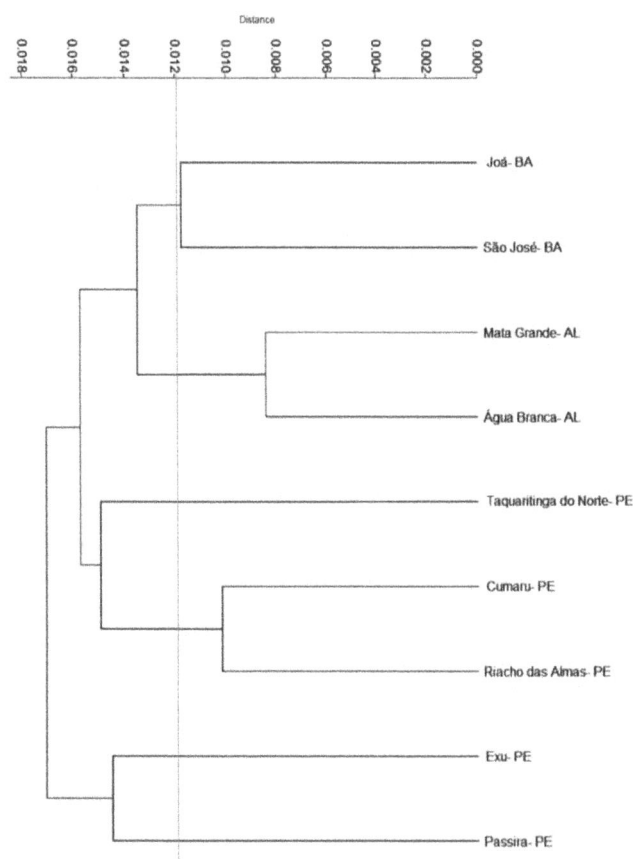

Fig 2 Dendrogram generated by UPGMA with the average morphometric distances between *Melipona subnitida* colonies from different localities.

We have previously found similar results when studying population diversity in *Melipona quadrifasciata anthidioides* Lepeletier from different regions of Bahia State (Nunes et al., 2008), observing that the populations that were located far from each other showed morphological divergence. Mendes et al. (2007) observed morphological differences between *Nannotrigona testaceicornis* (Lepeletier) groups from urban and rural areas in the municipality of Uberlândia, State of Minas Gerais, in the Southeast of Brazil. These authors associated these differences with selective pressures in the urban area and with a higher probability of genetic variability in the rural area.

Once the temperature is a factor that can influence the insect populations and distribution (Easterling et al., 2000; Damos & Savopoulou-Soultani, 2012), and considering the expected impacts of climate warming on stingless bees (Saraiva et al., 2012; Giannini et al., 2012), the characterization of population structures of *M. subnitida* will help to preserve its genetic diversity, increasing the likelihood of conserving lineages more adapted to the expected climatic changes, with potential impacts on pollination.

Melipona subnitida is narrowly distributed and it is currently found in fragmented and geographically-isolated environments (Zanella, 2000; Bonatti et al., 2014). The strong relationship between geographical distance and morphological variability found in our study indicates a lack of genetic flow among populations that are located far from each other or even among those that would be able to crossbreed if in favorable environmental conditions, corroborating the importance of conserving nesting sites and food sources. This finding will contribute to the formulation of management and conservation plans for *M. subnitida* in order to preserve its population diversity and, therefore, to collaborate to the generation of income in meliponiculture programs. In a broader sense, our results highlight the importance of characterizing the population structure of the stingless bees, because they help to identify differentiated populations, which need more human intervention as a way to decrease the rates of inbreeding.

Conclusion

There is a correlation between sampling site and wing shape in *M. subnitida* populations inhabiting the limit of the species distribution, indicating that the divergence among the sampled populations is due to geographical distance among the sampling sites and, hence, suggesting the formation of characterized groups of populations along the studied zone.

Acknowledgements

The authors thank Cherre S. B. Da Silva for the helpful suggestions for the manuscript, the beekeepers for providing samples of bees, and the anonymous reviewers for their criticisms and suggestions to the manuscript, the Empresa

Brasileira de Pesquisa Agropecuária (EMBRAPA) (Cod. SEG 02.11.01.029.00.00) for the financial support, Coordenação de Aperfeiçoamento de Pessoal de Nível Superior (CAPES) for the scholarship (C. B. da S. Lima) and the Conselho Nacional de Desenvolvimento Científico e Tecnológico (CNPq) (Proc. 305228/2013-7),.

References

Andrade-Lima, D. (1981). The caatingas dominium. Rev. Bras. Bot., 4: 149-153.

Bonatti, V., Simões, Z.L.P., Franco, F.F. & Francoy, T.M. (2014). Evidence of at least two evolutionary lineages in *Melipona subnitida* (Apidae, Meliponini) suggested by mtDNA variability and geometric morphometrics of forewings. Naturwissenschaften, 101: 17-24. doi:10.1007/s00114-013-1123-5.

Breuker, C.J., Patterson, J.S. & Klingenberg, C.P. (2006). A single basis for developmental buffering of *Drosophila* wing shape. PloS ONE 1(1) e7. doi:10.1371/journal.pone.0000007.

Brown, M.J.F. & Paxton, R.J. (2009). The conservation of bees: a global perspective. Apidologie, 40: 410–416. doi:10.1051/apido/2009019.

Camargo, J.M.F. & Pedro, S.R.M. (2007). Meliponini Lepeletier, 1836. In Moure, J.S., Urban, D. & Melo, G.A.R. (Orgs). Catalogue of bees (Hymenoptera, Apoidea) in the Neotropical region - online version. Available at http://www.moure.cria.org.br/catalogue. (accessed date: March 12, 2014).

CPRM, Site: http://www.cprm.gov.br/ (accessed date: February 25, 2014).

Cruz, D. de O., Freitas, B.M, Silva, L.A., Silva, E.M.S., & Bomfim, G.A. (2005). Pollination efficiency of the stingless bee *Melipona subnitida* on greenhouse sweet pepper. Pesq. Agropec. Bras., 40: 1197-1201.

Damos, P. & Savopoulou-Soultani, M. (2012). Temperature-driven models for insect development and vital thermal requirements. Psyche: J. Entomol., 2012: 1-13. doi: 10.1155/2012/123405.

Diniz-Filho, J.A.F. & Bini, L.M. (1994). Space-free correlation between morphometric and climatic data: a multivariate analysis of africanized honey bees (*Apis mellifera* L.) in Brazil. Global Ecol. Biogeogr., 4: 195-202.

Easterling, D.R., Meehl, G.A., Parmesan, C., Changnon, SA., Karl, T.R. & Mearns, L. O. (2000). Climate extremes: Observations, modeling, and impacts. Science, 289: 2068-2074. doi: 10.1126/science.289.5487.2068.

Ferreira, V.S., Aguiar, C.M.L., Costa, M.A. & Silva, J.G. (2011). Morphometric analysis of populations of *Centris aenea* Lepeletier (Hymenoptera: Apidae) from Northeastern Brazil. Neotrop. Entomol., 40: 97-102.

Freitas, B.M., Imperatriz-Fonseca, V.L., Medina, L.M., Kleinert, A. de M. P., Galetto, L., Nates-Parra, G. & Quezada-Euán, J.J.G. (2009). Diversity, threats and conservation of native bees in the Neotropics. Apidologie, 40: 332-346. doi: 10.1051/apido/2009012.

Francoy, T.M., Grassi, M.L., Imperatriz-Fonseca, V.L., May-Itza, W.J. & Quezada-Euan, J.J. (2011). Geometric mophometrics of the wing as a tool for assgning genetic lineages and geographic origin to Melipona beecheii (Hymenoptera: Meliponini). Apidologie, 42: 499-507. doi: 10.1007/s13592-011-0013-0.

Giannini, T.C., Acosta, A.L., Garófalo, C.A., Saraiva, A.M., Alves-dos-Santos, I., & Imperatriz-Fonseca. (2012). Pollination services at risk: Bee habitats will decrease owing to climate change in Brazil. Ecol. Model., 244:127–131

Heard, T.A. (1999). The role of stingless bees in crop plantation. Annu. Rev. Entomol., 44: 183-206.

Klingenberg, C.P. (2002). Morphometrics and the role of the phenotype in studies of the evolution of developmental mechanisms. Gene, 287: 3-10. PII:S0378-1119(01)00867-8.

Klingenberg, C,P. (2011). MorphoJ: In integrated software package for geometric morphometrics. Mol. Ecol. Resour., 11: 353–357.

Lima Junior, C.A., Carvalho, C.A.L., Nunes, L.A. & Francoy, T.M. (2012). Population divergence of Melipona scutellaris Latreille (Hymenoptera: Meliponina) in two Restricted Areas in Bahia, Brazil. Sociobiology, 59: 107-122.

Lyra, M.L.; Hatadani, L.M., De Azeredo-Espin, A.M.L & Klaczko L.B. (2010). Wing morphometry as a tool for correct identification of primary and secondary new world screwworm fly. Bul. Entomol. Res., 100: 19-26. doi: 10.1017/S0007485309006762.

Martins, C.F. (2002). Diversity of the bee fauna of the brazilian Caatinga. In: Kevan P & Imperatriz Fonseca VL (Eds), Pollinating bees: The conservation link between agriculture and nature (pp. 131-134). Brasília: Ministry of Environment.

Mendes, M.F.M., Francoy, T.M., Nunes-Silva, P., Menezes, C. & Imperatriz-Fonseca, V. L. (2007). Intra-populational variability of Nannotrigona testaceicornis Lepeletier, 1836 (Hymenoptera, Meliponini) using relative warp analysis. Biosci. J., 23: 147- 152.

Nunes, L.A., Araujo, E.D., Carvalho, C.A.L. & Waldschmidt, A.M. (2008). Population divergence of Melipona quadrifasciata anthidioides (Hymenoptera: Apidae) endemic to the semi-arid region of state of Bahia, Brazil. Sociobiology, 52: 81-93.

Nunes, L.A., Passos, G.B., Carvalho, C.A.L. & Araújo, E.D. (2013). Spatial variation of size and shape of the wing in Melipona quadrifasciata anthidioides Lepeletier, 1836 (Hymenoptera; Meliponini) assessed by geometric morphometrics. Braz. J. Biol., 73: 887-893. doi: 10.1590/S1519-69842013000400027.

Rohlf, F.J.; Marcus, L.F. (1993). A revolution in morphometrics. Trends in Ecol. Evol., 8: 129-132.

Rohlf, F.J. (1998). On applications of geometric morphometrics to studies of ontogeny and phylogeny. Systematic Biol., 47: 147–158.

Rohlf, F.J. (2008a).tpsUtil. Departament of Ecology and Evolution.State University of New York.

Rohlf, F.J. (2008b).tpsDig. Departament of Ecology and Evolution.State University of New York.

Saraiva, A.M., Acosta, A.L., Giannini, T.C., Carvalho, C.A.L. de, Alves, R.M. de O., Drummond, M.S., Blochtein, B., Witter, S., Alves-dos-Santos, I. & Imperatriz-Fonseca, V.L. (2012). Influencia das alterações climáticas sobre a distribuição de algumas espécies de Melipona no Brasil. In: Imperatriz- Fonseca, V.L., Canhos, D.A.L., Alves, D.A. & Saraiva, A.M. (Orgs.), Polinizadores do Brasil: Contribuição e perspectivas para a biodiversidade, uso sustentável, conservação e serviços ambientais (pp. 349-367). São Paulo: Editora da Universidade de São Paulo.

Slaa, E.J., Chaves, L.A.S., Malagodi-Braga, K.S. & Hofstede, F.E. (2006). Stingless bees in applied pollination: pratic and perspectives. Apidologie, 37: 293-315. doi: 10.1051/apido:2006022.

Wappler, T., Meulemeester, T. de., Aytekin, A.M., Michez, D. & Engel, M.S. (2012). Geometric morphometric analysis of a new Miocene bumble bee from the Randeck Maar of southwestern Germany (Hymenoptera: Apidae). Syst. Entomol., 37: 784–792. doi: 10.1111/j.1365-3113.2012.00642.x.

Webbee, (2014). Available in: http://www.webbee.org.br/jandaira/mapa.htm . Accessed date: March 5, 2014.

Zanella, F.C.V. (2000). The bees of the Caatinga (Hymenoptera, Apoidea, Apiformes): a species list and comparative notes regarding their distribution. Apidologie, 31: 579-592. doi: 10.1051/apido:2000148.

28

Waste management in the stingless bee *Melipona beecheii* Bennett (Hymenoptera: Apidae)

LAM Medina[1], AG Hart[2], FLW Ratnieks[3]

1 - Universidad Autónoma de Yucatán, Mérida, Yucatán, México
2 - University of Gloucestershire, Gloucestershire, U.K.
3 - University of Sussex, Brighton, U.K.

Keywords

Stingless bees, nest hygiene, task partitioning, division of labor age polyethism.

Corresponding author

Luis A. Medina Medina
Departamento de Apicultura, Facultad de Medicina Veterinaria y Zootecnia, Universidad Autónoma de Yucatán, Apartado Postal 4-116 Itzimná, C. P. 97100, Mérida, Yucatán, México.
E-Mail: mmedina@uady.mx

Abstract

Waste management is important in insect societies because waste can be hazardous to adults, brood and food stores. The general organization of waste management and the influence of task partitioning, division of labor and age polyethism on waste processing were studied in three colonies of the tropical American stingless bee *Melipona beecheii* Bennett in Yucatán, Mexico. Waste generated in the colony (feces, old brood cells, cocoons, dead adults and brood) was collected by workers throughout the nest and taken to specific waste dumps within the nest. During the day, workers based at the waste dumps formed waste pellets, which they directly transferred in 93% of cases, to other workers who subsequently removed them from the nest. This is an example of task partitioning and is hypothesized to improve nest hygiene as it has been found in leafcutting ants, *Atta*. To investigate division of labor and age polyethism we marked a cohort of 144 emerging workers. Workers forming waste pellets were on average 31.2±6.5 days old (X±SD, N= 40, range of 18-45 days). The life span of *M. beecheii* workers was 49.0±14.0 days (N= 144). There was no difference in the life span of workers which formed (52.2±11.6 days, N= 40) or did not form (49.9±11.5 days, N= 97) waste pellets, suggesting that waste work did not increase mortality. Although waste was probably not hazardous to adults and brood, because the dumps are located outside the brood chamber, its presence inside the nests can attract phorid flies and predators, which can harm the colony.

Introduction

Social living can cause problems which are not faced by non-social organisms, or exacerbates problems that are easily dealt with by solitary dwellers. In particular, waste poses a serious problem for social insects living within enclosed nests (Bot et al., 2001; Hart & Ratnieks, 2001, 2002a; Jackson & Hart, 2009). Localizing waste such as in a midden heap (e.g. *Atta colombica* Guérin-Méneville, Hart & Ratnieks, 2002a; *Messor barbarus* (Linnaeus), Anderson & Ratnieks, 2000), or removing waste regularly (e.g. defecating outside the nest and undertaking behavior in *Apis mellifera* Linnaeus, Visscher, 1983; Seeley, 1985) are common strategies to reduce waste building and its negative impact.

Waste management is often a combination of material transport and work organization and adaptations to either can improve effectiveness and efficiency (e.g. Hart & Ratnieks, 2001; Hart & Ratnieks, 2002a; Hart et al., 2002a). Work organization includes the organization of both tasks and the workforce performing them. Division of labor has proved to be an important component of work organization in social insect colonies (Robinson, 1992; Gordon, 1996).

Task partitioning (how tasks are divided into sub-tasks) is also proving to be a powerful organizational principle (Jeanne, 1986; Ratnieks & Anderson, 1999; Anderson & Ratnieks, 2000; Anderson et al., 2001; Hart & Ratnieks, 2002b; Hart et al., 2002a, b; Hart, 2013). Task partitioning describes situations where a single task, such as foraging, is divided into sequentially linked sub-tasks with a flow of material between them. It has been shown to be important in waste management in leafcutting ants *Atta*. Workers from "clean" parts of the nest, the fungus garden chambers, drop waste outside the "unclean" waste chambers where dedicated waste chamber workers collect it. The two-stage task partitioning (waste transporters coupled with waste chamber workers) with indirect transfer of waste, via caches outside the waste chambers, is hypothesized to help maintain nest hygiene by isolating the waste chambers (Hart & Ratnieks, 2001).

Waste management has not been extensively studied in social insects (Hart & Ratnieks, 2002a). One group which has received some attention are the stingless bees (Meliponini). Stingless bee waste comprises feces, old brood cells, cocoons and both dead adults and brood (Eltz et al., 2001) and waste handling strategies are varied. Workers of *Melipona favosa* (Fabricius) deposit waste in "waste dump areas" within the nest (where workers also defecate) before removing it (Sommeijer, 1984; Bruijn et al., 1989) whereas in some species, e.g. *Cephalotrigona capitata* (Smith) and *Trigona spinipes* (Fabricius), waste accumulates, eventually forming a large mass (scutellum), which can help to regulate nest temperature (Michener, 1974, 2000; Sakagami, 1982). In *Melipona compressipes* (Fabricius) and *Melipona scutellaris* Latreille, Kerr and Kerr (1999) reported that waste (cocoon + feces + wax) are removed from the nests and dropped at a distance of 1 to 45 m according to a Gaussian distribution which reduces the cost of energy and the attraction of predators and pests.

A detailed study of work organization in waste management (division of labor, task partitioning, spatial management and age polyethism) was carried out on the stingless bee *Melipona beecheii* Bennett. This species ranges from Mexico to Costa Rica (Ayala, 1999; Van Veen & Arce, 1999) and colonies typically contain 500 to 2500 workers with a single singly-mated queen (Van Veen & Arce, 1999; Paxton et al., 2001). Natural nests are built in tree hollows and have an entrance hole that connects to the nest cavity. Within the cavity, the brood area consists of multiple horizontal combs of wax cells covered by layers of involucrum (Van Veen & Arce, 1999). Outside that there are many egg-shaped pots made from cerumen (Wille & Michener, 1973; Van Veen & Arce, 1999) which are used for honey and pollen storage.

Material and Methods

Study site and study colonies

The study was carried out at the Department of Apiculture of the Campus of Biological and Animal Husbandry Sciences, Autonomous University of Yucatan, Merida, Mexico, from March to September 2001 during the main nectar flow of "Tsitsilche" (*Gymnopodium floribundum*

Rolfe, Polygonaceae) and many different Fabaceae species (Echazarreta et al., 1997).

Three healthy queenright colonies of *M. beecheii* were studied which originally were housed in log hives. Two months before the study, colonies were transferred into observation nests consisting of a box (30x25x8 cm) with wooden sides and glass top and bottom. This allowed observation of intranidal behavior. Colonies comprised c.900 workers of different ages, 2-4 brood combs and more than 20 honey and pollen pots surrounding the brood area.

General waste dump features

All colonies had two clear areas in which waste was dumped. The waste dump areas were 8-15 cm away from the brood area, outside the involucrum. No other waste dump areas were formed and the original dumps remained in the same place during the study period.

Task partitioning

One waste dump area in each colony was continually observed for 15 min before switching to the second waste dump area. Each dump was observed for a total of c. 170 h. Individuals that were molding waste into pellets were considered "waste dump workers". This behavior was characterized by the worker's body movements when trying to separate small portions of the waste material from the dump (Bruijn et al., 1989). Workers that went to a waste dump to perform a specific activity (e.g. defecation, grooming or dropping material), but did not manipulate the waste to form pellets for removal, were not considered waste dump workers (in contrast to Bruijn et al. (1989) who classified all bees performing activities at the dumps as waste workers).

Workers forming waste pellets were followed to determine the pellet's destination. There were only two destinations. A pellet was either transferred to another worker who removed it from the nest, or it was taken outside the nest directly by the worker that formed it. Task partitioning occurred when a pellet was transferred to another worker. Pellets weighing an average of 18 mg were carried in the mandibles of the workers which represents 25.7% of the weight of the workers (X=70 mg).

Division of labor, age polyethism and worker longevity

Three mature brood combs from one colony were taken from the original log nest and kept in an incubator. All workers that emerged from 28 May to 5 June were individually marked with a numbered tag (Opalithplättchen; 2.1 mg each tag which represents 3% of the weight of the workers) on the notum and introduced into an observation nest formed from the original colony. 144 workers in total were marked and introduced successfully. These marked bees allowed detailed data on the ages of workers performing waste–related activities to be collected and to determine whether waste-related activities affect worker longevity. They also enabled us to determine whether division of labor into waste workers and non-waste workers was occurring.

Individuals processing waste (e.g. collecting waste and dropping waste at the dumps to form waste caches, molding waste to form pellets) and workers releasing the waste pellets from the nests were recorded and their ages of workers performing these tasks and their longevity by a daily census were determined.

Data analysis

The proportions of pellets transferred to other workers, which then removed the pellets from the nest, were compared in the three colonies using a Chi-square contingency test (Zar, 1999). The life span and survival curves of workers which had or had not formed pellets were compared using Logrank Test (Machin et al., 2006).

All mean values are expressed ± one standard deviations (SD). Statistical analysis were performed using SPSS 11.0 for Windows (SPSS Inc., Norusis, 2002).

Results

General waste dump features

Waste management in *M. beecheii* was a three-stage process. First, workers collected waste from all over the nest, especially from the brood chamber, and deposited it in the waste dumps. This activity was observed during both day and night. Second, workers at the dumps molded waste into pellets during the day. Finally, and again only during the day, waste pellets were removed from the nest.

The average number of waste pellets formed per day in the study nests was 25.3±8.1 (X±SD, N = 40 days study), which represents 454.9±145.9 mg of waste (pellets weighed 18.0±11.1 mg, X±SD, N= 45 pellets).

Task partitioning

The destinations of 416 waste pellets were observed in the three study colonies (Nest 1, N = 112; Nest 2, N = 165; Nest 3, N = 109). In total, 386 workers (92.8%) observed forming pellets subsequently transferred them to other workers which removed the pellets from the nest. There were no significant differences in the occurrence of task partitioning between the study colonies (cross-colony comparison: Chi Squared = 1.17, df= 2, P= 0.56, Table 1). Overall, 92.8% of all waste pellets formed by waste dump workers were transferred directly (from mandibles to mandibles) to another worker, and 7.2% were taken outside by the waste dump worker without transfer. When waste dump workers took waste outside without transfer (i.e. when task partitioning was absent), it was

not because transfer partners were unavailable. In the marked cohort of workers, the only workers that formed pellets and subsequently removed them from the nest were workers that were close to switching tasks from pellet forming to flying outside the nest with waste (Figs 1b and c).

Table 1. Number (%) of waste pellets formed and transferred between waste dump workers and individuals removing pellets from the nest.

	Nest 1	Nest 2	Nest 3	Total
With transfer	112	165	109	**386**
	(91.0%)	(94.3%)	(92.4%)	**(92.8%)**
Without transfer	11	10	9	**30**
	(9.0%)	(5.7%)	(7.6%)	**(7.2 %)**

Cross-colony comparison: χ^2= 1.17, df= 2, P= 0.56

Division of labor, age polyethism and worker longevity

Workers that collected waste within the nest (mainly in the brood area) and then placed it at the waste dump areas were 20.7±5.7 days old (X±SD, N= 46) with a range of 10-31 days (Fig 1a). Workers engaged in this task were removing wax from old brood combs, destroying old brood cells and involucrum, removing dead brood and collecting dead adults. Workers then carried the waste to the dump areas where they dropped it to form waste caches. All workers performing duties inside the colonies defecated only at the two waste dumps areas, so feces do not have to be collected and carried out to these areas and other forms of waste were carried out to these places to form the wastes dumps. Thirty-three percent of the age-marked cohorts were observed collecting waste and taking it to the dump areas.

The age at which individuals first formed waste pellets was 31.2±6.5 days (X±SD, N= 40, range of 18-45 days). In total, 29.2% of the age-marked cohort engaged in this task (Fig 1b). Workers that started processing waste material at the dumps and forming waste pellets performed this task for about 2.5 days (range of 1-15 days) and were later observed flying outside the nest to remove the waste pellets at 34.5±5.9 days old (X±SD, N= 13) with a range of 29 to 59 days (Fig 1c). However, only 10% of the age-marked cohort removed waste pellets from the nest. These workers were very active, removing waste pellets during the day at a high rate (some handled 4-5 waste pellets within an hour) until all waste was removed from the nest.

Although we did not record the distance flown before waste pellets were released, total flight durations were short (X±SD = 21.7±11.1 s, N = 65 flights). The first foraging flights in the cohort of age-marked *M. beecheii* workers were performed at 41.7±7.1 days old (X±SD, N= 58) with a range of 22 to 63 days (Fig 1d).

The life span of the age-marked cohort was 49.0±14.0 days (X±SD, N = 144) with a range of 3 to 71 days (last marked-worker seen alive, Fig 1a). There was no significant difference (X^2 = 1.373, P= 0.354, Logrank Test) in the life span of workers which had (52.2 ± 11.6 days, N = 40) or had not (49.9±11.5 days, N= 97) formed pellets.

Discussion

Our results show that waste disposal in *M. beecheii* is subject to partitioned organization, encompassing spatial management, division of labor with age polyethism, and task partitioning. Waste management had three distinct stages connected by task partitioning with both direct and indirect transfer of waste between workers, and this pattern occurred consistently across all three study colonies. Younger workers collected waste around the nest and transferred it to the waste dump areas where it formed a cache (a case of task partitioning with indirect transfer; Ratnieks & Anderson 1999). The percentage (33%) of worker bees collecting waste and taking it to the dump areas indicates that this is a common but not universal activity in the worker age polyethism schedule.

Much of the waste in the dumps was feces, which was not transferred but deposited directly by defecating workers. Waste in the dump areas was then fashioned into discrete pellets by older workers. Most pellets, 93%, were transferred directly by a dump worker to another worker, who was on average even older, who flew from the nest to dump the pellet. The other pellets, 7%, were removed from the nest without transfer.

Waste management has not been extensively studied in social insects. However, the system found in *M. beecheii* is similar to that found in the leafcutting ant, *Atta cephalotes* Linnaeus, where the transfer of waste to disposal zones is also subject to task partitioning. In *A. cephalotes*, waste is stored in specialized chambers in the underground nest that are distinct from the chambers where the ants grow the fungus gardens that they depend on for food. Fungus garden workers deposit waste just outside the waste chambers, which they do not enter. Dedicated waste chamber workers retrieve the waste and carry it into the chamber. Thus, indirect transfer (coupled with division of labor and nest compartmentalization) enables effective isolation of hazardous waste (Hart & Ratnieks, 2001). Good hygiene is important in leafcutting ants (and other fungus growing ants) because of the presence of the vulnerable fungus on which the ants depend. In *M. beecheii*, waste may not be as hazardous. Possibly reflecting this, leafcutting ants working in waste chambers never become foragers, but in *M. beecheii* all waste dump workers eventually became foragers.

Task partitioning can have ergonomic advantages. In *M. beecheii*, de-coupling pellet making from pellet disposal allows workers to become specialized at each task (albeit only for a short period), which is likely to improve efficiency. This may be the advantage of task partitioning in this case. A potential cost of task partitioning with direct transfer is that the two task groups may become out of phase, such that members of one or other groups must wait for transfer partners. One way to reduce this problem is to introduce indirect transfer via a cache (Hart et al., 2002a). However, in *M. beecheii*, introducing

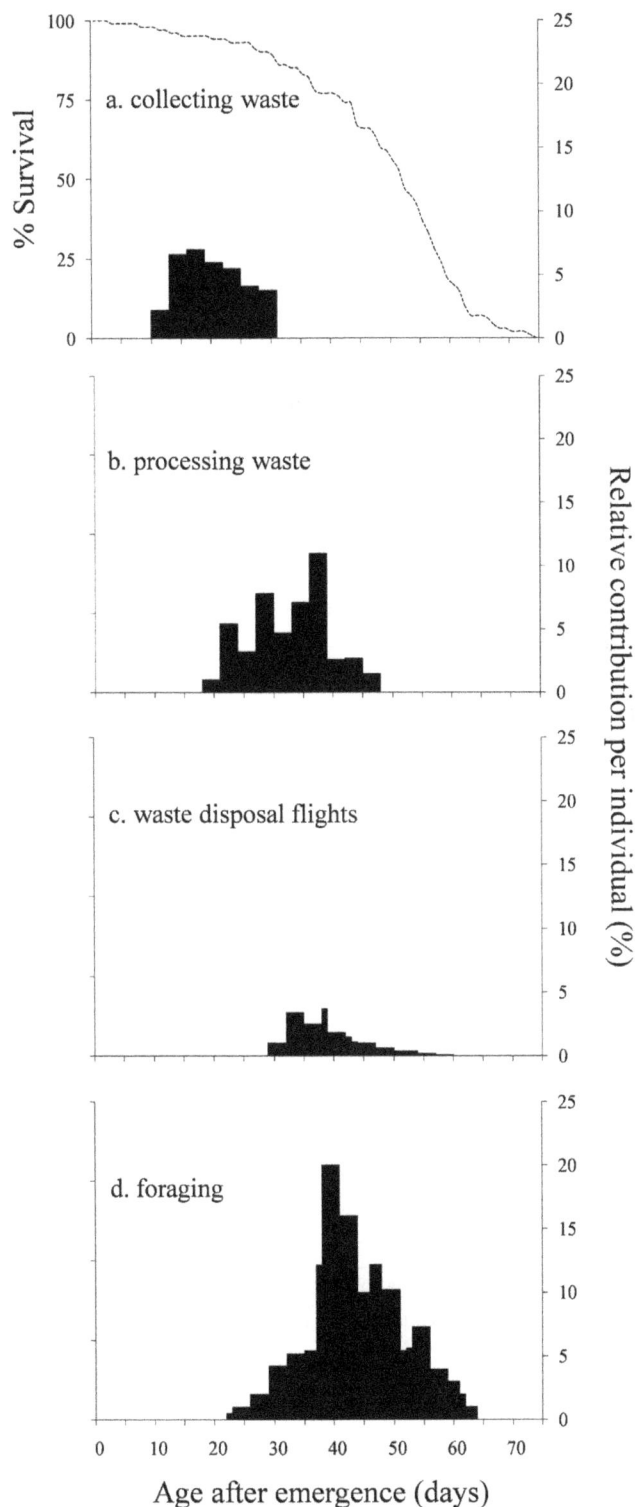

Fig 1. Survival curve (line, a) and frequency (%) of workers of *Melipona beecheii* performing tasks related with the collection of waste (a), processing (molding) the waste at dump areas (b), releasing waste outside the nest (c) and foraging (d). Age is given in days after emergence.

a further cache of waste pellets (in addition to the initial waste dump areas) is not likely to improve nest hygiene, since waste will be more distributed around the colony (increasing the area it could contaminate) and might not be collected. In order to limit the distribution of waste, workers must either transfer pellets directly to other workers who will subsequently remove the pellets from the nest or remove waste pellets themselves.

We did not record the destination of waste outside the nest, but return flights were short (mean 21.7 seconds) compared with nectar and pollen collecting flights (15.6 and 25.1 minutes respectively; Biesmeijer & Toth, 1998) and it is likely that waste was dropped at a distance comparable to other *Melipona* species (e.g. 18m in *M. compressipes* and 31m in *M. scutellaris*; Kerr & Kerr 1999). Dumping waste close to the nest is typical in stingless bees. Some species, like *Lestrimellita limao* (Smith) and *Lestrimellita niitkib* Ayala, simply drop waste in front of the nest entrance and in *Frieseomellita varia* (Lepeletier) waste is carried less than 1m from the nest (Kerr & Kerr, 1999). Carrying waste away from the nest has a number of possible advantages. Waste can reduce the space available for food storage or brood rearing and in species where waste accumulates in the nest, the thickness of waste increased with nest age (in *Trigona corvina* Cockerell a total thickness of 11-24 cm is found around the sides and base of the nest; Michener, 1974).

Additionally, waste removal may help to avoid attracting parasitic phorid flies (in *M. compressipes*, larvae and pupae of the phorid fly *Megaselia scalaris* (Loew) were found only in the waste (Kerr, 1996) and vertebrates (e.g. *Eira barbara* (Linnaeus) [Mustelidae]; Kerr & Kerr 1999), both of which can attack and destroy the colony. Waste may also provide a substrate on which potentially harmful micro-organisms can thrive.

Division of labour in waste management was age-dependent. Workers collecting waste were younger (mean age, 20.7 days old) than workers processing (31.2 days old) and releasing waste pellets (34.5 days old), and also performed brood chamber duties such as wax removal. This pattern of multi-tasking is found in other species. For example, in *M. favosa* (Sommeijer, 1984) and *Melipona bicolor* Lepeletier (Bego, 1983), young workers (<20 days) both collected waste and performed brood chamber tasks such as building and provisioning cells. In addition to temporal polyethism, the schedule of waste management tasks in *M. beecheii* exhibited a strong spatial pattern. Young workers first performed tasks within the brood chamber (waste collection), later graduating to tasks outside of the brood chamber (waste pellet forming) before finally performing tasks outside the nest (waste removal and foraging). This is an excellent example of the "conveyor belt" model (Schmid-Hempel, 1998), in which individuals move from safe tasks inside the nest to more hazardous tasks outside the nest as they age. It also mirrors the situation in honey bees, *A. mellifera* (Seeley, 1982) in which workers carry out a range of tasks that co-occur spatially in the nest. Seeley (1982) commented that stingless bees, with their distinctive nest layout, could be an important taxon to further investigate this.

Large societies, both human and insect, struggle with how to collect and dispose their waste (Doan, 1998; Dijkema et al., 2000; Bot et al., 2001; Hart & Ratnieks, 2002a). Effective waste management needs to encompass collection, processing and eventual disposal. In human society these together prevent waste building up, thereby reducing negative effects such as odor, unsightliness and disease-carrying vermin. Similar to human waste management, we found that in the nests of *M. beecheii*, waste was collected daily and taken to an intermediate processing site (the waste dump areas), where it was processed and then disposed of. This insect management system probably prevents waste accumulation and consequent attraction of "vermin" (e.g. phorid flies and predators) in the same way that human waste management does. Social organisms from widely different taxa often face similar problems for example, maintaining social hygiene and managing waste. Waste management in social insects is currently understudied and we suggest that further studies need to be carried out to investigate this potentially vital component of social life. The stingless bees provide *a priori* examples of waste management diversity and this, combined with the ease with which they can be cultured and studied in the laboratory, give them the potential to become a model group for waste management and nest hygiene in social insects studies.

Acknowledgements

L.M. Medina thanks H. Moo-Valle and WR Angulo for their personal support during fieldwork. The work of L.M. Medina was supported by PROMEP-Mexico PhD studentship. The study complies with the current laws of Mexico.

References

Anderson, C., Franks, N.R. & McShea, D.W. (2001). The complexity and hierarchical structure in insect societies. Anim. Behav. 62: 643-651. doi:10.1006/anbe.2001.1795

Anderson, C. & Ratnieks, F.L.W. (2000). Task partitioning in insect societies: novel situations. Insectes Soc. 47: 198-199. doi: 10.1007/PL00001702

Ayala, R. (1999). Revisión de las abejas sin aguijón de México (Hymenoptera; Apidae; Meliponini). Folia Entomol. Mex. 106: 1-123.

Bego, L.R. (1983). On some aspects of bionomics in *Melipona bicolor bicolor* Lepeletier (Hymenoptera, Apidae, Meliponinae). Braz. J. Biol. 27: 211-224.

Biesmeijer, J.C. & Tóth, E. (1998). Individual foraging, activity level and longevity in the stingless bee *Melipona beecheii* in Costa Rica (Hymenoptera, Apidae, Meliponinae). Insectes Soc. 45: 427-443.

Bot, A.N.M., Currie, C.R., Hart, A.G. & Boomsma, J.J. (2001).

Waste management in leafcutting ants. Ethol. Ecol. Evol. 13: 225-237. doi:10.1080/08927014.2001.9522772

Bruijn, de L.M.M., Sommeijer, M.J. & Dijkstra, E. (1989). Behaviour of workers on waste dumps in the nest of *Melipona favosa* (Apidae, Meliponini). Actes Colloq. Insectes Soc. 5: 31-37.

Dijkema, G.P.J., Reuter, M.A. & Verhoef, E.V. (2000). A new paradigm for waste management. Waste Manage. 20: 633-638. doi: 10.1016/S0956-053X(00)00052-0

Doan, P.L. (1998). Institutionalizing household waste collection: the urban environmental management project in Côte d'Ivoire. Habitat Int. 22: 27-39.

Echazarreta, C.M., Quezada-Euan, J.J.G., Medina, L.M. & Pasteur, K.L. (1997). Beekeeping in the Yucatan peninsula: development and current status. Bee World. 78: 115-127.

Eltz, T., Bruhl, C.A., Van der Kaars, S. & Insenmair, K.E. (2001). Assessing stingless bee pollen diet by analysis of garbage pellets: a new method. Apidologie 32: 341-353. doi: 10.1051/apido:2001134.

Gordon, D.M. (1996). The organization of work in social insect colonies. Nature 38: 121-124. doi:10.1038/380121a0.

Hart, A.G. (2013). Task partitioning: is it a useful concept? In: K. Sterenly, R. Joyce, & B. Calcott (Eds.), Cooperation and its Evolution (pp. 203-221). Cambridge: MIT Press.

Hart, A.G., Anderson, C. & Ratnieks F.L.W. (2002a). Task partitioning in leafcutting ants. Acta Ethol. 5: 1-11. doi:10.1007/s10211-002-0062-5.

Hart, A.G., Bot, A.N.M. & Brown, M.J.F. (2002b). A colony-level response to disease control in a leaf-cutting ant. Naturwissenschaften 89: 275-277. doi:10.1007/s00114-002-0316-0

Hart, A.G. & Ratniek, F.L.W. (2001). Task partitioning, division of labour and nest compartmentalisation collectively isolate hazardous waste in the leafcutting ant *Atta cephalotes*. Behav. Ecol. Sociobiol. 49: 387-392. doi:10.1007/s002650000312.

Hart, A.G. & Ratnieks, F.L.W. (2002a). Waste management in the leaf-cutting ant *Atta colombica*. Behav. Ecol. 13: 224-231. doi:10.1093/beheco/13.2.224.

Hart, A.G. & Ratnieks, F.L.W. (2002b). Task-partitioned nectar transfer in stingless bees: work organization in a phylogenetic context. Ecol. Entomol. 27: 163-168. doi:10.1046/j.1365-2311.2002.00411.x

Jackson, D.J. & Hart, A.G. (2009). Does sanitation facilitate sociality? Anim. Behav. 77: e1-e5. doi:10.1016/j.anbehav.2008.09.013.

Jeanne, R.L. (1986). The evolution of the organization of work in social insects. Monit. Zool. Ital. 20: 119-133.

Kerr, A.S. & Kerr, W.E. (1999). *Melipona* garbage bees release their cargo according to a gaussian distribution. Braz. Rev. Biol. 59: 119-123.

Kerr, W.E. (1996). Biologia e manejo da tiúba: a abelha do Maranhão. São Luís: Editora da Universidade Federal do Maranhão, 156 p.

Machin, D., Cheung, Y.B. & Parmar, M.K.B. (2006). Survival Analysis: a practical approach. West Sussex: John Wiley & Sons Press, 266 p.

Michener, C.D. (1974). The Social Behaviour of the Bees: a comparative study. Cambridge Mass: Harvard University Press, 404 p.

Michener, C.D. (2000). The Bees of the World. Baltimore: Johns Hopkins University Press, 913 p.

Norusis, M. (2002). SPSS 11.0 Guide to data analysis. New Jersey: Prentice Hall.

Paxton, R.J., Ruhnke, H., Shah, M., Bego, L.R., Quezada-Euan, J.J.G. & Ratnieks, F.L.W. 2001. Social evolution of stingless bees: are the workers or is the queen in control of male parentage? 2nd Stingless Bee Seminar, Merida, pp 104-107.

Ratnieks, F.L.W. & Anderson, C. (1999). Task partitioning in insect societies. Insectes Soc. 46: 95-108.

Robinson, G.E. (1992). Regulation of division of labor in insect societies. Annu. Rev. Entomol. 37: 637-665.

Sakagami, S.F. (1982). Stingless bees. In R.H. Hermann (Eds.) Social Insects III (pp. 361-423). Academic Press, London.

Schmid-Hempel, P. (1998). Parasites in Social Insects. New Jersey: Princeton University Press, 409 p.

Seeley, T.D. (1982). Adaptive significance of the age polyethism schedule in honeybee colonies. Behav. Ecol. Sociobiol. 11: 287-293.

Seeley T.D. (1985). Honeybee Ecology: A study of adaptation in social life. New Jersey: Princeton University Press, 201 p.

Sommeijer, M.J. (1984). Distribution of labour among workers of *Melipona favosa* F.: age polyethism and worker oviposition. Insectes Soc. 31: 171-184.

Van Veen, J.W. & Arce, H.G. (1999). Nest and colony characteristic of log-hived *Melipona beecheii* (Apidae: Meliponinae). J. Apicult. Res. 38: 43-48.

Visscher, P.K. (1983). The honey bee way of death: necrophoric behaviour in *Apis mellifera* colonies. Anim. Behav. 31: 1070-1076.

Wille, A. & Michener, C.D. (1973). The nest architecture of the stingless bees with special reference to those of Costa Rica. Rev. Biol. Tropic. 21: 1-278.

Zar, J. (1999). Biostatistical Analysis. New Jersey: Prentice Hall, 663 p.

29

Pollen Analysis of Food Pots Stored by *Melipona subnitida* Ducke (Hymenoptera: Apidae) in a Restinga area

RS Pinto, PMC Albuquerque, MMC Rêgo

Universidade Federal do Maranhão, São Luís, MA, Brazil

Keywords
Pollen, honey, nectariferous plants, floral resources, stingless bee, Meliponini.

Corresponding author
Rafael Sousa Pinto
Universidade Federal do Maranhão
Departamento de Biologia
Laboratório de Estudos Sobre Abelhas
Av. dos Portugueses, 1966, 65080-805
São Luís, MA, Brazil
E-Mail: rafael_spinto@hotmail.com

Abstract

The geographic distribution of *Melipona subnitida* Ducke covers the dry areas in the northeastern Brazil, where it plays an important role as pollinator of many wild plant species. In the current study, the botanical species this bee uses as pollen and nectar sources in a restinga area of Maranhão State, Brazil, were identified by analyzing pollen grains present in their storage pots in the nests. Samples were collected from five colonies bimonthly, from April 2010 to February 2011. In all the samples, 58 pollen types were identified; the families Fabaceae (8) and Myrtaceae (5) had the largest number of pollen types. In the pollen pots, 52 pollen types were identified; Fabaceae, Melastomataceae, Myrtaceae and Dilleniaceae species were dominant. In honey samples, 50 pollen types were found, with a predominance of nectariferous and polliniferous plant species. Out of the total of pollen types from nectariferous plants identified in honey, 20 pollen types contributed to the honey composition. *Humiria balsamifera* occurred in high frequency and was predominant in October. *Chrysobalanus icaco*, *Coccoloba* sp., *Cuphea tenella* and *Borreria verticillata* were also important for honey composition. The occurrence of a high number of minor pollen types indicated that *M. subnitida* visits many species in the locality; however, it was possible to observe that its floral preferences are very similar to those from other *Melipona* species.

Introduction

Stingless bees (Apidae: Meliponini) occur in most tropical or subtropical regions of the world and the *Melipona* genus is exclusively found in the Neotropical region (Camargo & Pedro, 2007). They have perennial colonies with hundreds to thousands of workers and require continuous foraging activity to meet their food requirements (Roubik, 1989).

Faced with the need to forage several food resources, the stingless bees have a generalist behavior concerning the plants visited, but a small number of plant species are most exploited in local communities (Ramalho et al., 1989). The study of the plant-pollinator interaction can be performed by sampling bees in flowers (Imperatriz-Fonseca et al., 2011) or indirectly by morphological identification of pollen loads transported on the workers' corbiculae or stored in food pots inside their nests (Barth, 2004).

Stingless bees usually store food resources in pots built with cerumen (the mixture of bees wax and resin) for the colony's future use. The pots that contain pollen or honey are irregularly distributed, using all the free spaces of the hollow or cavity where they are located, and have a completely random arrangement (Camargo, 1970). However, in general, the pollen pots are located closer to the brood combs and opposite to the honey pots.

Several studies have identified the plants species collected by stingless bees by analyzing the contents of pollen pots (Ramalho et al., 1989; Wilms & Wiechers, 1997; Pick & Blochtein, 2002) and pollen grains present in honey (Iwama & Melhem, 1979; Carvalho et al., 2001; Martins et al., 2011). Using these analyses, it is possible to define the floral preferences of the visitors; the most abundant pollen types have greater relevance for the bees' species.

The knowledge of the floral resources are necessary for the maintenance of bee communities in their habitats is crucial to understanding the mutualistic relationship between plants and bees and to developing management programs for pollinators, reforestation and environmental restoration

(Luz et al., 2007). For example, strategies for planting natural resources can be developed to supply food during periods of shortage that bees may face.

The aim of the current study was to identify the botanical species that are sources of pollen and nectar for *Melipona subnitida* Ducke in a restinga (coastal sandy plain) environment. *M. subnitida*, locally known as jandaíra, occurs in Brazil only in the northeast region and is very frequent in the Caatinga biome (Martins, 2002). In Maranhão state, this species occurs in a restinga area of the Lençóis Maranhenses Nacional Park (Rego & Albuquerque, 2006) and in the Parnaíba Delta (Silva et al., 2014). *M. subnitida* is of great importance for the pollination of the regional native flora (Ferraz et al., 2008) and cultivated plants (Cruz et al., 2004; Silva et al., 2005) and is traditionally reared for honey production, which has a high economic value.

Materials and Methods

The current study was conducted on a *M. subnitida* meliponary located in the municipality of Barreirinhas, in the Lençóis Maranhenses National Park (Parque Nacional dos Lençóis Maranhenses; 2°58'12"S, 42°79'56"W), Maranhão state, Brazil. The climate in the study region is classified as tropical megathermal (Aw' type, according to the Köppen classification). The average annual temperature is approximately 27°C, and the annual precipitation is approximately 2,000 mm. There are two well-defined seasons: a rainy season from January to July and a dry season from August to December (Brazilian Institute of Environment and Renewable Natural Resources - IBAMA, 2002).

The vegetation of the Lençóis Maranhenses National Park covers an area of 453.28 km², of which 405.16 km² are predominately composed of restinga vegetation. The rest of the vegetation consists of mangroves and riparian forests. The restinga area has plant species that are specific to this type of vegetation and plants characteristic of Cerrado (Brazilian savanna), Caatinga (semi-arid) and rainforest. Shrub species are dominant, and herbaceous communities are also present in large areas surrounding lakes (Brazilian Institute of Environment and Renewable Natural Resources - IBAMA, 2002).

Samples of the pollen and honey pots were collected in April, June, August, October and December 2010 and February 2011 from five randomly chosen nests. In each nest, 2-3 g of pollen and 10 ml of honey were extracted with a spatula and a disposable syringe, respectively. In total, 30 pollen samples and 30 honey samples were collected (5 nests x 6 months). Pollen and honey samples were acetolysed by Erdtman's method (1960) to facilitate the observation of the outer pollen wall (exine). However, the honey samples (10 ml) were divided into two test tubes before they were subjected to the acetolysis process (Erdtman, 1960). Distilled water (10 ml) was added to each tube, and the tubes were centrifuged for 5 min at 2,000 rpm (Louveaux et al., 1978). The supernatant was discarded, and the pellet was subjected to the acetolysis process.

Slides were then prepared using glycerin jelly for optical microscopy analysis. The pollen grains were separated into pollen types according to their morphology and were photographed under a Zeiss Primo Star optical microscope. The identification of the pollen grains was performed by comparing them with both a pollen collection from the regional flora and the literature (Roubik & Moreno, 1991; Carreira & Barth, 2003; Silva et al., 2010). The classification system adopted for the level of family was the APG III (2009).

In total, 2,000 and 1,000 pollen grains from each pollen and honey pot were counted, respectively. The monthly means of five pollen and honey samples were calculated. The quantitative results were classified as frequency classes (Louveaux et al., 1978): Predominant pollen 'P' (more than 45% of the grains counted), Secondary pollen 'S' (15% to 45%), Important minor pollen 'I' (3% to 15%) and Minor pollen 'M' (less than 3%).

Based on the literature and floristic surveys performed in the study region, the pollen types identified in honey that were considered to be from nectariferous plants were separately analyzed to determine which species actually contributed to the honey composition.

Minitab®15 software was used to generate the dendrograms of percentage similarity of the two types of pots analyzed.

Results

Fifty-eight pollen morphospecies were observed in the *M. subnitida* food samples. In the qualitative analysis of the pollen pots, 52 pollen types belonging to 29 families, 40 genera and 29 species were recorded. Only two pollen types were not identified (Table 1). Fifty pollen types were present in the honey samples, which were grouped into 29 families, 37 genera and 28 species. Four pollen types were indeterminate (Table 1).

The botanical families with greatest species diversity in the samples' pollen spectra were Fabaceae (8), Myrtaceae (5), Malpighiaceae (3) and Melastomataceae (3). Most plants species occurred occasionally (minor pollen) in the samples. Of the pollen types identified in the pollen pots, only 12 (23.07%) were identified with percentages higher than 3% in any given month. In the honey, 17 pollen types (34%) were considered at least once as important minor, secondary or predominant pollens.

Figure 1 shows the percentage frequencies of the main species identified in the pollen (Fig 1A) and honey samples (Fig 1B) throughout the study. In pollen pots, *Mimosa misera* Benth. (Fabaceae) (Fig 2A), *Chamaecrista ramosa* (Vogel) H.S. Irwin & Barneby (Fabaceae) (Fig 2B), *Mouriri guianensis* Aubl. (Melastomataceae) (Fig 2C), *Doliocarpus* sp. (Dilleniaceae) (Fig 2D), *Myrcia obtusa* Schauer (Myrtaceae) (Fig 2E), *Comolia lythrarioides* Naudin (Melastomataceae) (Fig 2F), *Myrcia* sp. 3 (Myrtaceae), *Tibouchina* sp. (Melastomataceae), *Myrcia sylvatica* (Myrtaceae) and *Eugenia* sp. (Myrtaceae) were particularly abundant. In April, the presence of *Orbignya phalerata* Mart. (Arecaceae) was also noteworthy.

Table 1. Occurrences and frequencies of pollen types in pollen and honey pots of *Melipona subnitida* in a restinga area, Maranhão state, Brazil. Classification: P – Predominant Pollen (more than 45% of the grains counted), S – Secondary Pollen (15 to 45%), I – Important Minor Pollen (3 to 15%) and M – Minor Pollen (less than 3%). Apr (April), Jun (June), Aug (August), Oct (October), Dec (December), Feb (February).

Family	Pollen Type	Pollen Pots						Honey Pots					
		Apr	Jun	Aug	Oct	Dec	Feb	Apr	Jun	Aug	Oct	Dec	Feb
Amaranthaceae	*Alternanthera* sp.								M		M		
Anacardiaceae	*Anacardium microcarpum*		M	M	M				M	M	M		
Arecaceae	Arecaceae type		M	M	M					M	M	M	M
	Orbignya phalerata	I	M	M	M	M			M	M	I	M	M
Asteraceae	*Wulffia baccata*				M			M	M				M
Bignoniaceae	*Arrabidaea dispar*	M	M	M									
	Bignoniaceae type					M					M		
Boraginaceae	*Heliotropium* sp.			M					M				
Burseraceae	*Protium heptaphyllum*	M						M	M	M	M		
Caryocaraceae	*Caryocar brasiliense*					M				M		M	
Chrysobalanaceae	*Chrysobalanus icaco*	M	M	M	M	M	M	M	M	I	I	I	I
	Licania sp.	M					M						
Clusiaceae	*Clusia grandiflora*	M	M		M					M	M	M	
Dilleniaceae	*Doliocarpus* sp.	M	I	M	M	M	P	I	M	I	M	M	S
Euphorbiaceae	*Croton* sp.						M					M	M
	Mabea pohliana		M	M	M				M				
Fabaceae	*Abarema cochleata*				M				M				
	Centrosema sp.	M	M	M	M						M		
	Chamaecrista ramosa	M	S	S	S	M	M	I	I	M	M	M	M
	Copaifera sp.		M	M	M				M	M	M		
	Mimosa caesalpiniifolia				M			M	M	M	M		
	Mimosa misera	S	P	S	P	M	M	S	S	I	I	M	M
	Senna sp.				M			M	M	M	M		
	Stryphnodendron adstringens	M	M		M	M			M	M	M	M	M
Gentianaceae	*Irlbachia pratensis*										M		
Humiriaceae	*Humiria balsamifera*	M	M	M	M	M	M	I	I	I	P	I	I
Loranthaceae	*Phthirusa pyrifolia*		M	M		M				M		M	M
Lythraceae	*Cuphea tenella*	M	M	M	M		M	I	M	M	M	M	M
Malpighiaceae	*Byrsonima chrysophylla*	M	M	M	M						M		
	Byrsonima crassifolia	M		M	M	M	M	M		M	M	M	M
	Mascagnia sp.				M							M	M
Malvaceae	*Sida* sp.				M								
Melastomataceae	*Comolia lythrarioides*	I		I	I	M	M	I	I	P	M	M	I
	Mouriri guianensis	S	M	I	I	S	P	I	I	I	I	S	I
	Tibouchina sp.	I	M	I	M	I		I	M	M	M	I	
Meliaceae	Meliaceae type			M					M				
Menyanthaceae	*Nymphoides indica*	M		M	M				M		M		
Myrtaceae	*Eugenia* sp.	I	M	M		M		I	I	M	M	M	
	Myrcia obtusa	M	M	M	M	S	M	M	M	M	M	I	I
	Myrcia sylvatica	I	M	I	M			I	M	I	S		
	Myrcia sp.3	M	M	M	M	S	M					M	I
	Myrcia sp.4			M	M							M	I
Ochnaceae	*Ouratea racemiformis*	M			M							M	M
Passifloraceae	*Passiflora* sp.					M							
	Turnera ulmifolia	M	M					M	M	M	M		
Poaceae	Poaceae type	M	M	M	M					M	M		M
Polygonaceae	*Coccoloba* sp.	I	M			M	M	M	I	M		I	I
Rubiaceae	*Borreria latifolia*		M	M	M								
	Borreria verticillata	M	M	M	M	M	M	I	I	M	M	M	M
Rutaceae	*Citrus* sp.						M						
Sapindaceae	*Paullinia pinnata*	M											
Sapotaceae	*Pouteria* sp.	M	M	M						M	M		
Xyridaceae	*Xyris paraensis*							M		M		M	M
Not Identified	Type 01									M			
	Type 02	M	M		M				M				
	Type 03										M		
	Type 04									M			
	Type 05	M	M	M	M								

The major pollen types identified in the honey samples were *M. misera*, *Humiria balsamifera* Aubl. (Humiriaceae) (Fig 2G), *M. guianensis*, *C. lythrarioides*, *Doliocarpus* sp., *Chrysobalanus icaco* L. (Chrysobalanaceae) (Fig 2H) and *Coccoloba* sp.. Other species also stood out in a few of the months, such as *O. phalerata*, *C. ramosa*, *Cuphea tenella* Hook. & Arn. (Lythraceae), *Tibouchina* sp., *Borreria verticillata* (L.) G. Mey. (Rubiaceae) and all the species of the family Myrtaceae.

In the similarity dendrogram of the pollen pots, the samples from April, June, August and October were grouped together with 86.89% of similarity; the samples from June and October had 98.43% of similarity. Conversely, February and December were the least similar months compared to the other months (Fig 3A). In the similarity dendrogram of the honey samples, December and February had 75.70% of similarity. The samples collected in August had 73.55% of similarity with the samples collected in April and June. Samples collected in April and June had the highest similarity (95.87%). The samples collected in October showed less similarity with those collected in the other months (Fig 3B).

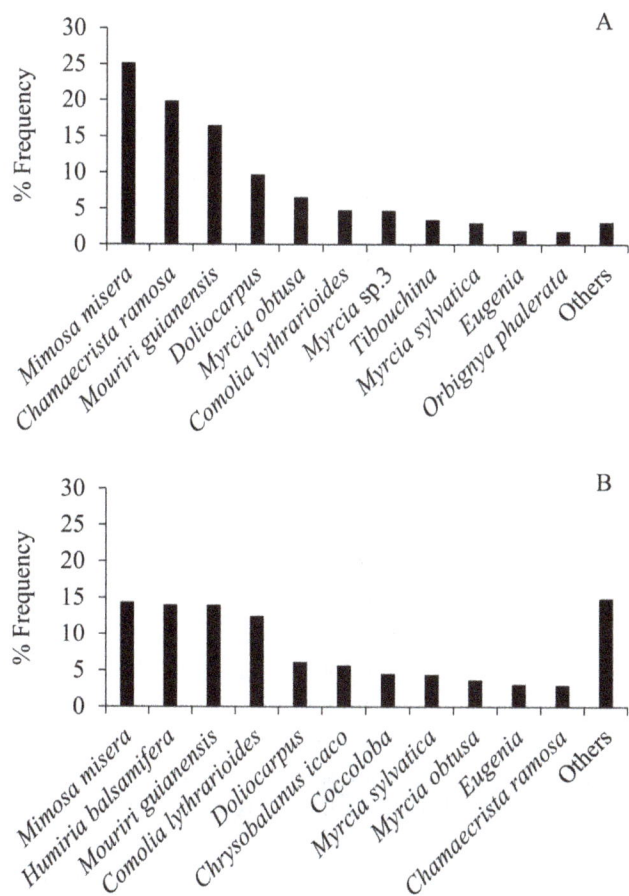

Fig 2. Photomicrographs of pollen types observed in food pots of *Melipona subnitida*. (A) *Mimosa misera*, (B) *Chamaecrista ramosa*, (C) *Mouriri guianensis*, (D) *Doliocarpus*, (E) *Myrcia obtusa*, (F) *Comolia lythrarioides*, (G) *Humiria balsamifera*, and (H) *Chrysobalanus icaco*. Bar scale – 10 μm.

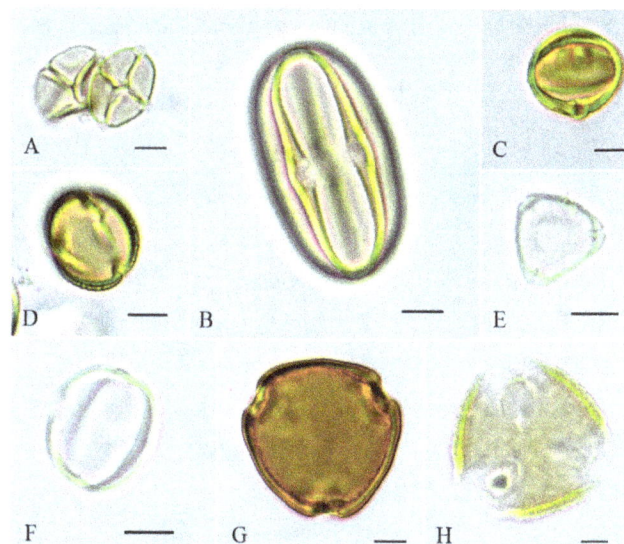

Fig 1. Percentages of pollen types identified in the samples from (A) pollen pots and (B) honey pots of *Melipona subnitida* in a restinga area, Maranhão state, Brazil.

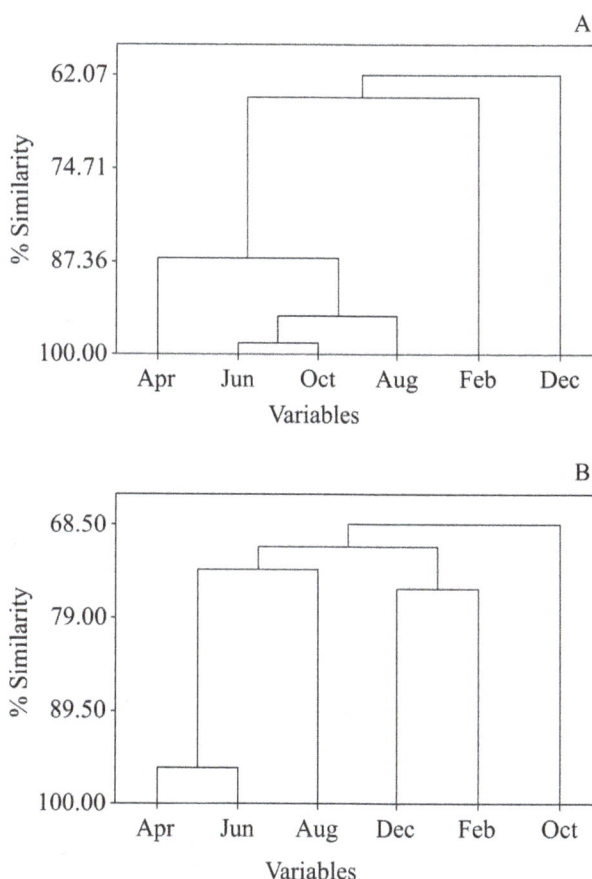

Fig 3. Similarity dendrogram based on the percentage data of (A) the 52 pollen types in pollen pots and (B) the 50 pollen types in honey pots of *Melipona subnitida*.

Based on floristic surveys performed in the region and specialized literature, we separated the genera and species of nectariferous plants present in honey. Of the 50 pollen types recorded in the honey samples, 20 were from plants that contributed to the production of *M. subnitida* honey (Table 2).

Assessing the number of pollen quantified in nectariferous plants only, we calculated the percentage frequencies of those pollen types in each month. *Humiria balsamifera* occurred as

Table 2. Frequencies (%) of pollen types of nectariferous species from honey samples produced by *Melipona subnitida* in a restinga area, Maranhão state, Brazil.

Family	Pollen type	Honey Pots					
		Apr	Jun	Aug	Oct	Dec	Feb
Anacardiaceae	*Anacardium microcarpum*		6.67	2.09	0.73		
Asteraceae	*Wulffia baccata*	0.12	0.75				0.09
Boraginaceae	*Heliotropium* sp.		0.03				
Burseraceae	*Protium heptaphyllum*	7.72	1.95	3.40	1.95		
Caryocaraceae	*Caryocar brasiliense*			0.15		0.04	
Chrysobalanaceae	*Chrysobalanus icaco*	3.13	5.44	24.56	13.32	41.84	27.80
Euphorbiaceae	*Mabea pohliana*		6.84				
Fabaceae	*Centrosema* sp.				0.09		
	Copaifera sp.		0.36	0.52	0.09		
	Mimosa caesalpiniifolia	0.72	2.93	0.42	1.08		
	Stryphnodendron adstringens		0.39	2.72	0.05	1.04	1.03
Gentianaceae	*Irlbachia pratensis*				0.09		
Humiriaceae	*Humiria balsamifera*	28.33	32.58	43.63	82.14	31.59	21.11
Loranthaceae	*Phtirusa pyrifolia*			0.05		0.40	0.13
Lythraceae	*Cuphea tenella*	20.74	0.81	4.23	0.14	3.39	0.16
Menyanthaceae	*Nymphoides indica*		0.33		0.06		
Passifloraceae	*Turnera ulmifolia*	5.55	6.05	3.34	0.05		
Polygonaceae	*Coccoloba* sp.	11.75	12.47	13.37		20.62	41.97
Rubiaceae	*Borreria verticillata*	21.94	22.40	1.15	0.14	1.08	7.71
Sapotaceae	*Pouteria* sp.			0.37	0.07		

secondary pollen in five of the six samples; in October, it was the predominant nectariferous species, with 82.14% of the total grains counted. *Chrysobalanus icaco* was identified as secondary pollen in August, December and February. *Coccoloba* sp. was considered secondary pollen in December and February. Two other species worth noting were *Borreria verticillata* and *Cuphea tenella* (Table 2).

Discussion

In the current study, we observed that *M. subnitida* depends on a variety of plant species to obtain pollen and nectar, but this specie prefers a particular plant spectrum. In the restinga region of the Maranhão eastern coast, *M. subnitida* collected pollen preferably from Fabaceae, Melastomataceae, Myrtaceae and Dilleniaceae species, which are frequently observed in palynological studies of pollen of the genus *Melipona* (Kerr et al., 1986/1987; Ramalho et al., 1989; Marques-Souza, 1996; Wilms & Wiechers, 1997; Luz et al., 2011).

In the Brazilian semi-arid region, where *M. subnitida* is abundant, the floral preferences of this species also agree with our study. Silva et al. (2006), in Paraíba state, observed the dominance of *Mimosa* and other Fabaceae species in pollen stored by *M. subnitida*. In addition, Maia-Silva et

al. (2012) in a botanical survey on the border of Ceará and Rio Grande do Norte states, also noted that the *M. subnitida* visited many Fabaceae species. This seems to fit the pattern of phylogenetically close species that interact with a similar set of species in a wide geographic range, with highly conserved associations (Thompson, 2005; Rezende et al., 2007).

Fabaceae and Myrtaceae families had the greatest number of observed pollen types, a finding that is closely related to the large richness of Fabaceae and Myrtaceae species and specimens in restinga areas of northeastern Brazil (Freire & Monteiro, 1993; Santos-Filho et al., 2013).

Three Melastomataceae species were very important for *M. subnitida*. This plant family is one of the most representative in South America (Ramalho et al., 1990). Given that many species of this family bloom sequentially throughout the year (Wilms & Wiechers, 1997), a high frequency of Melastomataceae in the food samples would be expected because of the intrinsic preference of *Melipona* for this plant family.

The preference of *M. subnitida* for species of Melastomataceae and *Chamaecrista ramosa* must be attributed to the fact that the flowers of those plants have poricidal anthers, which restrict the number of floral visitors. This feature is important for the foraging activity of *Melipona*, the only highly eusocial genus that performs buzz pollination – the capacity to vibrate flowers with poricidal anthers (Buchmann, 1983; Nunes-Silva et al., 2013). However, pollen collection is not restricted to this flower type, and the behavior of vibrating the thoracic flight muscles also occurs when visiting non-poricidal flowers, such as several Myrtaceae species and *Mimosa* (Nunes-Silva et al., 2010), which were abundant in the study samples.

The predominance of *Doliocarpus* in the February pollen samples is likely due to the great supply of this resource in that month. All the plant species visited have flowering peaks – periods with the highest number of flowers available, which is crucial for attracting foraging bees. Therefore, consecutive months tend to be more similar in regards to the plants visited, as observed by the similarity analyses of bimonthly samples.

Orbignya phalerata (the babassu, a native species of the transition zone between the Cerrado and the South Amazonian open forests (Albiero et al., 2007)) was recorded in the pollen pots in April, and it was also a relevant species among the studied samples, although to a lower degree. Several species of family Arecaceae are visited by bees due to the plants prolonged flowering and pollen abundance in their inflorescences (Oliveira et al., 2009).

Regarding the honey analysis, there was a high richness of pollen types in the samples, which usually hinders the determination of the nectariferous species. The polliniferous species tend to be greatly represented in honey and are considered contaminants because either the pollen grains were attached to the bee's body or the flower of the nectariferous species was contaminated by pollen from anemophilous plants that attached to the nectaries (Barth, 1989). The pollen types of plants that do not produce nectar but are present in honey are important because they help to broaden the knowledge of the flora where the colonies are located.

Mimosa misera was the most representative plant in the pollen pots and was abundant in the honey samples collected in April and June. *Mimosa misera* is a polliniferous species (Novais et al., 2009) and was therefore excluded from the honey analysis because it was considered a contaminant. Nevertheless, some authors consider *Mimosa* species to be nectar sources for *M. subnitida* (Almeida-Muradian et al., 2013; Silva et al., 2013). However, in general, the *Mimosa* genus consists of plants that produce high amounts of pollen and relatively little nectar, as they are over-represented in the pollen spectra. To be considered free of these contaminating plants, the honey samples need to have more than 98% of pollen from nectariferous species (Barth, 1989).

It is noteworthy that the polliniferous *Chamaecrista ramosa* and Melastomataceae and Myrtaceae species were abundant in the honey samples but did not contribute to its production. According to Buchmann (1983), plants adapted to buzz pollination, such as species of Melastomataceae and *C. ramosa*, usually experience secondary loss of floral nectar. The true nectariferous importance of many Myrtaceae species

is also doubtful because their flowers supply high amounts of pollen grains as their main resource but produce small amounts of nectar (Wilms & Wiechers, 1997; Gressler et al., 2006). In floristic surveys performed in the study region, Myrtaceae species were only considered as pollen sources.

Of the more abundant nectariferous species in the honey samples, *Humiria balsamifera* was present in all samples, since it blooms throughout the entire year (Machado et al., 2007). On average, each specimen of this plant has 50,000 small flowers (Viana & Kleinert, 2006), and they produce a small amount of nectar (Machado et al., 2007). Thus, *M. subnitida* foragers may compensate for the small of amount of nectar offered by the plants by continually visiting the abundant flowers.

Humiria balsamifera was considered predominant in October; therefore, honey sampled in October can be characterized as unifloral honey, due to the pollen dominance of this species (Barth, 2004). Although *H. balsamifera* was less frequent in other months, Ohe et al. (2004) noted that honey samples may be underrepresented in the pollen from plants that provided the nectar. For example, the occurrence of 2% to 42% of *Citrus* pollen in a honey sample is enough to characterize it as unifloral.

The nectariferous species *Chrysobalanus icaco*, *Coccoloba* sp. and *Borreria verticillata*, which were greatly important in different samples, are examples of sub-representations, not reflecting the real importance of those plants for *M. subnitida*. The possibility cannot be ruled out that at certain periods of the year, some honeys are produced solely from one of those plants.

The only representative of Lythraceae family (*Cuphea tenella*) was observed throughout the entire study in honey samples, but the percentage was only high in April. This finding may be related to increased flowering in that period and the bees' preference for certain food sources, depending on the ease of food gathering, the quantity and quality of trophic resources and interactions with competitors (Cortopassi-Laurino & Ramalho, 1988). In the coastal dunes of Abaeté, *Cuphea brachiata* Mart. ex Koehne produces nectar constantly but with low sugar concentrations (Viana & Kleinert, 2006), which can also occur with *C. tenella*, thus making it less attractive to bees.

In the pollen and honey samples, most of the pollen types was considered as minor pollen, indicating the polylectic nature of this bee. Plants with low pollen frequencies may have reduced pollen production or may be present in the samples due to the bees' indirect collection behavior. It is also known that different pollen types may be either stored for prolonged periods or be immediately offered to brood; therefore, minor pollens are poorly represented in the analyses of pollen pots (Malagodi-Braga & Kleinert, 2009). It must be considered that less frequent plants complement the colony's needs, ensuring nutritional balance in environments where the floral resource supply constantly changes due to seasonality.

The spectrum of floral sources visited by *M. subnitida* indicates its benefit for the pollination of several native species of the Lençóis Maranhenses restinga, as the bees clearly have a strong mutualistic relationship with both pollen-supplying and nectar-producing plants. In general, *M. subnitida* has botanical affinities that are very similar to those of *Melipona* genus.

Acknowledgments

The authors thank Dr. Léa Carreira of the Museu Paraense Emílio Goeldi, Msc. Angela Corrêa and Dr. Maria Amélia Cruz of the Botanical Institute of São Paulo (Instituto de Botânica de São Paulo) for their cooperation in identifying pollen types. The authors also thank Irene Aguiar, Mr. Emídio and Mrs. Maria, the owners of the jandaíra meliponary, and the Research and Scientific Development Foundation of Maranhão (Fundação de Amparo à Pesquisa e Desenvolvimento Científico do Maranhão – FAPEMA) for financial support.

References

Albiero, D., Maciel, A.J.S., Lopes, A.C. & Gamero, C.A. (2007). Proposta de uma máquina para colheita mecanizada de babaçu (*Orbignya phalerata* Mart.) para a agricultura familiar. Acta Amazon. 37: 337-346. doi: 10.1590/S0044-59672007000300004

Almeida-Muradian, L.B., Stramm, K.M., Horita, A., Barth, O.M., Freitas, A.S. & Estevinho, L.M. (2013). Comparative study of the physicochemical and palynological characteristics of honey from *Melipona subnitida* and *Apis mellifera*. Int. J. Food Sci. Tech. 48: 1698-1706. doi: 10.1111/ijfs.12140

APG III. (2009). An update of the Angiosperm Phylogeny Group classification for the orders and families of flowering plants: APG III. Bot. J. Linn. Soc. 161: 105-121. doi: 10.1111/j.1095-8339.2009.00996.x

Barth, M.O. (1989). O pólen no mel brasileiro. Rio de Janeiro: Gráfica Luxor, 45 p.

Barth, M.O. (2004). Melissopalynology in Brazil: A review of pollen analysis of honey, propolis and pollen loads of bees. Sci. Agric. 61: 342-350. doi: 10.1590/S0103-90162004000300018

Buchmann, S.L. (1983). Buzz pollination in angiosperms. In C.E. Jones & R.J. Little (Eds.), Handbook of experimental pollination biology (pp. 73-113). New York: Van Nostrand Reinhold.

Camargo, J.M.F. (1970). Ninhos e biologia de algumas espécies de meliponíneos (Hymenoptera: Apidae) da região de Porto Velho, território de Rondônia, Brasil. Rev. Biol. Trop. 16: 203-239.

Camargo, J.M.F. & Pedro, S.R.M. (2007). Meliponini Lepeletier, 1836. In J.S. Moure, D. Urban & G.A.R. Melo (Eds.), Catalogue of bees (Hymenoptera, Apoidea) in the Neotropical region (pp. 272-578). Curitiba: Sociedade Brasileira de Entomologia.

Carreira, L.M.M. & Barth, O.M. (2003). Atlas de pólen da vegetação de canga da Serra de Carajás. Belém: Museu Paraense Emílio Goeldi, 112 p.

Carvalho, C.A.L., Moreti, A.C.C.C., Marchini, L.C., Alves, R.M.O. & Oliveira, P.C.F. (2001). Pollen spectrum of honey "uruçu" bee (*Melipona scutellaris* Latreille, 1811). Rev. Bras. Biol., 61: 63-67. doi: 10.1590/S0034-71082001000100009

Cortopassi-Laurino, M. & Ramalho, M. (1988). Pollen harvest by Africanized *Apis mellifera* and *Trigona spinipes* in São Paulo botanical and ecological views. Apidologie 19: 1-24. doi: 10.1051/apido:19880101

Cruz, D.O., Freitas, B.M., Silva, L.A., Silva, E.V.S. & Bomfim, I.G.A. (2004). Adaptação e comportamento de pastejo da abelha jandaíra (*Melipona subnitida* Ducke) em ambiente protegido. Acta Sci. Anim. Sci. 26: 293-298. doi: 10.4025/actascianimsci.v26i3.1777

Erdtman, G. (1960). The acetolysis method. Sven. Bot. Tidskr. 54: 561-564.

Ferraz, R.E., Lima, P.M., Pereira, D.S., Freitas, C.C.O. & Feijó, F.M.C. (2008). Microbiota fúngica de *Melipona subnitida* Ducke (Hymenoptera: Apidae). Neotrop. Entomol. 37: 345-6. doi: 10.1590/S1519-566X2008000300017

Freire, M.C.C.C. & Monteiro, R. (1993). Florística das praias da Ilha de São Luís, Estado do Maranhão (Brasil): Diversidade de espécies e suas ocorrências no litoral brasileiro. Acta Amazon. 23: 125-140.

Gressler, E., Pizo, M.A. & Morellato, L.P.C. (2006). Polinização e dispersão de sementes em Myrtaceae do Brasil. Rev. Bras. Bot. 29: 509-530. doi: 10.1590/S0100-84042006000400002

Imperatriz-Fonseca, V.L., Alves-dos-Santos, I., Santos-Filho, P.S., Engels, W., Ramalho, M., Wilms, W., Aguilar, J.B.V., Pinheiro-Machado, C.A., Alves, D.A. & Kleinert, A.M.P. (2011). Checklist of bees and honey plants from São Paulo State, Brazil. Biota Neotrop. 11: 1-25. doi: 10.1590/S1676-06032011000500029

Instituto Brasileiro do Meio Ambiente e dos Recursos Naturais Renováveis (IBAMA). (2002). Parque Nacional dos Lençóis Maranhenses - Plano de Manejo.

Iwama, S. & Melhem, T.S. (1979). The pollen spectrum of the honey of *Tetragonisca angustula* Latreille (Apidae, Meliponinae). Apidologie 10: 275-295. doi: 10.1051/apido:19790305

Kerr, W.E., Absy, M.L. & Marques-Souza, A.C. (1986/1987). Espécies nectaríferas e poliníferas utilizadas pela abelha *Melipona compressipes fasciculata*, no Maranhão. Acta Amazon. 16/17: 145-156.

Louveaux, J., Maurizio, A. & Vorwohl, G. (1978). Methods of melissopalynology. Bee World 59: 139-157.

Luz, C.F.P., Fernandes-Salomão, T.M., Lage, L.G.A., Resende, H.C., Tavares, M.G. & Campos, L.A.O. (2011). Pollen sources for *Melipona capixaba* Moure & Camargo: An endangered Brazilian stingless bee. Psyche 2011: 1-7. doi: 10.1155/2011/107303

Luz, C.F.P., Thomé, M.L. & Barth, O.M. (2007). Recursos tróficos de *Apis mellifera* L. (Hymenoptera, Apidae) na região de Morro Azul do Tinguá, Estado do Rio de Janeiro. Rev. Bras. Bot. 30: 29-36. doi: 10.1590/S0100-84042007000100004

Machado, C.G., Coelho, A.G., Santana, C.S. & Rodrigues, M. (2007). Beija-flores e seus recursos florais em uma área de campo rupestre da Chapada Diamantina, Bahia. Rev. Bras. Ornitol. 15: 267-279. Retrieved from: http://inot.org.br/wp-content/uploads/ Beija-flores_e_seus_recursos_florais_em_uma_%C3%A1rea_ de_campo_rupestre_da_Chapada_Diamantina_Bahia.pdf

Maia-Silva, C., Silva, C.I., Hrncir, M., Queiroz, R.T. & Imperatriz-Fonseca, V.L. (2012). Guia de Plantas visitadas por abelhas na Caatinga. http://www.mma.gov.br/estruturas/203/_ arquivos/livro_203.pdf (accessed date: 11 March, 2013).

Malagodi-Braga, K.S. & Kleinert, A.M.P. (2009). Comparative analysis of two sampling techniques for pollen gathered by *Nannotrigona testaceicornis* Lepeletier (Apidae, Meliponini). Genet. Mol. Res. 8: 596-606. doi: 0.4238/vol8-2kerr014

Marques-Souza, A.C. (1996). Fontes de pólen exploradas por *Melipona compressipes manaosensis* (Apidae: Meliponinae), abelha da Amazônia Central. Acta Amazon. 26: 77-86.

Martins, A.C.L., Rêgo, M.M.C., Carreira, L.M.M. & Albuquerque, P.M.C. (2011). Espectro polínico de mel de tiúba (*Melipona fasciculata* Smith, 1854, Hymenoptera, Apidae). Acta Amazon. 4: 183-190. doi: 10.1590/S0044-59672011000200001

Martins, C.F. (2002). Diversity of the bee fauna of the Brazilian Caatinga. In P.G. Kevan, & V.L. Imperatriz-Fonseca (Eds.), Pollinating bees – the conservation link between agriculture and nature (pp. 131-134). Brasília: Ministério do Meio Ambiente.

Novais, J.S., Lima, L.C.L. & Santos, F.A.R. (2009). Botanical affinity of pollen harvested by *Apis mellifera* L. in a semi-arid area from Bahia, Brazil. Grana 48: 224-234. doi: 10.1080/00173130903037725

Nunes-Silva, P., Hrncir, M. & Imperatriz-Fonseca, V.L. (2010). A polinização por vibração. Oecol. Aust. 14: 140-151. doi: 10.4257/oeco.2010.1401.07

Nunes-Silva, P., Hrncir, M., Silva, C.I., Roldão, Y.S. & Imperatriz-Fonseca, V.L. (2013). Stingless bees, *Melipona fasciculata*, as efficient pollinators of eggplant (*Solanum melongena*) in greenhouses. Apidologie 44: 537-546. doi: 10.1007/s13592-013-0204-y

Ohe, W.V.D., Oddo, L.P., Piana, M.L., Morlot, M. & Martin, P. (2004). Harmonized methods of melissopalynology. Apidologie 35: S18-S25. doi: 10.1051/apido:2004050

Oliveira, F.P.M., Absy, M.L. & Miranda, I.S. (2009). Recurso polínico coletado por abelhas sem ferrão (Apidae, Meliponinae) em um fragmento de floresta na região de Manaus – Amazonas. Acta Amazon. 39: 505-518. doi: 10.1590/S0044-59672009000300004

Pick, R.A. & Bloctein, B. (2002). Atividades de coleta e origem floral do pólen armazenado em colônias de *Plebeia saiqui* (Holmberg) (Hymenoptra, Apidae, Meliponinae) no Sul do Brasil. Rev. Bras. Zool. 19: 289-300. doi: 10.1590/S0101-81752002000100025

Ramalho, M. (1990). Foraging by the stingless bees of the genus *Scaptotrigona* (Apidae, Meliponinae). J. Apicult. Res. 29: 61-67.

Ramalho, M., Kleinert-Giovannini, A. & Imperatriz-Fonseca, V.L. (1989). Utilization of floral resources by species of *Melipona* (Apidae, Meliponinae): floral preferences. Apidologie 20: 185-195. doi: 10.1051/apido:19890301

Rêgo, M.M.C & Albuquerque, P.M.C. (2006). Redescoberta de *Melipona subnitida* Ducke (Hymenoptera: Apidae) nas Restingas do Parque Nacional dos Lençóis Maranhenses, Barreirinhas, MA. Neotrop. Entomol. 35: 416-417. doi: 10.1590/S1519-566X2006000300020

Rezende, E.L., Lavabre, J.E., Guimarães Jr, P.R., Jordano, P. & Bascompte, J. (2007). Non-random coextinctions in phylogenetically structured mutualistic networks. Nature 448: 925-929. doi:10.1038/nature05956

Roubik, D.W. (1989). Ecology and Natural History of Tropical Bees. New York: Cambridge University Press, 526 p.

Roubik, D.W. & Moreno, J.E. (1991). Pollen and Spores of Barro Colorado Island. St. Louis: Missouri Botanical Garden, 268 p.

Santos-Filho, F.S., Almeida Jr, E.B. & Zickel, C.S. (2013). Do edaphic aspects alter vegetation structures in the Brazilian restinga? Acta Bot. Bras. 27: 613-623. doi: 10.1590/S0102-33062013000300019

Silva, C.I., Ballesteros, P.L.O., Palmero, M.A., Bauermann, S.G., Evaldt, A.C.P. & Oliveira, P.E. (2010). Catálogo polínico: Palinologia aplicada em estudos de conservação de abelhas do gênero *Xylocopa* no Triângulo Mineiro. Uberlândia: EDUFU, 154 p.

Silva, E.M.S., Freitas, B.M., Silva, L.A., Cruz, D.O. & Bomfim, I.G.A. (2005). Biologia floral do pimentão (*Capsicum annuum*) e a utilização da abelha jandaíra (*Melipona subnitida* Ducke) como polinizador em cultivo protegido Rev. Ciênc. Agron. 36: 386-390. http://www.ccarevista.ufc.br/seer/index.php/ccarevista/ article/view/256/ 251 (accessed date: 11 March, 2013).

Silva, G.R., Souza, B.A., Pereira, F.M., Lopes, M.T.R., Valente, S.E.S. & Diniz, F.M. (2014). New molecular evidence for fragmentation between two distant populations of the threatened stingless bee *Melipona subnitida* Ducke (Hymenoptera, Apidae, Meliponini). J. Hymenopt. Res. 38: 1-9. doi: 10.3897/JHR.38.7302

Silva, T.M.S., Camara, C.A., Lins, A.C.S., Barbosa-Filho, J.M., Silva, E.M.S., Freitas, B.M. & Santos, F.A.R. (2006). Chemical composition and free radical scavenging activity of pollen loads from stingless bee *Melipona subnitida* Ducke. J. Food Comp. Anal. 19: 507-511. doi:10.1016/j.jfca.2005.12.011

Silva, T.M.S., Santos, F.P., Evangelista-Rodrigues, A., Silva,

E.M.S., Silva, G.S., Novais, J.S., Santos, F.A.R. & Camara, C.A. (2013). Phenolic compounds, melissopalynological, physicochemical analysis and antioxidant activity of jandaíra (*Melipona subnitida*) honey. J. Food Comp. Anal. 29: 10-18. doi: 10.1016/j.jfca.2012.08.010

Thompson, J.N. (2005). The geographic mosaic of coevolution. Chicago: University of Chicago Press, 443 p.

Viana, B.F. & Kleinert, A.M.P. (2006). Structure of bee-flower system in the coastal sand dune of Abaeté, Northeast of Brazil. Rev. Bras. Entomol. 50: 53-63. doi: 10.1590/S0085-56262006000100008

Wilms, W. & Wiechers, W. (1997). Floral resource partitioning between native *Melipona* bees and the introduced Africanized honey bee in the Brazilian Atlantic rain forest. Apidologie 28: 339-355. doi: 10.1051/apido:19970602

30

Pollen and nectar foraging by *Melipona quadrifasciata anthidioides* Lepeletier (Hymenoptera: Apidae: Meliponini) in natural habitat

C Oliveira-Abreu[1], SD Hilário[2], CFP Luz[3], I Alves-dos-Santos[2]

1 - Universidade de São Paulo, Ribeirão Preto, Brazil
2 - Universidade de São Paulo/IBUSP, São Paulo, Brazil
3 - Instituto de Botânica de São Paulo, São Paulo, Brazil

Keywords
Stingless bees, foraging behavior, nectar, pollen.

Corresponding author
Sergio Dias Hilário
Departamento de Ecologia, Instituto de Biociências, Universidade de São Paulo
Rua do Matão, trav. 14, 321
Cidade Universitária
05508-090, São Paulo, SP, Brazil
E-Mail: sedilar@usp.br

Abstract
This study shows the influence of meteorological factors on the collection of nectar and pollen by *Melipona quadrifasciata anthidioides* Lepeletier foragers in their natural habitat. Five *M. quadrifasciata anthidioides* colonies were studied in Parque das Neblinas, Mogi das Cruzes district (23º44'52"S/46º09'46"W), from October 2009 to September 2010. The foraging activity of the worker bees was observed monthly and the temperature and relative humidity were registered and the pollen grains types from the pollen loads were identified. The peaks of pollen collection occurred between 08:30-09:50 am, while nectar was gathered along the day. The relation between resources sampling and environmental temperature is best described with a polynomial function, while in relation to relative humidity the curves of foraging activity is slightly asymmetric to left. A total of 24 pollen types were identified and the most frequents were Myrtaceae (*Eucalyptus*, *Myrcia*), Melastomataceae and Solanaceae. The tolerance to the environmental conditions is discussed, as well as the plants explored for pollen sources.

Introduction

Flight activity in stingless bees have been studied for many species in Brazil, such as: *Melipona quadrifasciata quadrifasciata* Lepeletier, *Melipona marginata marginata* Lepeletier, *Melipona obscurior* Moure, *Melipona bicolor bicolor* Lepeletier (review in Hilário et al., 2000), *Trigona hyalinata* Lepeletier (Contrera et al., 2004*)*, *Tetragona clavipes* Fabricius (Rodrigues et al., 2007), *Melipona bicolor schencki* Gribodo (Ferreira-Junior et al., 2010), *Melipona rufiventris* Lepeletier (Fidalgo & Kleinert, 2010*)*, *Melipona eburnea* Friese (Nates-Parra & Rodríguez, 2011), *Plebeia remota* Holmberg (Hilário & Imperatriz-Fonseca, 2012), *Scaptotrigona depilis* Moure (Figueiredo-Mecca et al., 2013) and *Geotrigona mombuca* Smith (Gobatto & Knoll 2013). In the majority of these studies the flight activity of the workers was associated with environmental conditions like temperature, barometric pressure, relative humidity, wind speed and even time of the day. However, any of these conditions can interact with biotic factors such as bees' physiology or biological clocks (Hilário et al., 2003; Gouw & Gimenes, 2013), influencing the performance of the individuals. In fact, the amplitude of tolerance to many ecological factors (with the minimum and maximum limits of tolerance) may have synergistic effects.

Besides the environmental conditions, the availability of food resources also play an important role on the limits where the species can live, grow and reproduce. In stingless bees, the essential food resources are pollen and nectar, which are stored in pots inside the nest (Michener, 1974; Roubik, 1989). The health and size of the colonies may reflect the availability of the floral resources in the field. Thus, the number of individuals in the nest and the colony survival along the years are dependent, among other factors, on the amount of food collected and stored. On the other hand, the population size of each nest characterizes the colony as small (or weak), medium or large (or strong) and has a direct relationship with the foraging activities among other factors (Kleinert-Giovannini & Imperatriz-Fonseca, 1986; Hilário et al., 2000; Hilário et al., 2003; Hilário & Imperatriz-Fonseca, 2009).

Melipona quadrifasciata Lepeletier, has a relative broad geographic distribution along eastern Brazil, reaching the Misiones region in Argentina and Paraguay (Camargo & Pedro, 2008; Batalha-Filho et al., 2009). This species is commonly associated with the Atlantic Rainforest habitats in Southern and Southeastern Brazil. The colony of this species is less numerous than other stingless bees, approximately 300-400 individuals (Tóth et al., 2004). In nature the nest is constructed inside trees hollows, but it is easily kept and maintained into artificial hives. Concerning the plants visited by M. quadrifasciata this species is considered generalist, but several species of Myrtaceae, Asteraceae, Melastomataceae and Solanaceae are among the preferred floral resources (Ramalho et al., 1989; Wilms & Wiechers, 1997; Antonini et al., 2006a; Antonini et al., 2006b).

The aim of this study was to evaluate the pattern of pollen and nectar collection by M. quadrifasciata and their relationship with environmental factors like the temperature and relative humidity under natural conditions. This species has an economic importance for the beekeeper and can be used as pollinator in greenhouse, in tomato production for example (Del Sarto et al., 2005; Bispo dos Santos et al., 2009; Bartelli et al., 2014; Bartelli & Nogueira-Ferreira, 2014).

Material and Methods

Study area

This study was conducted in the Parque das Neblinas, located in the Mogi das Cruzes district in Sao Paulo state, Brazil (23°44'52"S/46°09'46"W; 700 to 1100m a.s.l.). The park has 2,788.15 hectares and the main vegetation is the Atlantic Rainforest. The climate of the region according to Köppen is *Af*, considered as tropical and always humid with the average temperature above 18°C during the summer. Annual precipitation varies between 1600 and 2000 mm. There is no winter dry season and only a decrease of precipitation is registered.

External Activities of the Bees

Five colonies of M. quadrifasciata were maintained in free-foraging wooden nest boxes at the Park. During the first month the colonies were fed every 2 weeks with a sugar syrup (1:1 sugar and water).

The flight activity of the bees was observed monthly

between October 2009 and September 2010. For each month, the number of bees returning from the field to the nest was registered along two consecutive days, during five minutes per hour, between 05:30 am to 04:30 pm, which corresponded to 1430 observations. Pollen was easily observed on the corbiculae of foragers, due to the color and texture of the load. On the other hand, foragers returning to the colonies, without load on the legs, were considered as nectar foragers considering that nectar is stored and transported inside the honey stomach (Roubik, 1989; Pierrot & Schlindwein, 2003; Souza et al., 2006). Other materials, like resin and mud were not considered in the analysis.

During the observations the temperature and relative humidity were registered every 30 minutes, using dataloggers *(HOBO Pro RH/Temp)*, which were installed in the area and later the data were transferred to the computer.

Pollen analysis

One day after the field observations, samples of the honey collected from the pots and the pollen collected from the corbiculae of the foragers were taken from the five nests from December 2009 to September 2010. For that, the entrance of the colony was closed for five minutes and five arriving foragers were caught with entomological net. The pollen loads were extracted from their corbiculae and kept on separated plastic tubes (eppendorff) and the bees were released. For each colony samples of 10mL of honey from three new storage pots were sampled with syringes and kept separated into sterilized vials. In February honey and pollen were not sampled due to excessive rainfall.

The honey samples and pollen pellets were prepared fresh according to the standard European protocol (Maurizio & Louveaux, 1965) for palynological studies. Three microscopic slides with glycerine jelly were prepared for each sample and were sealed with paraffin.

Pollen grains identification was carried out using literature data (e.g., Barth, 1989; Roubik & Moreno, 1991) and the reference of pollen grains previously identified and maintained in slides in the collection from the Research Center of Palynology, Botanical Institute, Department of Environment of São Paulo State. The pollen grains were identified using the term "pollen type" that means pollen from a single plant species as well as a group of species or higher taxa presenting similar pollen morphology.

Data analyses

Since our data had a non-normal distribution, we performed nonparametric correlation Spearman's rho (Zar, 1999) to verify if there was relationship between the meteorological factors and the pattern of nectar and pollen collection. For statistical analysis the SPSS for Windows release 8.0 was used.

Results

The temperature range during the experiment was between 5.8° and 37.0°C, and the relative humidity range was between 15% and 100%. Foraging flights were registered within a temperature range of 13.7°-36.8°C, while relative humidity (RH) oscillated between 15.7% and 100%.

The number of pollen and nectar foragers was positively correlated to the air temperature, and negatively with the relative humidity (Table 1). These correlations were weak, since that there was a peak in intermediates temperature and relative humidity ranges (Figs 1 and 2).

Table 1. Spearman's correlation coefficients (rho) for pollen and nectar collection by *Melipona quadrifasciata anthidioides* foragers and meteorological factors.

		rho	p
Pollen	Temperature	0.174	0.001 (S)
	Relative Humidity	-0.057	0.030 (S)
Nectar	Temperature	0.208	0.001 (S)
	Relative Humidity	-0.063	0.017 (S)

N=1430; S= Significat, NS=Non-Significat; p=probability.

The relation of resources sampling and environmental temperature was best described with a polynomial function, showing an optimal interval between ca. 17.0°C -29.0°C, which corresponded to an intense foraging activity for nectar and pollen (Fig 1). It is also possible to observe that there is a limit of tolerance in temperature below 13.7°C, and above 35.0°C. Peaks of pollen and nectar collection occurred at the temperature intervals of 20.1°-23.0°C and 17.1°-20.0°C, respectively (Fig 1).

In relation to relative humidity the tendency to polynomial function is not clear and the curves of resources sampling are slightly asymmetric to left (Fig 2). Foraging activity increased with 40% relative humidity. Both peaks of pollen and nectar collection occurred at the relative humidity intervals of 60.1-70.0% (Fig 2).

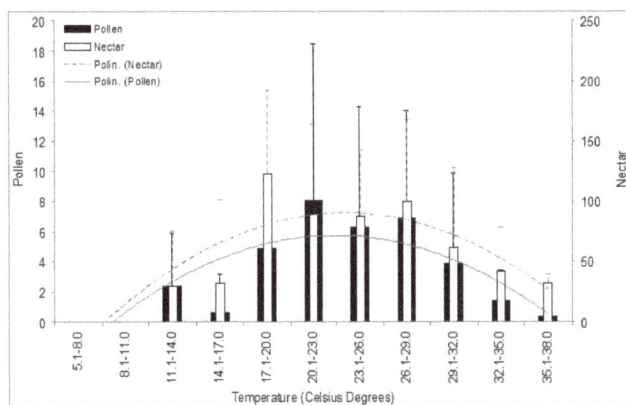

Fig 2. Pollen and nectar collection by *Melipona quadrifasciata anthidiodes* related to the relative humidity intervals. Solid and dashed bars represent standard deviation of pollen and nectar collection, respectively.

The number of bees collecting resource increased until 09:50 am, with a peak of nectar sample between 08:30-08:50 am and pollen gather between 09:30-09:50 am. After this period the income of pollen decreased gradually, while the bees remain collecting nectar for the rest of the day (Fig 3).

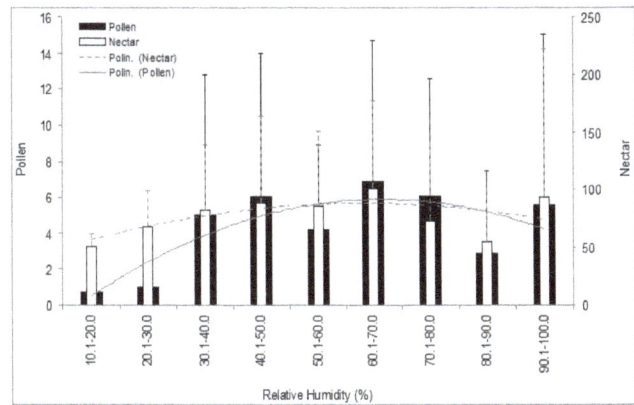

Fig 1. Pollen and nectar collection by *Melipona quadrifasciata anthidioides* related to the environmental temperature intervals. Solid and dashed bars represent standard deviation of pollen and nectar collection, respectively.

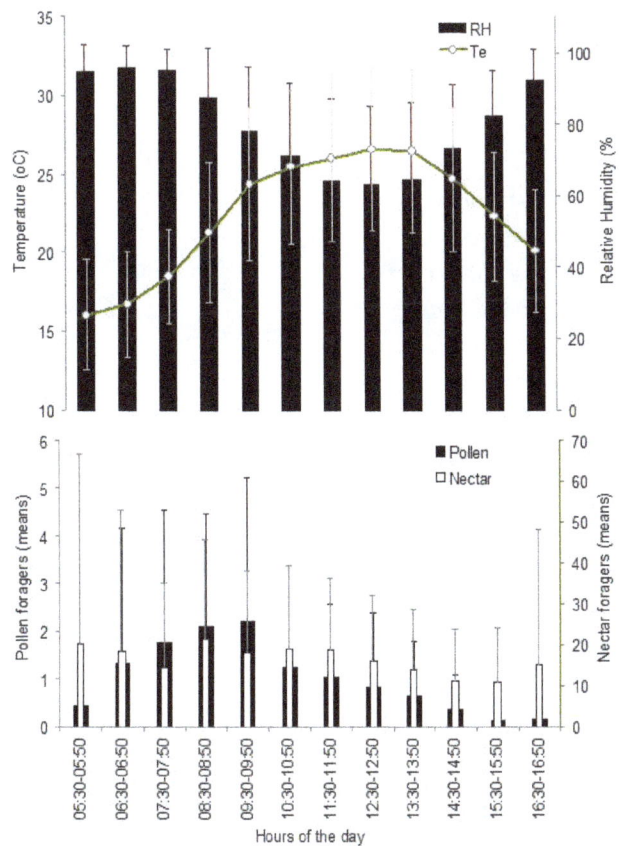

Fig 3. Hourly variation of pollen and nectar collection related to the meteorological factors (temperature and relative humidity). Solid and dashed bars represent standard deviation of pollen and nectar collection, respectively.

The number of pollen types from honey pots and corbiculae was moderately correlated to the number of bees entering inside the nest with pollen and nectar (rho = 0.410; p = 0.009; Fig 4).

In samples of the honey pots and pollen of the corbiculae a total of 24 pollen types were recorded, recognizing 18 genera, 16 families. The most frequent pollen types in samples of honey and pollen of were Myrtaceae (*Eucalyptus*, *Myrcia*), Melastomataceae and Solanaceae, which were also presented along all the year (Table 2; Fig 5).

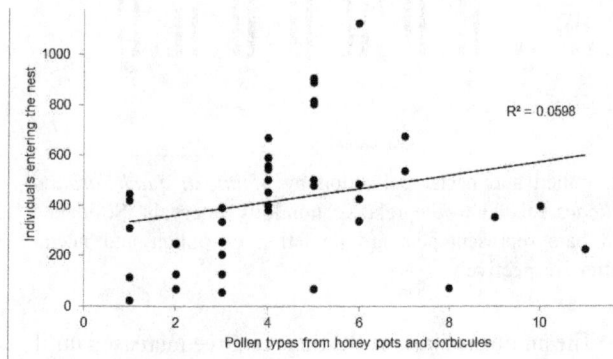

Fig 4. Relationship between the number of pollen grain types (from corbiculae loads and honey from pots) with the total numbers of pollen and nectar foragers returning to the nest.

Discussion

The concept of niche proposed by Hutchison (Hutchinson, 1957) refers to the way that the tolerance and needs of the many conditions interact with the resources used by a species. The variables analyzed in this work are part of the ecological niche of *M. quadrifasciata anthidioides* in their natural environment, since we recorded the ideal temperatures and humidity for the foraging activity. In other words, environmental conditions influences the adaptive value in this species. In addition another important component of the niche, which are the plants explored for pollen sources, was also recorded.

According to Cobert et al., (1993) the temperature is the environmental factor that most affects the flight activity on bees. The extreme values of temperature affect the individual thermoregulation prerequisite to flight behavior (Heinrich & Esch, 1994).

Our results demonstrate that *M. quadrifasciata anthidioides* reduces the collection of resources with temperatures below 13.7°C and over 29.0°C when the foraging activity decreases considerably, in the studied area. The lowest value is very similar to temperatures found to this species in an urban

Table 2. Pollen types present on the corbiculae of the workers (*) and honey pots (+) in 5 *Melipona quadrifasciata* nests, sampled during 9 months (December 2009 to September 2010) in the Parque das Neblinas.

Pollen types/ Months	Dec	Jan	Mar	Apr	Mai	Jun	Jul	Ago	Sep
Aegiphila (+)		X							
Alchornea (+)			X						
Baccharis (*) (+)	X	X	X		X		X	X	
Begonia (*)							X		
Eucalyptus (*) (+)	X	X	X	X	X	X	X	X	X
Euterpe/Syagrus (+)		X			X				
Hibiscus (+)		X	X						
Inga (+)		X							
Machaerium (+)	X	X	X						
Maranthaceae (+)		X							
Melastomataceae (*) (+)	X	X	X	X	X	X	X	X	X
Mimosa caesalpiniaefolia (+)			X	X		X			
Mimosa scabrella (*)							X		
Monocot (+)		X		X					
Myrcia (*) (+)	X	X	X	X	X	X	X	X	X
Schefflera (+)				X					
Serjania (+)		X	X		X	X			
Solanaceae (*) (+)	X	X	X		X	X	X	X	X
Sorocea (+)				X					
Struthanthus (+)		X							
Stylosanthes (+)							X		
Triumfetta (+)			X						
Vernonia (*) (+)		X						X	X
Not identified (+)		X							

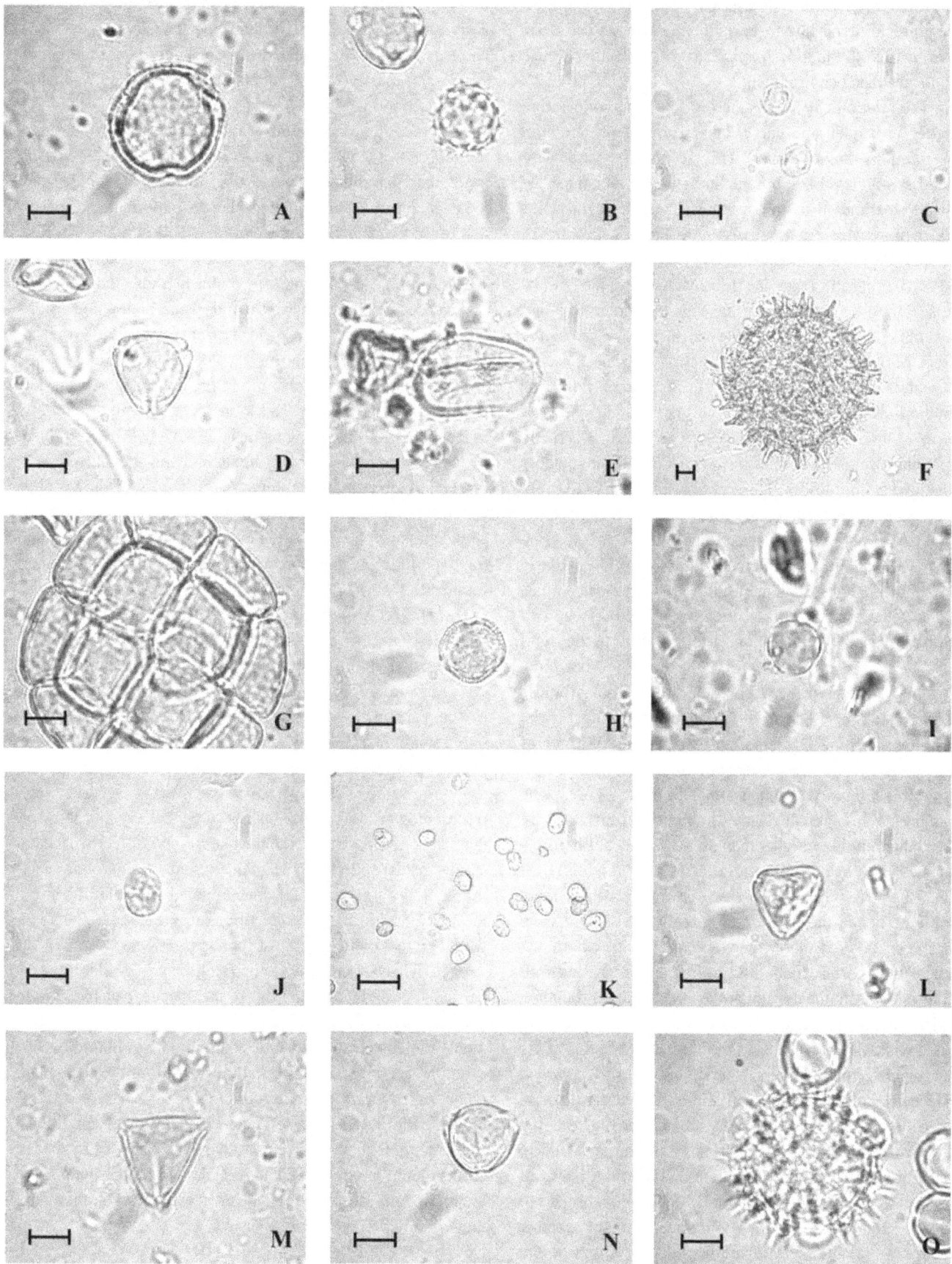

Fig 5. Light microscopic (LM) pollen grain types micrographs observed in both honey and pollen pellets samples of *Melipona quadrifasciata* in Parque das Neblinas, Mogi das Cruzes district, São Paulo, Brazil. Figure A. *Aegiphila*. B. *Baccharis*. C. *Begonia*. D. *Eucalyptus*. E. *Euterpe/Syagrus*. F. *Hibiscus*. G. *Inga*. H. *Machaerium*. I. Melastomataceae. J. *Mimosa caesalpiniaefolia*. K. *Mimosa scabrella*. L. *Myrcia*. M. *Serjania*. N. Solanaceae. O. *Vernonia*. Scales = 10µ.

area (Guibu & Imperatriz-Fonseca, 1984), but the highest temperature limits seems to be higher in our studied area, comparing also to other *Melipona*. This may occurs because of the local shaded environment.

The relation between gathering food and temperature fits better in a quadratic curve, demonstrating an optimum range and limits of tolerance. This seems to be truth for many other stingless bees, but the limits vary a lot among the species. Sometimes the limits are tested, for instance when a sudden mass flowering provoke a copious collection of pollen outside the foraging pattern (Pierrot & Schlindwein, 2003). In the subtribe Trigonina (tribe Meliponini) the optimal range seems to be shorter, starting with temperature higher. For example Mouga (1984) observed that for *Paratrigona subnuda* Moure the greatest flight activity occurs between temperatures of 24.0°C and 25.0°C; in *Plebeia saiqui* Friese in southern Brazil, the collection of pollen had a higher intensity in the range of 18.0° to 19.0°C (Pick & Blochtein, 2002). Some *Melipona* species shows the optimal range wider and the lowest limit lower, like *M. marginata* (19.0°-30.0°C), *M. bicolor bicolor* (16.0°-26.0°C) and *M. rufiventris* (16.0°-30.0°C) for the activity of related pollen and nectar collection (Kleinert-Giovannini & Imperatriz-Fonseca, 1986; Hilário et al., 2000; Fidalgo & Kleinert, 2007). In Cruz das Almas, Bahia, northeastern Brazil, Nascimento et al., (2012) recorded a significant negative correlation of temperature to the flight activity of *M. quadrifasciata*, and the peaks of bee flow occurred in temperature between 21.2° and 23.3°C. All these mentioned bee species are of different size, and probably the body dimension plays an important role on the tolerance of most suitable climatic factors for foraging.

Another important dimension of the niche for *Melipona* bees is the environmental relative humidity. In this case we recorded that humidity higher than 30-40% is ideal for pollen gathering, while nectar collection occurred in a wide range of RH. But, it is important to consider that pollen is more available earlier in the day, when humidity is higher, while nectar in general is available the whole day. Furthermore, the habitat of the coastal Atlantic Rainforest is moist almost all over the year, due to the dynamic of the trade winds producing precipitation or fog daily. More than $^1/_3$ of our data were recorded with RH above 90%, in the studied area. Silva et al., (2011) suggested that foragers from large species of *Melipona* use RH as indicator of pollen availability in the environment.

As it was expected for *Melipona* bees the foraging activities are concentrated early in the morning and tend to decrease after midday. But in the studied area the peaks of pollen and nectar collection are slightly delay, probably due to the fogs formed at dawn and early in the morning. For many *Melipona* species the peaks of pollen harvest occurs before 08:00 am (Bruijn & Sommeijer, 1997; Fidalgo & Kleinert, 2007) for example.

The moderate correlation among the number of pollen types and bees entering the nest suggest that an increasing in flowering plants elicited an increasing in foraging intensity. But we need to consider that pollen type diversity not necessary correspond to floral abundance. It is know that Meliponini species use massive flowering plants (Roubik, 1989; Wilms & Wiechers, 1997).

According to Antonini et al., (2006a) *M. quadrifasciata anthidioides* is considered a generalist species, since it visited more than ten plant species. But, there is no doubt that Myrtaceae (*Eucalyptus*) pollen play an important role in the diet of *M. quadrifasciata anthidioides*. This was also verified in other areas (Ramalho et al., 1989; Wilms & Wiechers, 1997; Antonini et al., 2006b). Similarly, more than 75% of Myrtaceae pollen grains was found in the honey and in the larval food from five species of *Melipona* in the Amazon forest (Cortopassi-Laurino et al., 2007). Luz et al., (2011) and Serra et al., (2012) also found preference for Myrtaceae pollen in *Melipona capixaba* Moure and Camargo. According to Roubik (1989) the foraging behavior of a bee is determined by the resources it has access, being influenced by the quality, dispersion, quantity and competition for these resources plus the environmental conditions. In the study area, *Eucalyptus* is abundant, since it was used in the park reforestation program for a period, and so represents no competition with other foragers.

Hilgert-Moreira et al., (2014) demonstrated that regardless of the landscape characteristics, *M. obscurior* foragers were able to collect pollen from *Eucalyptus* throughout the year. According to these authors, this persistence is due to the capability of *Eucalyptus* in supporting more than one foraging species, which results in low or inexistent competition for the resources.

In samples of the honey pots many species were represented just once during the year. We believe that some of these species (like the first of the list: *Aegiphila*, Verbenaceae) are just nectar sources for *M. quadrifasciata anthidioides*, and probably the foragers were contaminated with the pollen on the body during nectar collection. An alternate explanation would be that their bloom just for short period.

The environmental conditions and availability of particular resources define where the species can live, grow and reproduce. Furthermore, we have to remember that the niche dimensions (temperature and relative humidity for example) act together. Under natural conditions the ecological niche of *M. quadrifasciata anthidioides* is composed of multiple dimensions, i.e. tolerance to many other conditions and requirements that were not measured in this study, but it seems that in the studied area, *M. quadrifasciata anthidioides* are finding the good requirements since the five colonies kept the reproduction the whole year.

Acknowledgment

We are grateful for the financial support of CAPES, FAPESP, CNPq and the permission and help in the field of the Instituto Ecofuturo. The authors express their gratitude to Roberto Shimizu for his assistance in data analysis, Denise de Araujo Alves and Carlos Eduardo Pinto that helped in the discussion of the data and the two anonymous referees for their valuable comments and suggestions.

References

Antonini, Y., Costa, R.G. & Martins, R.P. (2006a). Floral preferences of a Neotropical stingless bee, *Melipona quadrifasciata* Lepeletier (Apidae: Meliponina) in an urban Forest fragment. Braz. J. Biol., 66: 463-471. doi: 10.1590/S1519-69842006000300012.

Antonini, Y., Soares, S.M. & Martins, R.P. (2006b). Pollen and nectar harvesting by the stingless bee *Melipona quadrifasciata anthidioides* (Apidae: Meliponini) in an urban forest fragment in Southeastern Brazil. Stud. Neotrop. Fauna Environ. 41: 209-215. doi: 10.1080/01650520600683088#.VHPJl9LF-No.

Bartelli, B.F., Santos, A.O.R & Nogueira-Ferreira, F.H. (2014). Colony performance of *Melipona quadrifasciata* (Hymenoptera, Meliponina) in a Greenhouse of *Lycopersicon esculentum* (Solanaceae). Sociobiology 61: 60-67. doi: 10.13102/sociobiology.v61i1.60-67

Bartelli, B.F. & Nogueira-Ferreira, F.H. (2014). Pollination services provided by *Melipona quadrifasciata* Lepeletier (Hymenoptera, Meliponina) in greenhouses with *Solanum lycopersicum* L. (Solanaceae). Sociobiology 61: 510-516. doi: 10.13102/sociobiology.v61i4.510-516

Barth, O.M. (1989). O pólen no mel brasileiro. Editora Luxor: Rio de Janeiro, 151p.

Batalha-Filho, H., Melo, G.A.R, Waldschmidt, A.M., Campos, L.A.O. & Fernandes-Salomão, T.M. (2009). Geographic distribution and spatial differentiation in the color pattern of abdominal stripes of the Neotropical stingless bee *Melipona quadrifasciata* (Hymenoptera, Apidae). Zoology, 26: 213-219. doi: 10.1590/S1984-46702009000200003.

Bispo dos Santos, S. A., Roselino, A. C., Hrncir, M. & Bego, L. R. (2009). Pollination of tomatoes by the stingless bee *Melipona quadrifasciata* and the honey bee *Apis mellifera* (Hymenoptera, Apidae). Genet. Mol. Res., 8: 751-757.

Bruijn, L.L.M. & Sommeijer, M.J. (1997). Colony foraging in different species of stingless bees (Apidae, Meliponinae) and the regulation of individual nectar foraging. Insectes Soc., 44: 35-47. doi: 10.1007/s000400050028.

Camargo, J.M.F. & Pedro, S.R.M. (2008). Meliponini Lepeletier, 1836. *In* J.S. Moure, D. Urban & G.A.R. Melo (Orgs). Catalogue of Bees (Hymenoptera, Apoidea) in the Neotropical Region - online version, 2008. Available at http://www.moure.cria.org.br/catalogue. Accessed Set/28/2014.

Contrera, F.A.L., Imperatriz-Fonseca, V.L. & Nieh, J.C. (2004). Temporal and climatological influences on flight activity in the stingless bee *Trigona hyalinata* (Apidae, Meliponini). Rev. Tecn. Amb., 10: 35-43.

Corbet S.A., Fussell, M., Ake, R., Fraser, A., Gunson, C., Savage, A. & Smith, K. (1993). Temperature and the pollination activity of social bees. Ecol. Entomol., 18: 17-30. doi: 10.1111/j.1365-2311.1993.tb01075.x

Cortopassi-Laurino, M., Velthuis, H.H.W. & Nogueira-Neto, P. (2007). Diversity of stingless bees from the Amazon forest in Xapuri (Acre), Brazil. Proc. Neth. Entomol. Soc. Meet., 18: 105-114.

Del Sarto, M.C.L., Peruquetti, R.C. & Campos, L.A.O. (2005). Evaluation of the neotropical stingless bee *Melipona quadrifasciata* (Hymenoptera: Apidae) as pollinator of greenhouse tomatoes. J. Econ. Entomol., 98: 260-266. doi: 10.1603/0022-0493-98.2.260.

Ferreira-Junior, N.T., Blochtein, B. & Moraes, J.F. (2010). Seasonal flight and resource collection patterns of colonies of the stingless bee *Melipona bicolor schencki* in an Araucaria forest area in southern Brazil. Rev. Bras. Entomol., 54: 630-636. doi: 10.1590/S0085-56262010000400015.

Fidalgo, A.O. & Kleinert, A.M.P. (2007). Foraging behavior of *Melipona rufiventris* Lepeletier (Apinae; Meliponini) in Ubatuba, SP, Brazil. Braz. J. Biol., 67: 137-144. doi: 10.1590/S1519-69842007000100018.

Fidalgo, A.O. & Kleinert, A.M.P. (2010). Floral preferences and climate influence in nectar and pollen foraging by *Melipona rufiventris* Lepeletier (Hymenoptera: Meliponini) in Ubatuba, São Paulo State, Brazil. Neotrop. Entomol., 39: 879-884. doi:10.1590/S1519-566X2010000600005.

Figueiredo-Mecca. G., Bego, L.R. & Nascimento, F.S. (2013). Foraging behavior of *Scaptotrigona depilis* (Hymenoptera, Apidae, Meliponini) and its relationship with temporal and abiotic factors. Sociobiology, 60: 277-282. doi: 10.13102/sociobiology.v60i3.277-282.

Gobatto, A.R. & Knoll, F.R.N. (2013). Influence of seasonal changes in daily activity and annual life cycle of *Geotrigona mombuca* (Hymenoptera, Apidae) in a Cerrado habitat, São Paulo, Brazil. Iheringia, Sér. Zool., 103: 367-373. doi: 10.1590/S0073-47212013000400006.

Gouw, M.S. & Gimenes, M. (2013). Differences of the daily flight activity rhythm in two Neotropical stingless bees (Hymenoptera, Apidae). Sociobiology, 60: 183-189. doi: 10.13102/sociobiology.v60i2.183-189.

Guibu, L. & Imperatriz Fonseca, V.L. (1984). Atividade externa de *Melipona quadrifasciata* Lepeletier (Hymenoptera, Apidae, Meliponinae). Ciên. e Cult., 36(7): 623.

Heinrich, B. & Esch, H. (1994). Thermoregulation in Bees.

Amer. Sci., 82: 164-170.

Hilário, S.D., Imperatriz Fonseca, V.L. and Kleinert & A.M.P. (2000). Flight activity and colony strength in the stingless bee *Melipona bicolor bicolor* (Apidae, Meliponinae). Rev. Bras. Biol., 60: 299-306. doi: 10.1590/S0034-71082000000200014.

Hilário, S.D., Gimenes, M. & Imperatriz Fonseca, V.L. (2003). The influence of colony size in diel rhythms on flight activity of *Melipona bicolor* Lepeletier, 1836 (Hymenoptera, Apidae, Meliponini). In G.A.R. Melo & I. Alves dos Santos (Eds.), Apoidea Neotropica: Homenagem aos 90 anos de Jesus Santiago Moure (pp. 191-197). Criciúma: UNESC.

Hilário, S.D. & Imperatriz Fonseca, V.L. (2009). Pollen foraging in colonies of *Melipona bicolor* (Apidae, Meliponini): effects of season, colony size and queen number. Genet. Mol. Res., 8: 664-671.

Hilário, S.D. & Imperatriz Fonseca, V.L. (2012). Can climate shape flight activity patterns of *Plebeia remota* (Hymenoptera, Apidae)? Iheringia, Sér. Zool., 102: 269-276. doi: 10.1590/S0073-47212012000300004.

Hilgert-Moreira, S.B., Fernandes, M.Z., Marchett, C.A. & Blochtein, B. (2014). Do different landscapes influence the response of native and non-native bee species in the *Eucalyptus* pollen foraging, in southern Brazil? Forest Ecol. Manag., 313: 153-160. doi: 10.1016/j.foreco.2013.10.049.

Hutchinson, G.E. (1957). Concluding remarks. Cold Spr. Harb. Symp. Quant. Biol., 22: 415-427.

Kleinert-Giovannini, A. & Imperatriz Fonseca, V.L. (1986). Flight activity and responses to climatic conditions of two subspecies of *Melipona marginata* Lepeletier (Apidae, Meliponinae). J. Apic. Res, 25: 3-8.

Luz, C.F.P., Fernandes-Salomão, T.M., Lage, L.G.A., Resende, H.C., Tavares, M.G. & Campos. L.A.O. (2011). Pollen sources for *Melipona capixaba* Moure & Camargo: an endangered Brazilian stingless bee. Psyche, 2011, Article ID 107303. doi: 10.1155/2011/107303.

Maurizio, A. & Louveaux, J. (1965). Pollens de plantes mellifères d'Europe. Un. Des. Group. Apic. Franç., Paris.

Michener, C.D. (1974). The Social Behavior of the Bees: A Comparative Study, Cambridge, Massachussets: The Belknap Press of Harvard University Press, 404p.

Mouga, D.M.D.S. (1984). Coleta de pólen e néctar em *Paratrigona subnuda* e Atividade externa de *Paratrigona subnuda*. Cienc. Cult., 36: 696-697.

Nascimento, A.S., Pereira, L.L., Carvalho, C.A.L., Machado, C.S., Oda-Souza, M. & Souza, B.A. (2012). Flight activity of the eusocial bee *Melipona quadrifasciata anthidioides* (Hymenoptera: Apidae, Meliponini). Magistra, 24: 112-118.

Nates-Parra, G. & Rodríguez, A. (2011). Forrajeo en colonias de *Melipona eburnea* (Hymenoptera: Apidae) en el piedemonte llanero (Meta, Colombia). Rev. Colomb. Entomol., 37: 121-127.

Pick, R.A. & Blochtein, B. (2002). Atividades de coleta e origem floral do pólen armazenado em colônias de *Plebeia saiqui* (Holmberg) (Hymenoptera, Apidae, Meliponinae) no sul do Brasil. Rev. Bras. Zool., 19: 289-300. doi: 10.1590/S0101-81752002000100025.

Pierrot, L.M & Schlindwein, C. (2003). Variation in daily flighty and foraging patterns in colonies of uruçu-*Melipona scutellaris* Latreille (Apide, Meliponini). Rev. Bras. Zool., 20: 565-571. doi: 10.1590/S0101-81752003000400001.

Ramalho, M., Kleinert-Giovannini, A. & Imperatriz-Fonseca, V.L. (1989). Utilization of floral resources by species of *Melipona* (Apidae, Meliponinae): floral preferences. Apidologie, 20: 185-195. doi: 10.1051/apido:19890301.

Rodrigues, M., Santana, W.C., Freitas, G.S. & Soares, A.E.E. (2007). Flight activity of *Tetragona clavipes* (Fabricius, 1804) (Hymenoptera: Apidae) at the São Paulo University Campus in Ribeirão Preto. Biosc. J., 23: 118-124.

Roubik, D.W. (1989). Ecology and natural history of tropical bees. New York: Cambridge Univ. Press, 514p.

Roubik, D.W. & Moreno, J.E.P. (1991). Pollen and spores of Barro Colorado Island. St Louis: Miss. Botan. Gard. Press, Monogr. Syst. Bot., 36: 268.

Serra, B.D.V, Luz, C.F.P. & Campos, L.A.O. (2012). The use of polliniferous resources by *Melipona capixaba*, an endangered stingless bee species. J. Insect Sci., 12(148): 1-14. doi: 10.1673/031.012.14801.

Silva, M.D., Ramalho, M. & Rosa, J. (2011). Por que *Melipona scutellaris* (Hymenoptera, Apidae) forrageia sob alta umidade relativa do ar? Iheringia, Sér. Zool., 101: 131-137. doi: 10.1590/S0073-47212011000100019.

Souza, B.A., Carvalho, C.A.L. & Alves, R.M.O. (2006). Flight activity of *Melipona asilvai* Moure (Hymenoptera: Apidae). Braz. J. Biol., 66: 731-737. doi: 10.1590/S1519-69842006000400017.

Tóth, E., Queller, D.C., Dollin, A. & Strassman, J.E. (2004). Conflict over male parentage in stingless bees. Insectes Soc., 51: 1-11. doi: 10.1007/s00040-003-0707-z.

Wilms, W. & Wiechers, B. (1997). Floral resource partitioning between native *Melipona* bees and the introduced Africanized honeybee in the Brazilian Atlantic rain Forest. Apidologie, 28: 339-355.

Zar, J.H. (1999). Biostatistical Analysis. Fourth ed. Upper Saddle River, New Jersey: Prentice-Hall, 664p.

31

Foraging Distance of *Melipona subnitida* Ducke (Hymenoptera: Apidae)

AG Silva[1], RS Pinto[1], FAL Contrera[2], PMC Albuquerque[1], MMC Rêgo[1]

1 - Universidade Federal do Maranhão (UFMA), São Luís, MA, Brazil

2 - Universidade Federal do Pará (UFPA), Belém, PA, Brazil

Keywords

Artificial feeding, Capture-recapture, Distance, Foragers, Meliponini, Stingless bees.

Corresponding author

Albeane Guimarães Silva
Universidade Federal do Maranhão - UFMA
Departamento de Biologia, Laboratório de Estudos Sobre Abelhas.
Av. dos Portugueses, 1966, 65080-805
São Luís, MA, Brazil
E-Mail: albeaneguimaraes@hotmail.com

Abstract

The current study aimed at estimating the maximum foraging distance of the stingless bee *Melipona subnitida* Ducke by comparing the efficacy of two methods: training of workers with an artificial feeding source and the capture-recapture technique, which consisted at marking bees that were released at different distances from the nest, after which the number of bees that returned to the colony was recorded. Under the training method, the mean foraging distance of the three colonies studied was 1,120 m and maximum foraging distance of 1,160 m. Yet the number of recruits and reactivated foragers for each colony were quantified, the average maximum distance until recruitment occurred was 886,66 m. In the capture-recapture method, the maximum flight distance of captured foragers ranged from 3,600 to 4,000 m, which was 2,700 m farther than the maximum flight distance recorded using the artificial feeding method. Therefore, we verified that *M. subnitida* is a species that can travel long distances in search for food. Our results also suggest that an abundance of resources near the nest can reduce its foraging area.

Introduction

Bees are primary pollinators in most regions of the world (Bawa, 1990; Silberbauer-Gottsberger & Gottsberger, 1988). Their flight range strongly influences the sexual reproduction of most flowering plants and can further determine the genetic structure of plant populations (Campbell, 1985; Waser et al., 1996). The distance that bees travel in search of a resource can directly affect agricultural crops, given that bee pollination is necessary to generate 30% of the human food supply (Slaa et al., 2006).

To increase the efficiency of collection and exploitation of good resources (Kerr, 1994; Contrera & Nieh, 2007), eusocial bees developed a sophisticated communication system that allows foragers to recruit other bees to profitable food sites (Lindauer & Kerr, 1960; Kerr, 1969; Wille, 1983; Jarau et al., 2000; Nieh, 2004; Aguilar et al., 2005). To save time and energy, bees do not forage over long distances unnecessarily (Frisch, 1967; Seeley, 1994). For example, in the summer, when there is a decrease in the food supply near the nest, bees of the genus *Apis* may use an area 6 to 22-fold larger than the area used during spring or fall (Couvillon et al., 2014).

The methods that have been most commonly used for the study of the maximum flight range of honey bees in general, and stingless bees in particular, include training foragers to feed from artificial feeders, and the marked forager capture-recapture method (Zurbuchen et al., 2010).

The use of artificial feeders to train forager workers makes it possible to train them to the maximum distance that a species may travel in search of food. In this test, the workers that reach the feeder to collect a food sample are marked with a specific color or number combination, usually on the thorax or abdomen (Seeley, 1995). Measuring the maximum foraging distance for each stingless bee species provides information related to the communication and recruitment techniques used by stingless bees to obtain food in response to their environment (Contrera & Nieh, 2007; Kuhn-Neto et al., 2009).

The capture-recapture method involves the release of previously marked workers at a given distance from the mother colony, and then counting the number of marked bees that return to the colony thereafter (Roubik & Aluja, 1983). According to Kuhn-Neto et al. (2009), this method may be less efficient when the workers are released in remote areas that they are not familiar with. Despite a potential loss of foragers that may get lost in unfamiliar release spots, the number of bees that do return to the nest, it is possible to estimate the species' flight radius, which could possibly be greater than obtained with others methods, and provide more reliable information about the distances that bees travel in the community away from their colony to search for resources (van Nieuwstadt & Iraheta, 1996).

The current study aimed to estimate the maximum foraging distance of *Melipona subnitida* Ducke, a species distributed in the Brazilian States of Alagoas, Bahia, Ceará, Maranhão, Paraíba, Pernambuco, Piauí, Rio Grande do Norte and Sergipe (Camargo & Pedro, 2013), while comparing the efficacy of two experimental methods: (1) training workers to visit an artificial feeder; and (2) capture-recapture (adapted from Roubik & Aluja, 1983). Foraging distance information is critical for understanding the scale at which bee populations respond to the landscape, assessing the role of bee pollinators in affecting plant population structure and planning conservation strategies for plants (Greenleaf et al., 2007).

Materials and Methods

Study area and colony selection

This study was conducted at "Ponta do Mangue" village (2°58'12"S; 42°79'56"W), which is located in the municipality of Barreirinhas along the eastern coast of the state of Maranhão, Brazil. The village is located within the National Park of Lençóis Maranhenses. The park covers a total area of 155,000 ha, of which 453.28 km² are covered by vegetation (Brazilian Institute of Environment and Renewable Natural Resources - IBAMA, 2002).

Within the area covered by vegetation, 405.16 km² are predominately composed of Restinga, term that represents a set of physiognomically distinct plant communities under marine influence. The species that colonize these areas are mainly from other ecosystems, but with phenotypic variations that differ from those expressed in their original environments (Freire, 1990). The restinga area has shrub species dominance, and herbaceous communities are also present in large areas surrounding lakes (Brazilian Institute of Environment and Renewable Natural Resources - IBAMA, 2002).

To test the efficacy of using artificial feeder and the capture-recapture methods to determine the flight range of *M. subnitida*, four colonies were selected from a meliponary consisting of natural nests and nest boxes kept at Ponta do Mangue village. Two *M. subnitida* colonies were used for

both methods (i.e., "colony 1" and "colony 2"), while the other two colonies were used individually for each method (i.e., "colony 3" - artificial feeder and "colony 4" - capture-recapture). Strong colonies with a large number of workers (mean 890 foragers) and sufficient stored food were selected (mean of 115 pots food).

Artificial feeder method – Foraging distance

The experiments were performed in two different months: March 2013 and August 2013. In March, only the workers from colony 1 visited the artificial feeder. After the study with colony 1, which lasted five days, the same methodology was attempted with the other three colonies, though without success. The study continued in August with two other colonies (colony 2 and colony 3), thus totaling three colonies using the artificial feeder method. The weather in the days of study (March and August) did not differ (without rain), thus allowing for a comparison of the replicates.

An artificial feeder consisting of a flat acrylic disc with grooves from the center to the edge was used to evaluate foraging distances. A jar containing food (2.5M, 60%: 40% sugar: water concentration), with the opening facing downwards so that the food would drain, was placed in the center of the disc. Vanilla extract (9 μl/l) was added to the food to simulate floral odor and to increase the attractiveness of workers to the artificial food (Fig 1). The feeder was placed atop a tripod, which allowed for the adjustment of the height and horizontal inclination of the food supply (method described by Frisch, 1967; Nieh et al., 2003).

The first step of the experiment was to train the bees to visit the feeder. The feeder was placed right outside of the hive entrance and then moved a few cm away from the nest (thus, creating a food trail), so that when bees cleared the colony entrance, they tried the food and then followed the trail to reach the tripod with the artificial food source. Each worker that visited the feeder was marked with a dab of acrylic paint on their thorax with one color or a combination of specific

Fig 1. Feeder used to artificially train *Melipona subnitida* to a food source.

colors to differentiate them foragers from each other, as described in Seeley (1995). Foragers were classified as recruits or reactivated foragers. Recruits were those foragers that had never visited the feeder before and found the feeder on their own, while reactivated foragers were those that had fed from the feeder when it was placed in a previous location, for some reason had stopped visiting the feeder temporarily, but found it again later (Kuhn-Neto et al., 2009).

During the training process, when a minimum of 10 bees had visited the feeder and recruitment was in progress (i.e. when new foragers were recruited), the feeder was gradually moved away from the nest. To facilitate the forager's memorization of the new location, the feeder was moved only when workers were feeding. The feeder was always moved only after the first bee recorded at a specific distance left the feeder probably to unload her food sample at the nest, and later returned to the feeder for more food. The number of workers that visited the feeder was recorded every 20 m. If the total number of foragers decreased by 50% from the maximum number recorded previously, the feeder was not moved throughout the rest of the day. The next morning, the feeder was placed at the final location reached on the previous day, and remained at this location until the same number of foragers arrived as was recorded the day before. If the same number of foragers did not arrive within two hours, the feeder was moved to the next distance.

The experiments were performed daily during the period of natural foraging activity for *M. subnitida* (6 am – 5 pm) and ended when the workers stopped visiting the feeder. The experiments with colonies 1, 2, and 3 respectively lasted 56 h over 6 days, 33 h over 3 days, and 32 h over 3 days. Trial duration was determined by the maximum distance to which the colony could be trained.

The full path of the foraging range of each studied colony was classified as three distances: close foraging distance, average foraging distance and maximum foraging distance. The close foraging distance was defined as the distance from the colony's nest to the point where the number of foragers remained above the overall mean of visiting workers along the path during the experiment ($\bar{X}_{VF} = \frac{Tvw}{Ttp}$), where T_{vw} is the total number of visits by workers along the path, and T_{tp} is the total number of points where bees were recorded along the path, dividing these values, we obtain X_{VF} (the mean number of bees that visited the feeder). The average foraging distance was defined as the distance at which the number of workers fell below average to the last distance where workers were recruited. The maximum foraging distance was defined as the last point where occurred the foraging activities at the feeder.

Capture-recapture method – Flight radius

The capture-recapture experiment was performed in August 2013. Three *M. subnitida* colonies were used, two of which were also used for the feeder test (2.2). Workers (n = 150) were captured at the nest entrance and marked on the thorax with nontoxic paint of different colors, totaling 15 groups of 10 workers in each nest. Initially, 10 linear points were marked every 200 m starting at the colony's nest entrance, reaching a total distance of 2,000 m. From the 2,000-meter mark onwards, five more points were marked at a distance of 400 m apart from each other, reaching a total distance of 4,000 m. The distances at which the bees were released were measured from the colony entrance, and the points were located using a GPS device. The bees were released between 9 and 10 am, and the number of bees that returned to the colony was recorded to calculate the percent of successful return.

Workers were recaptured at the colony entrance by observing the bees that arrived. To do this, one researcher released the bees at the established distances away from the nest, while another researcher stood at the nest entrance, observing and quantifying the bees that returned to the nest until 6 pm. The nest was kept closed during the experiments, and was only opened when a new worker approached the nest to prevent the erroneous recounting of bees that had already entered the nest.

Statistical analysis

We performed linear regression between the total distribution of workers visiting the experimental feeders and the distance from the feeder to the colony to evaluate the effect of distance on foraging. We also made a linear regression on the number of recruits and reactivated foragers for each colony and distance (Zar, 1999). In addition, for each colony, the distances at which we measured 75%, 95% and 100% of foraging activity were estimated to reflect the percentage of active workers relative to the total number of workers within a certain distance.

To determine whether the number of workers that returned to the nest (as obtained through the capture-recapture method) differed between colonies, we used the non-parametric Kruskal-Wallis test. The relationship between the percentage of bees that successfully return to the nest and the distance at which they were released was assessed using a simple regression (Zar, 1999). We also used the Mann-Whitney test to assess the significance of the difference between the two methods used in the current study (artificial feeder vs. capture-recapture). All statistical analyses were performed using the Statistica 7.0 software with a critical P-value of 0.05 (Statsoft Inc., 2004).

Results

Foraging distance

Overall, *M. subnitida* foragers achieved a maximum foraging distance of 1,080 - 1,160 m and a maximum recruitment distance of 940 m. The mean foraging distances of the three colonies were as follows: close distance of 653 m, average distance of 863 m and maximum distance of 1,120 m (Table 1). The mean values for 75%, 95% and 100% of foraging activity were 566 m, 833 m and 1,120 m, respectively (Table 2). The number of workers found at each foraging distance is given in Fig 2 for each of the three colonies studied. Overall, the number of foragers trained and recruited to the artificial feeder decreased with distance away from the feeder.

The linear regression analysis of the relationship between the number of bees that visited the feeder and distance showed a strong negative correlation for the three studied colonies (R = 0.67; p < 0.0001), i.e., the greater the distance, the lower the number of foragers that visited the artificial feeder

Table 1. Foraging distances (m) calculated for the three colonies evaluated by the artificial feeder method. Data shown as averages and ± SD.

	Colony 1	Colony 2	Colony 3	Average/ SD
Close distance	740	600	620	653.33 ± 75.71
Average distance (Maximum recruitment distance)	900	940	820	886.66 ± 61.10
Maximum distance	1,160	1,120	1,080	1,120 ± 40.00

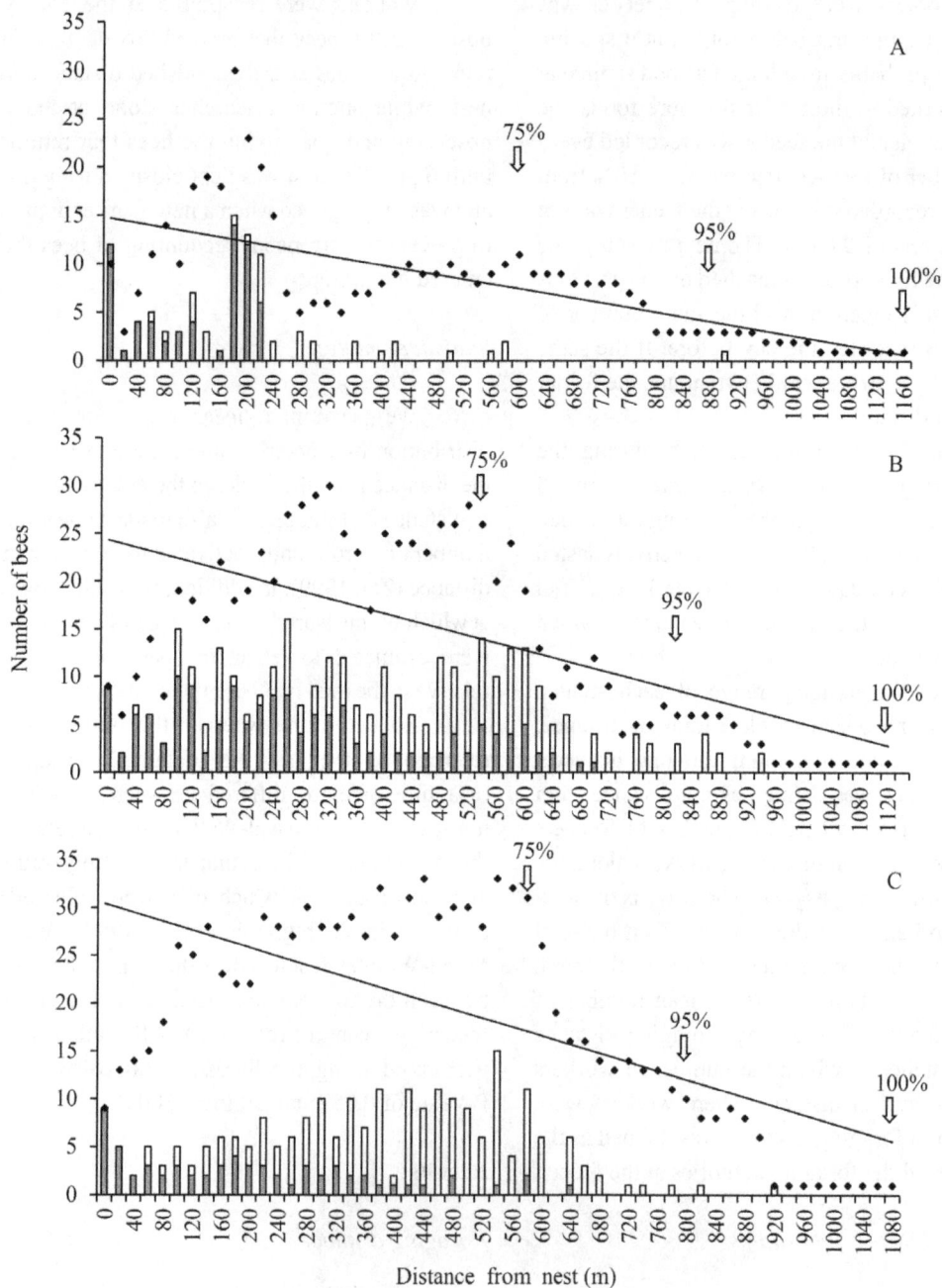

Fig 2. Total distribution of the number of *Melipona subnitida* foragers (black diamonds), recruits (grey bars) and reactivated foragers (white bars) that were observed at artificial feeders to which the bees were trained over time. Arrows indicate the distances defining the areas of 75%, 95% and 100% foraging activity (see Methods for details).

(Table 2). The correlation between the number of recruits and reactivated foragers and distance were also evaluated by linear regression for each colony. In all three colonies, both the number of recruits and the number of reactivated bees were negatively correlated with distance away from the nest (Table 2). Therefore, there was a decline in the number of recruits and reactivated foragers at more distant sites, compared to sites near the nest (Fig 2).

Table 2. Linear relationships between distance (m) and the number of *Melipona subnitida* workers that visited the feeder. In addition, distances that delimited the areas of 75%, 95% and 100% cumulative foraging activity. Data shown as averages and ± SD. A – total number of bees, B – recruits, C – reactivated workers.

A	Colony	Linear relationship between distance and the total number of bees that visited the feeder	R	75%	95%	100%
	1	y = 14.7203 - 0.0121x	0.70*	600	880	1,160
	2	y = 24.3987 - 0.0193x	0.66*	540	820	1,120
	3	y = 30.4169 - 0.023x	0.67*	560	800	1,080
	Avarege/		0.67*	566,66	833,33	1,120
	SD		± 0.02	± 30.55	± 41.66	± 40

B	Colony	Linear relationship between distance and the number of recruits	R
	1	y = 3.5068 - 0.0042x	0.48*
	2	y = 6.9074 - 0.0074*x	0.80*
	3	y = 7.0013 - 0.0076*x	0.76*

C	Colony	Linear relationship between distance and the number of reactivated bees	R
	1	y = 1.3186 - 0.0013x	0.44*
	2	y = 4.4289 - 0.0025x	0.27*
	3	y = 4.3045 - 0.0021x	0.30*

*p < 0.05 (significant)

Flight radius

The number of workers released by the capture-recapture method and the number of workers that returned to the nest were significantly different in all three colonies studied (Kruskal-Wallis n = 350; p = 0.0175). Thus, the data for the percentage of bees that returned to the nest were analyzed separately. The percentages of released bees that returned from different distances from the nest are shown in Fig 3. The correlation between the percentage of bees that successfully returned to the nest and the distance traveled, as evaluated by linear regression, was highly negative for all three colonies. Therefore, the number of bees that returned to the nest gradually decreased as the distance increased (Table 3). The maximum flight distance of *M. subnitida* measured by the capture-recapture method was 3,600 m for colonies 2 and 3, and colony 1 reached 4,000 m (Fig 3).

Table 3. Linear relationship between the distance (m) at which the *Melipona subnitida* workers were released by the capture-recapture method and the percentage of return to the nest. Data shown as averages and ± SD.

Colony	Linear relationship between distance and % of return	R	Flight radius
1	y = 9.7306 - 0.0022x	0.95*	4,000
2	y = 9.9095 - 0.0024x	0.89*	3,600
3	y = 11.0524 - 0.0022x	0.83*	3,600
Average/ SD		0.89 ± 0.06	373.33 ± 230.94

*p < 0.001 (significant)

Comparison of methods

The results obtained for the foraging distance of *M. subnitida* workers were very different between the two methods used. There was a highly significant difference between the maximum foraging distances obtained by the two methods (U < 0.0001; p = 0.04; Fig 4), with the maximum flight distance recorded in the capture-recapture method being approximately 2,700 m further than that observed in the artificial feeder method.

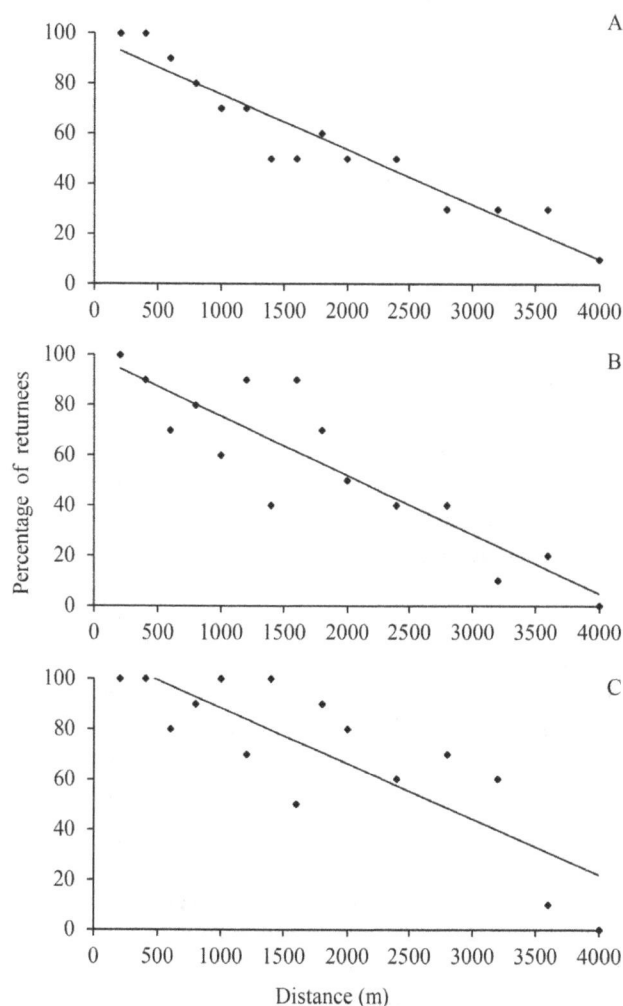

Fig 3. Percentage of returnees for each distance (m) at which the workers were released by the capture-recapture method: A – Colony 1, B – Colony 2, C – Colony 3.

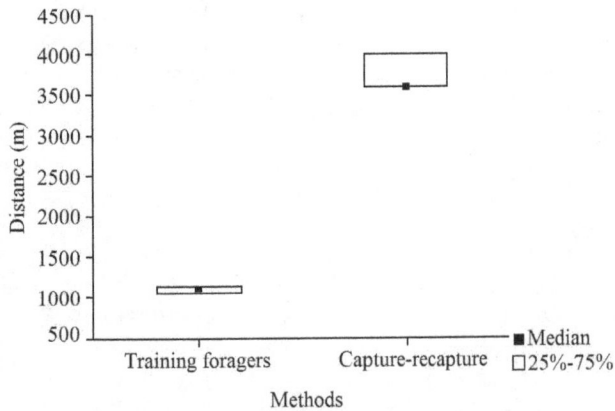

Fig 4. Comparison of the efficacy of two methods to determine the foraging distance of *Melipona subnitida* foragers: (a) training foragers to an artificial feeder, and (b) the capture-recapture method.

Discussion

Our results suggest that the mean foraging distance of *M. subnitida*, as obtained from the artificial feeder experiment, was approximately 1,120 m. This value was lower that reported by Kerr (1987) for *Melipona fasciculata* Smith (approximately 2,470 m) and that reported by Kuhn-Neto et al. (2009) for *Melipona mandacaia* Smith (1,800 m).

Interestingly, the three classifications of *M. subnitida* foraging distances revealed that the close foraging distance was approximately 653 m, indicating that at this distance, *M. subnitida* colonies have a greater ability to maintain a large number of foragers and a facilitated recruitment capacity. The close foraging distance is the distance within which the bees truly dominate a food resource and therefore have a higher probability of pollinating the plants where foragers (van Nieuwstadt & Iraheta, 1996). Considering the distance at which it is possible to recruit workers, which in our study we found to be 863 m on average, it is still likely to establish an effective control of the food source. Beyond this recruitment distance, it seems that it is possible for the foragers to find food, but they would probably have to explore it by themselves and potentially the exploitation of the food resource would be less efficient (Kuhn-Neto et al., 2009).

According to van Nieuwstadt and Iraheta (1996), more than 75% of a stingless bee colony's food-searching activity usually occurs within 40% of the maximum foraging distance. For *M. mandacaia*, 75% of the foraging activity (893.33 ± 11.54 m) occurred within corresponded to 50% of the maximum foraging distance (Kuhn-Neto et al., 2009). This is similar to what we observed for *M. subnitida*, in which 75% of the food-searching activity (566.66 ± 30.55 m) occurred within 51% of the maximum foraging distance in their Restinga environment. Therefore, it can be stated that the farther the colony is from the resource, the lower the number of interested or available foragers is, and consequently, the lower the recruitment of new workers to that food source.

One factor that may have been crucial to the smaller flight radius of *M. subnitida* is the large amount of flowering plants near the colonies. *M. subnitida* workers visit flowers of *Humiria balsamifera* (Aubl.) A. St.-Hil. and *Chrysobalanus icaco* L. to collect nectar. These plant species bloom almost all year round in the studied region; they exhibited high flowering when colony 1 was trained but less flowering during the experiments with colonies 2 and 3. Because other colonies did not follow the feeder in March, the experiment was resumed five months later. The results obtained in March and August were very similar. Therefore, it can be inferred that an abundant supply of a profitable nectar source near the nest decreased foraging and recruitment to an artificial site located at more distant locations.

The large food supply near the experimental colonies in March certainly influenced the number of bees of colony 1 that were recruited to the feeder. However, the foraging distance did not differ among the colonies. Although it was not possible to verify such influence (abundant food sources), workers could potentially be "discouraged" in the search for a farther food source simply because there were other abundant food sources near their nest. A study looking at the foraging ecology of *M. mandacaia*, a native bee from the Caatinga biome (Kuhn-Neto et al. 2009), found a larger foraging distance in that species than what we found in *M. subnitida*. The longer foraging distance of *M. mandacaia* may be in part because there is a marked reduction in the food supply in the Caatinga biome during the prolonged dry season, compared to the constant nectar flow in the Restinga region where our study was conducted.

In the current study, the distances traveled by *M. subnitida* were measured without analyzing other factors. For instance, the distances traveled by bees responsible for foraging activities depend on several factors, including density and seasonality of the food supply, species (Dornhaus et al., 2006), physiology, and body size (Imperatriz-Fonseca et al., 1985). Furthermore, other factors, separately or combined, may also affect flight radius, such as the internal colony conditions and climatic factors (Hilário et al., 2000). However, the days upon which the study was performed did not have any heavy rain and were thus favorable for the experiment because it is known that rain prevents or reduces bee activity (Hilário et al., 2007).

Furthermore, in Apoidea bees, body size may act as a limiting factor to the maximum flight capacity and therefore, maximum foraging distance (Araújo et al., 2004; Greenleaf et al., 2007; Kuhn-Neto et al., 2009; Zurbuchen et al., 2010). However, it is likely that many species actually exercise a lower flight capacity (i.e., occupy a smaller space) depending on other variables such as foraging, specialization in resource searching, navigation, abundance of food resources, and availability of nesting sites.

By using the capture-recapture method, the percentage of bees that returned to the nest was 80% if they were released within 1,000 m of their colonies. Roubik and Aluja (1983), released bees of the genus *Melipona* at different distances from the nest, and found that the mean flight radius of these bees is

2,100 m from the nest. The authors estimated the maximum flight radius for this genus using regression tests and concluded that it could be up to 2,400 m. Taking into account that Roubik and Aluja (1983) used the same method as that used in our study our results in the capture-recapture experiment exceeded the maximum distance conjectured by Roubik and Aluja (1983). This difference could be attributable to the fact that, in the current study, we waited until the end of the day for the return of the workers, while Roubik and Aluja (1983) waited just a few hours.

When comparing the two methodologies used in the current study, we observed that each technique has benefits and limitations. In the artificial feeder test, the foragers' flight range could have been underestimated because some bees got lost in the course of the experiment, or because the artificial food is less attractive than a natural flower, which may discourage the search for a food supply at long distances (van Nieuwstadt & Iraheta, 1996). In the capture-recapture test, by contrast, an overestimation of the actual flight distance could be possible because the released bees only needed to fly back to the nest. Alternatively, there may have been a smaller number of bees that returned to the nest after traveling greater distances, potentially because they became lost due to unfamiliarity with the environment where they were released and because the energy costs associated with orientation are high (van Nieuwstadt & Iraheta, 1996).

The advantage of the artificial feeder method over the capture-recapture method is that it can assess the distance that workers can travel, the number of bees they recruit at each distance as well as the exact gradual reduction in the number of workers along the path, which would be closer to the distance travelled by the bees under natural conditions (Kuhn-Neto et al., 2009). Nevertheless, even with the capture-recapture method, we can determine the distance at which the workers begin to return less successfully.

When an environment is fragmented, numerous aspects of the landscape ecology are affected, as the reduction of dispersion and potential colonization of plant species (Lovejoy et al., 1986; Bierregaard et al., 1992). We can infer that such fragmentation of the landscape may have interfered with our results given that most of the vegetation in the study site is concentrated near the meliponary and the region is surrounded by dunes. According to Roubik (1989), the behavior of bees is likely to be adapted to their environment and will also be determined by the "resource landscape," which includes aspects such as resource quality. Bees require a large area of vegetation to obtain food throughout the whole year and to nest in regions beyond the mother colony. As the Lençóis Maranhenses region is an area in constant flux due to the strong winds and the presence of dunes, the area covered by vegetation tends to decline (Gonçalves et al., 2003), thus resulting in an increase in the distance that M. subnitida must fly in search of food. The information related to the foraging distance of M. subnitida reported in the current study can thus help to promote strategies for the conservation of this species.

Acknowledgments

The authors thank the owners of the Jandaíra meliponary, Irene Aguiar, Mr. Emídio and Mrs. Maria; the Research and Scientific Development Foundation of Maranhão (Fundação de Amparo à Pesquisa e Desenvolvimento Científico do Maranhão – FAPEMA) for financial support.

References

Aguilar, I., Fonseca, A. & Biesmeijer, J.C. (2005). Recruitment and communication of food source location in three species of stingless bees (Hymenoptera, Apidae, Meliponini). Apidologie 36: 313-324. doi: 10.1051/apido:2005005.

Araújo, E.D., Costa, M., Chaud-Netto, J. & Fowler, H.G. (2004). Body size and flight distance in stingless bees (Hymenoptera: Meliponini): inference of flight range and possible ecological implications. Braz. J. Biol. 64: 563-568. doi: 10.1590/S1519-69842004000400003.

Bawa, K.S. (1990). Plant–pollinator interactions in tropical rain forests. Annu. Rev. Ecol. Syst. 21: 399-422.

Bierregaard, R.O., Lovejoy, T.E., Kapos, V., Santos, A.A. & Hutchings, W. (1992). The biological dynamics of tropical rain forest fragments. BioScience 42: 859-866.

Camargo, J.M.F. & Pedro, S.R.M. (2013). Meliponini Lepeletier, 1836. In J.S. Moure, D. Urban & G.A.R. Melo (Orgs.), Catalogue of Bees (Hymenoptera, Apoidea) in the Neotropical Region - online version. Available at http://www.moure.cria.org.br/catalogue. (accessed date: 27 November, 2014).

Campbell D.R. (1985). Pollen and gene dispersal - the influences of competition for pollination. Evolution 39: 418-431.

Contrera, F.A.L. & Nieh, J.C. (2007). Effect of forager-deposited odors on the intra-patch accuracy of recruitment of the stingless bees Melipona panamica and Partamona peckolti (Apidae, Meliponini). Apidologie 38: 584-594. doi: 10.1051/apido:2007054.

Couvillon, M.J., Schürch, R. & Ratnieks, F.L.W. (2014). Waggle dance distances as integrative indicators of seasonal foraging challenges. PLoS One 9 (4): 1-7. doi: 10.1371/journal.pone.0093495.

Dornhaus, A., Klügl, F., Oechslein, C., Puppe, F. & Chittka, L. (2006). Benefits of recruitment in honey bees: effects of ecology and colony size in an individual based model. Behav. Ecol. 17: 336-344. doi:10.1093/beheco/arj036.

Freire, M.S.B. (1990). Levantamento florístico do Parque Estadual das Dunas de Natal. Acta Bot. Bras. 4: 41-59.

Frisch, K. von. (1967). The dance language and orientation of bees. Cambridge: Harvard University Press, 592 p.

Gathmann, A. & Tscharntke, T. (2002). Foraging ranges of solitary bees. J. Anim. Ecol. 71: 757-764. doi: 10.1046/j.1365-2656.2002.00641.x.

Gonçalves, R.A., Lehugeur, L.G.O., Castro, J.W.A. & Pedroto, A.E.S. (2003). Classificação das feições eólicas dos Lençóis Maranhenses - Maranhão – Brasil. Mecator 3: 99-112.

Hilário, S.D., Imperatriz-Fonseca, V.L. & Kleinert, A.M.P. (2000). Flight activity and colony strength in the stingless bee *Melipona bicolor bicolor* (Apidae, Meliponinae). Rev. Bras. Biol. 60: 299-306. doi: 10.1590/S0034-71082000000200014.

Hilário, S.D., Ribeiro, M.F. & Imperatriz-Fonseca, V.L. (2007). Impacto da precipitação pluviométrica sobre a atividade de vôo de *Plebeia remota* (Holmberg, 1903) (Apidae, Meliponini) [Effect of rain on the flight activity of *Plebeia remota* (Holmberg, 1903) (Apidae, Meliponini)]. Biota Neotrop. 7: 135-143. doi: 10.1590/S1676-06032007000300016.

Instituto Brasileiro do Meio Ambiente e dos Recursos Naturais Renováveis (IBAMA). (2002). Parque Nacional dos Lençóis Maranhenses - Plano de Manejo.

Imperatriz-Fonseca, V.L., Kleinert-Giovannini, A. & Pires, J.T. (1985). Climate variations influence on the flight activity of *Plebeia remota* Holmberg (Hymenoptera, Apidae, Meliponinae). Rev. Bras. Entom. 29: 427-434.

Jarau, S., Hrncir, M., Zucchi, R. & Barth, F.G. (2000). Recruitment behavior in stingless bees, *Melipona scutellaris* and *M. quadrifasciata*. I. Foraging at food sources differing in direction and distance. Apidologie 31: 81-91. doi: 10.1051/apido:2000108.

Kerr, W.E. (1960). Evolution of communication in bees and its role in speciation. Evolution 14: 386.

Kerr, W.E. (1969). Some aspects of the evolution of social bees. Evol. Biol. 3: 119-175.

Kerr, W.E. (1987). Biologia, manejo e genética de *Melipona compressipes fasciculata* Smith (Hymenoptera, Apidae). PhD thesis (Head Professor). São Luiz: Universidade Federal do Maranhão.

Kerr, W.E. (1994). Communication among *Melipona* workers (Hymenoptera: Apidae). J. Insect Behav. 7: 123-128.

Kerr, W.E. & Rocha, R. (1988). Comunicação em *Melipona rufiventris* e *Melipona compressipes*. Ciênc. Cult. 40: 1200-1202.

Kuhn-Neto, B., Contrera, F.A.L., Castro, M.S. & Nieh, J.C. (2009). Long distance foraging and recruitment by a stingless bee, *Melipona mandacaia*. Apidologie 40: 472-480. doi: 10.1051/apido/2009007.

Lindauer, M. & Kerr, W.E. (1960). Communication between the workers of stigless bees. Bee World, 41: 29-41, 65-71.

Lovejoy, T.E., Bierregaard, R.O., Rylands, A.B., Malcolm, J.R., Uintela, C.E., Harper, L.H., Brown, K.S., Powell, G.V.N.,

Schubart, H.O.R. & Hay, M.B. (1986). Edge and other effects of isolation on Amazon forest fragments. In M.E. Saule (Ed.), Conservation biology. Massachusetts: Sinauer Press, pp. 257-285.

Michener, C.D. (2000). The bees of the world. Baltimore and London: Johns Hopkins University Press, 913 p.

Nieh, J.C. (2004). Recruitment communication in stingless bees (Hymenoptera, Apidae, Meliponini). Apidologie 35: 159-182. doi: 10.1051/apido:2004007.

Nieh, J.C., Contrera, F.A.L., Ramírez, S. & Imperatriz-Fonseca, V.L. (2003). Variation in the ability to communicate three-dimensional resource location by stingless bees from different habitats. Anim. Behav, 66: 1129-1139. doi: 10.1006/anbe.2003.2289.

Roubik, D.W. (1989). Ecology and natural history of tropical bees. Cambridge: University Press, 514 p.

Roubik, D.W. & Aluja, M. (1983). Flight ranges of *Melipona* and *Trigona* in tropical forest. J. Kans. Entomol. Soc. 56: 217-222.

Seeley, T.D. (1994). Honey bee foragers as sensory units of their colonies. Behav. Ecol. Sociobiol. 34: 51-62.

Seeley, T. D. (1995) The Wisdom of the Hive. Cambridge: Harvard University Press, 295 p.

Silberbauer-Gottsberger, I. & Gottsberger, G. (1988). A polinização de plantas do cerrado. Rev. Bras. Biol. 48: 651–663.

Slaa, E.J., Sànchez Chaves, L.A., Malagodi-Braga, K.S. & Hofstede, F.E. (2006). Stingless bees in applied pollination: practice and perspectives. Apidologie 37: 293-315. doi: 10.1051/apido:2006022.

Statsoft Inc. (2004) Statistica (data analysis software system), version 7.

van Nieuwstadt, M.G.L. & Iraheta, C.E.R. (1996). Relation between size and foraging range in stingless bees (Apidae, Meliponinae). Apidologie 27: 219-228. doi: 10.1051/apido:19960404.

Waser, N.M., Chittka, L., Price, M.V., Williams, N.M. & Ollerton, J. (1996). Generalization in pollination systems and why it matters. Ecology 77: 1043-1060.

Wille, A. (1983) Biology of the stingless bees. Annu. Rev. Entomol. 28: 41-64. doi: 10.1146/annurev.en.28.010183.000353.

Zar, J.H. (1999). Biostatistical analysis. New Jersey: Prentice Hall, 663 p.

Zurbuchen, A., Landert, L., Klaiber, J., Müller, A., Hein, S. & Dorn, S. (2010). Maximum foraging ranges in solitary bees: only few individuals have the capability to cover long foraging distances. Biol. Conserv. 143: 669-676. doi: 10.1016/j.biocon.2009.12.003.

32

Recognition and Aggression of conspecific and heterospecific worker in *Acromyrmex subterraneus subterraneus* (Forel) (Hymenoptera: Formicidae)

TG Pikart[1,2], PG Lemes[1], WC de C Morais[1], JC Zanuncio[1], TMC Della Lucia[1]

1 - Universidade Federal de Viçosa, Viçosa, MG, Brazil
2 - Universidade do Estado de Santa Catarina, Lages, SC, Brazil

Keywords
behavior, competition, defense, etogram, heterospecifics, leaf-cutting ant.

Corresponding author
Terezinha Maria Castro Della Lucia
Departamento de Biologia Animal
Universidade Federal de Viçosa
36570-000 Viçosa, Minas Gerais, Brazil
E-Mail: tdlucia@ufv.br

Abstract
Aggressive behavior is important for social insects because it makes possible for the colony to defend itself and the offspring from the action of invasive species. We studied the recognition and aggressiveness of the leaf-cutting ant *Acromyrmex subterraneus subterraneus* (Forel) to co-specific workers from other nest and heterospecific workers of *Acromyrmex subterraneus molestans* Santschi, *Acromyrmex subterraneus brunneus* (Forel) and *Acromyrmex niger* (Smith); and queens of their social parasite *Acromyrmex ameliae* De Souza, Soares and Della Lucia. Workers of other species were placed in contact with those of *A. subterraneus subterraneus* for three minutes and during this period the behavioral interactions were quantified. The aggressiveness index (AI) for each agonistic encounter was obtained. *Acromyrmex subterraneus subterraneus* workers exhibited greater aggressiveness against heterospecific than against conspecific competitors. Aggressiveness is connected to differences in the chemical profiles, which are larger in heterospecifics colonies.

Introduction

An important feature of social species is the existence of elaborate recognition systems that facilitate cooperation among members of a group, because they help to maintain the integrity of the society, reducing the negative impact caused by predators, competitors and social parasites (Crosland, 1990; Fishwild & Gamboa, 1992; Crowley et al., 1996; Wiley, 2013). Ants maintain coherence within the colony through chemical recognition of nestmates, allowing workers to differentiate other individuals between friend and foe through specific chemical signatures of each colony (Lenoir et al., 2001; Akino, 2008; Guerrieri et al., 2009).

Leaf-cutting ants (Hymenoptera: Formicidae: Myrmicinae: Attini) are dominant herbivores in the Neotropics (Hölldobler & Wilson, 1990). *Acromyrmex* Mayr, with 62 species and nine subspecies (Forti et al., 2006; Souza et al., 2007; Brandão et al., 2011), causes economic losses in agriculture and is one of the most important pests of forest plantations (Boulogne et al., 2012).

These ants are models for behavioral studies because of their elaborate social organization and interactions between individuals of the colony and with other organisms (Camargo et al., 2006). Workers of different sizes exhibit different behaviors and division of labor (polyethism), which maximizes foraging (Hart et al., 2002). However, there are activities of the colony that are still poorly studied, such as some interactions between the workers of a nest.

Social insects exhibit aggressive behavior towards individuals from other nests and invading organisms (Hölldobler & Michener, 1980). This behavior allows the nests to store resources (Hamilton, 1972) and protect their workers and offspring from external threats (Pollock & Rissing, 1989; Sakagami, 1993; Mori et al., 2000; Allon et al., 2012).

The recognition of nestmates occurs through a specific "label" called "colony odor" (Crozier & Dix, 1979). This label is made of cuticular hydrocarbons in social insects (Lenoir et al., 2001a) and allows ants to recognize nestmates and distinguish them from intruders (Hölldober & Michener,

1980). Soil characteristics, food source and other nest materials participate in the formation of this label (Heinze et al., 1996). Diet is more important than the influence of the queen and their genetic origin in the formation of odor in nests of *Acromyrmex subterraneus subterraneus* (Forel) (Hymenoptera: Formicidae) and, consequently, in the recognition of the ants (Richard et al., 2004). The fungi recognition behavior (Viana et al., 2001) and trophallaxis (Moreira et al., 2006) were studied in this subspecies. However, several types of behavior of *A. subterraneus subterraneus* workers such as aggressiveness have not yet been evaluated. We hypothesized that workers of *A. subterraneus subterraneus* exhibit greater aggressiveness against heterospecifics than against conspecific ones.

The objective of this study was to describe and compare the behavior of recognition and aggression of *A. subterraneus subterraneus* workers in relation to workers of: (1) *A. subterraneus subterraneus* from different colonies; (2) *Acromyrmex subterraneus molestans* Santschi; (3) *Acromyrmex subterraneus brunneus* (Forel); (4) *Acromyrmex niger* (Smith); and their social parasite queens (5) *Acromyrmex ameliae* De Souza, Soares & Della Lucia.

Material and Methods

Study site and species

The study was conducted at the Laboratório de Formigas Cortadeiras (LFC) of the Universidade Federal de Viçosa (UFV) in Viçosa, Minas Gerais State, Brazil in November 2011. *Acromyrmex subterraneus subterraneus*, *A. subterraneus molestans*, and *A. niger* naturally co-occur in the region of Viçosa. The social parasite *A. ameliae* occurs only in nests of *A. subterraneus subterraneus* and *A. subterraneus brunneus* in another region of Minas Gerais State, at about 300 km from Viçosa. All individuals used in the experiment were obtained from a total of ten colonies maintained in the LFC at $24 \pm 2°C$ and $\pm 75\%$ RH; and fed with *Ligustrum japonicum* Thunb. (Oleaceae) and *Acalipha wilkesiana* Müll. Arg. (Euphorbiaceae) leaves. One of the colonies of *A. subterraneus subterraneus* was collected on the UFV campus and kept under the same conditions as the others, in a separate room.

Recognition and aggressiveness evaluation

The aggressiveness and recognition among workers were assessed in a neutral arena, a plastic container (250 mL) with the inner upper half covered with talc. Five workers of *A. subterraneus subterraneus* were randomly selected and introduced into the arena and, after 10 minutes, another individual was introduced as the following treatments: nestmate worker (Control), co-specific worker of nest collected in the field (ASS); worker of *A. subterraneus molestans* (ASM); worker of *A. subterraneus brunneus* (ASB); winged queens of *A. ameliae* (AA); and worker of *A. niger* (AN).

We avoided the use of *A. subterraneus subterraneus* workers that were weak, sick or injured in the experiment, because the use of non-healthy individuals may produce inconsistent results (Roulston et al., 2003). Agonistic encounters were observed for 3 minutes and the evaluated behavioral interactions were divided into six different levels of aggression (modified from Suarez et al., 1999 and Velasquez et al., 2006): 0 - ignore; 1 - quick or repeated antennation on another individual; 2 - retreat towards contrary direction after contact; 3 - intimidation by opening the mandibles; 4 - biting and gaster flexing; and 5 - hold/dominate the other individual and try to remove it from the arena. The latency, the time between the release of the stimulus and the first reaction of the workers of *A. subterraneus subterraneus*, was also recorded. Each arena was washed with distilled water and neutral detergent and then wiped with 70% alcohol after each evaluation to remove any substance that could influence the ants' behavior. Ten replicates were conducted for each treatment.

The frequency of each behavior was determined for each treatment and an aggressiveness index (AI) was calculated (Velasquez et al., 2006): $AI = \sum_{i=1}^{5} \frac{OB_i * F_i}{N}$ where OB_i is the observed behavior i, F_i is the frequency of each behavior during three minutes of observation and N is the total number of interactions observed during the period.

Statistical analysis

The difference in total frequencies in each agression level per treatment was determined using the total number of behaviors per treatment. The average proportion of each independent aggression level and the aggressiveness index were compared among treatments using analysis of variance (ANOVA) and the Fisher test (LSD) with the STATISTICA 7.0 software (Statsoft Inc., Tulsa, USA).

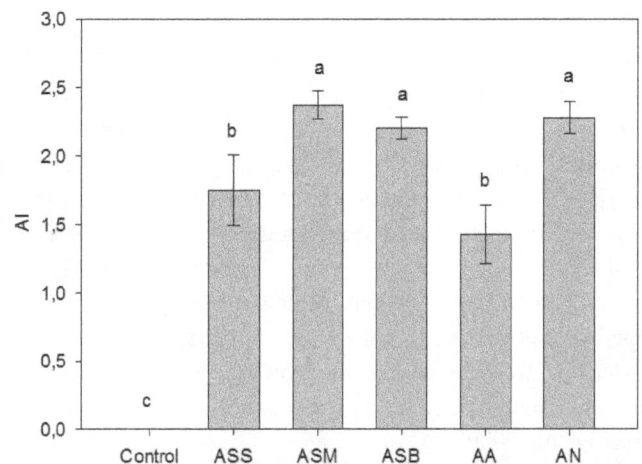

Fig 1. Aggressiveness index (AI) resulting from observed behaviors during interactions among workers of *A. subterraneus subterraneus* and conspecific nestmate workers (Control), conspecifics collected from a colony in the field (ASS), workers of *A. subterraneus molestans* (ASM), *A. subterraneus brunneus* (ASB), queens of *A. ameliae* (AA) or workers of *A. niger* (AN) with their standard errors. Different letters indicate significant differences by *post hoc* Fisher test (LSD).

Results

The aggressiveness index (AI) was higher in encounters between workers of *A. subterraneus subterraneus* and those of *A. subterraneus molestans*, *A. subterraneus brunneus* and *A. niger* (Fig 1) compared to the other treatments. *Acromyrmex subterraneus subterraneus* workers showed only antennation and ignored their nestmates (Table 1).

The workers in control treatment showed agression levels between 0 and 1, while all other treatments showed all aggression levels (Fig 2). Level 0 was more common in the control, followed by treatment with the parasite queen *A. ameliae* (Fig 2). Aggression levels 1, 2 and 5 did not differ among treatments (Fig 2). Levels 3 and 4 were more frequent in treatments with workers of *A. subterraneus brunneus* and *A. niger*, respectively.

Table 1. Observed behaviors during the interactions among workers of *A. subterraneus subterraneus* and conspecific nestmates (Control), conspecifics of colony collected in the field (ASS), workers of *A. subterraneus molestans* (ASM), *A. subterraneus brunneus* (ASB), queens of *A. ameliae* (AA) or workers of *A. niger* (AN).

Behavior	Control	ASS	ASM	ASB	AA	AN
Ignore	74	14	6	2	49	2
Antennation	79	103	120	130	107	123
Retreat	0	6	19	16	6	10
Intimidation	0	37	80	79	41	56
Bite	0	28	49	24	13	47
Gaster flexing	0	20	23	20	2	35
Grab/Dominate	0	7	10	8	1	7
Remove	0	0	1	2	5	2
Total	153	215	308	281	224	282

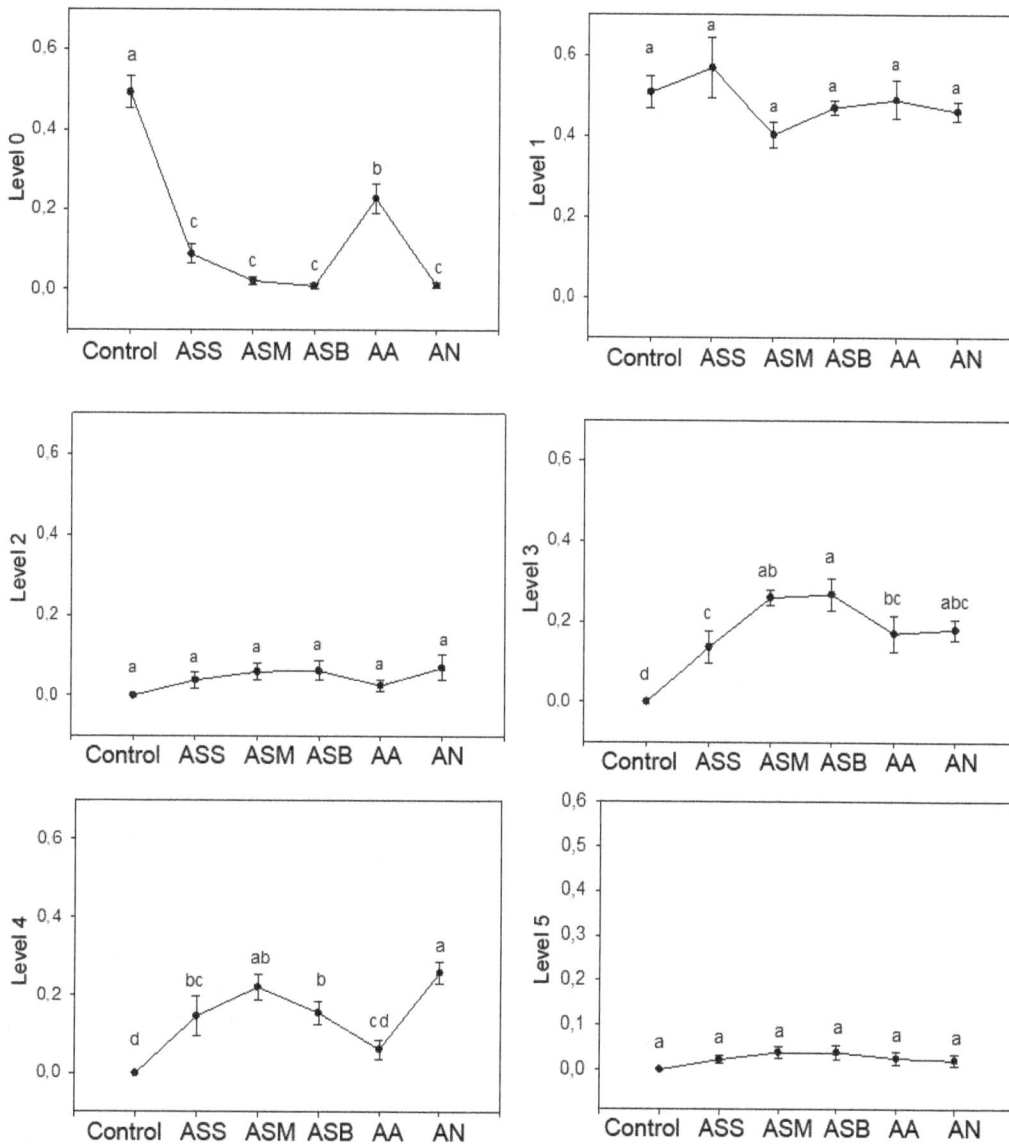

Fig 2. Average ratio of six aggression levels (0-5) in six treatments: conspecific nestmate workers (Control); conspecific from colony collected in the field (ASS), workers of *A. subterraneus molestans* (ASM), *A. subterraneus brunneus* (ASB), queens of *A. ameliae* (AA) or workers of *A. niger* (AN) with their standard errors. Different letters indicate significant differences by *post hoc* Fisher test (LSD).

Discussion

The presence of workers of *A. subterraneus molestans*, *A. subterraneus brunneus* and *A. niger* induced more aggression than the other intruder workers. Workers of *A. subterraneus subterraneus* made distinction between conspecific nestmates and non-nestmates, being more aggressive when in contact with non-nestmates, unlike the observed for *A. subterraneus molestans* (Souza et al., 2006). This is probably related to the ants diet. Colonies of *A. subterraneus subterraneus* maintained in the laboratory and the field had different food sources; this was the condition reported by Souza et al. (2006). Although the diet of colonies of *A. subterraneus subterraneus*, *A. subterraneus molestans*, *A. subterraneus brunneus* and *A. niger* kept in the laboratory was the same, the high aggressiveness of workers on these heteroepecifics is a sign that other factors, such as genetic influence, interfere with the composition of the "colony odor" (Vanzweden et al., 2010; Krasnec & Breed, 2013).

The low aggressiveness of workers of *A. subterraneus subterraneus* when in contact with winged queens of *A. ameliae* differs from that observed for workers of the genus *Temnothorax* (Hymenoptera: Formicidae), which responded more aggressively to the presence of workers of the social parasite *Protomognathus americanus* (Emery) (Hymenoptera: Formicidae) than to conspecifics or heterospecifics (Pamminger et al., 2011; Scharf et al., 2011). The tolerance observed in this experiment may be related to the phenomenon called "chemical insignificance". Just after emergence, immature ants are devoid of cuticular chemicals, acquiring the odor of the colony and integrating to other workers later (Lenoir et al., 1999). The weak signal of young workers allows them to be more easily accepted in other colonies than older workers (Lenoir et al., 2001a), similar to what can have occurred with the queens of *A. ameliae*. Adaptations in the morphology of these insects and the production of similar chemicals are possibly linked to non-aggression towards these queens (Martin et al., 2010; Bauer et al., 2009). Chemical camouflage, which occurs when the social parasite acquires the odor by direct contact with the host when entering the host colony, may also be occurring (Lenoir et al., 2001b). Males of *Bombus vestalis vestalis* (Fourcroy) (Hymenoptera: Apidae) lack the morphological and chemical adaptations that females have to infiltrate the host colony, but they produce a repellent odor that prevents them from being attacked by workers of the host species (Lhomme et al., 2012). It should be investigated if a similar phenomenon occurs with winged queens of *A. ameliae*.

Leaf-cutting ants defend their territories against conspecific and heterospecific intruders, protecting food resources and their offspring. The species *A. subterraneus subterraneus*, *A. subterraneus molestans*, *A. subterraneus brunneus* and *A. niger* co-exist in their natural range, besides having habits that make them potential competitors for feeding resources. On the other hand, the possible similarity of the cuticular chemical profile of different colonies of *A. subterraneus subterraneus* tested

and the strategy of the parasite *A. ameliae* to infiltrate in the host colony may explain the lower intensity of aggression suffered by the latter.

Workers of *A. subterraneus subterraneus* are more aggressive when in contact with heterospecifics competitors compared to the co-especifics and the social parasite *A. ameliae*. The mechanisms involved in the reduced aggression towards these treatments have to be evaluated to have a better understanding of the ecological process between competitors and social parasites in these ants.

Acknowledgments

We are greateful to the laboratory technician Manoel Ferreira for his help in performing the experiments. We also thank "Conselho Nacional de Desenvolvimento Científico e Tecnológico (CNPq)", "Coordenação de Aperfeiçoamento de Pessoal de Nível Superior (CAPES)" and "Fundação de Amparo à Pesquisa do Estado de Minas Gerais (FAPEMIG)" for financial support.

References

Akino, T. (2008). Chemical strategies to deal with ants: A review of mimicry, camouflage, propaganda, and phytomimesis by ants (Hymenoptera: Formicidae) and other arthropods. Myrmecological News, 11: 173-181.

Allon, O., Pascual-Garrido, A. & Sommer, V. (2012). Army ant defensive behaviour and chimpanzee predation success: field experiments in Nigéria. Journal of Zoology, 288: 237-244.

Bauer, S., Witte, V., Böhm, M. & Foitzik, S. (2009). Fight or flight? A geographic mosaic in host reaction and potency of a chemical weapon in the social parasite *Harpagoxenus sublaevis*. Behavioral Ecology and Sociobiology, 64, 45-56.

Boulogne, I., Ozier-Lafontaine, H., Germosén-Robineau, L., Desfontaines, L. & Loranger-Merciris, G. (2012). *Acromyrmex octospinosus* (Hymenoptera: Formicidae) Management: Effects of TRAMILs Fungicidal Plant Extracts. Journal of Economic Entomology, 105: 1224-1233.

Brandão, C.R.F., Mayhé-Nunes, A.J. & Sanhudo, C.E.D. (2011). Taxonomia e filogenia das formigas-cortadeiras. In T.M.C. Della Lucia (Ed.), Formigas-cortadeiras: da bioecologia ao manejo (pp. 27-48). UFV, Viçosa.

Camargo, R.S., Forti, L.C., Lopes, J.F.S. & Andrade, A.P.P. (2006). Brood care and male behavior in queenless *Acromyrmex subterraneus brunneus* (Hymenoptera: Formicidae) colonies under laboratory conditions. Sociobiology, 48: 717-726.

Crosland, M.W.J. (1990). Variation in ant aggression and kin discrimination ability within and between colonies. Journal of Insect Behavior, 3: 359-379.

Crowley, P.H., Provencher, L., Sloane, S., Dugatkin, L.A., Spohn, B., Rogers, L. & Alfieri, M. (1996). Evolving cooperation:

the role of individual recognition. Biosystems, 37: 49-66.

Crozier, R.H. & Dix, M.W. (1979). Analysis of two genetic models for the innate components of colony odor in social Hymenoptera. Behavioral Ecology and Sociobiology, 4: 217-224.

Fishwild, T.G. & Gamboa, G.J. (1992). Colony defence against conspecifics: caste-specific differences in kin recognition by paper wasps, *Polistes fuscatus*. Animal Behavior, 43: 95-102.

Forti, L.C., Andrade, M.L., Andrade, A.P.P., Lopes, J.F.S. & Ramos, V.M. (2006). Bionomics and identification of *Acromyrmex* (Hymenoptera: Formicidae) through an illustrated key. Sociobiology, 48: 135-153.

Guerrieri, F.J., Nehring, V., Jorgensen, C.G., Nielsen, J., Galizia, C.G., D'Ettore, P. (2009). Ants recognize foes and not friends. Proceedings of the Royal Society B, 276: 2461-2468.

Hamilton, W.D. (1972). Altruism and related phenomena mainly in the social insects. Annual Review of Ecology and Systematics, 3: 193-232.

Hart, A.G., Anderson, C. & Ratnieks, F.L.W. (2002). Task partitioning in leafcutting ants. Acta Ethologica, 5: 1-11.

Heinze, J., Foitzik, S., Hippert, A. & Hölldobler, B. (1996). Apparent dear-enemy phenomenon and environment-based recognition cues in the ant *Leptothorax nylanderi*. Ethology, 102: 510-522.

Hölldobler, B. & Michener, C.D. (1980). Mechanisms of identification and discrimination in social hymenoptera. In H. Markl (Ed.), Evolution of social behavior: hypotheses and empirical tests (pp. 35-38). Weinheim, Chemie Gmbh.

Hölldobler, B. & Wilson, E.O. (1990). *The ants*. Cambridge: Harvard University Press, 732 p.

Krasnec, M.O. & Breed, M.D. (2013). Colony-specific cuticular hydrocarbon profile in *Formica argentea* ants. Journal of Chemical Ecology, 39: 59-66.

Lenoir, A., Fresneau, D., Errard, C. & Hefetz, A. (1999). The individuality and the colonial identity in ants: the emergence of the social representation concept. In O. Detrain, J.L Deneubourg & J. Pasteels (Eds.), Information processing in social insects (pp. 219-237). Basel, Switzerland, Birkhauser.

Lenoir, A., D'ettorre, P., Errard, C. & Hefetz, A. (2001a). Chemical ecology and social parasitism in ants. Annual Review of Entomology, 46: 573-599.

Lenoir, A., Cuisset, D. & Hefetz, A. (2001b). Effects of social isolation on hydrocarbon pattern and nestmate recognition in the ant *Aphaenogaster senilis* (Hymenoptera, Formicidae). Insectes Sociaux, 48: 101-109.

Lhomme, P., Ayasse, M., Valterova, I., Lecocq, T. & Rasmont, P. (2012). Born in an alien nest: how do social parasite male offspring escape from host aggression? PLoS ONE, **7**, e43053. doi:10.1371/journal.pone.0043053

Martin, S.J., Carruthers, J.M., Williams, P.H. & Drijfhout,

F.P. (2010). Host specific social parasites (*Psithyrus*) reveal evolution of chemical recognition system in bumblebees. Journal of Chemical Ecology, 36: 855-863.

Moreira, D.D.O., Erthal, M., Carrera, M.P., Silva, C.P. & Samuels, R.I. (2006). Oral trophallaxis in adult leaf-cutting ants *Acromyrmex subterraneus subterraneus* (Hymenoptera, Formicidae). Insectes Sociaux, 53: 345-348.

Mori, A., Grasso, D.A. & Le Moli, F. (2000). Raiding and foraging behaviour of the blood-red ant, *Formica sanguinea* Latr. (Hymenoptera, Formicidae). Journal of Insect Behavior, 13: 421-438.

Pamminger, T., Scharf, I., Pennings, P.S. & Foitzik, S. (2011). Increased host aggression as an induced defense against slavemaking ants. Behavioral Ecology, 22: 255-260.

Pollock, G.B. & Rissing, S.W. (1989). Intraspecific brood raiding, territoriality, and slavery in ants. American Naturalist, 133: 61-70.

Roulston, T.H., Buczkowski, G. & Silverman, J. (2003). Nestmate discrimination in ants: effect of bioassay on aggressive behavior. Insectes Sociaux, 50: 151-159.

Richard, F.J., Hefetz, A., Christides, J.-P. & Errard, C. (2004). Food influence on colonial recognition and chemical signature between nestmates in the fungus-growing ant *Acromyrmex subterraneus subterraneus*. Chemoecology, 14: 9-16.

Sakagami, S.F. (1993). Ethology of the robber bee *Lestrimelitta limao* (Hymenoptera: Apidae). Sociobiology, 21: 237-277.

Scharf, I., Pamminger, T. & Foitzik, S. (2011). Differential response of ant colonies to intruders: attack strategies correlate with potential threat. Ethology, 117: 731-739.

Souza, D.J., Della Lucia, T.M.C. & Barbosa, L.C.A. (2006). Discrimination between workers of *Acromyrmex subterraneus molestans* from monogynous and polygynous colonies. Brazilian Archives of Biology and Technology, 49: 277-285.

Souza, D.J., Soares, I.M.F & Della Lucia, T.M.C. (2007). *Acromyrmex ameliae* sp.n. (Hymenoptera: Formicidae): a new social parasite of leaf-cutting ants Brazil. Insect Science, 14: 251-257.

Suarez, A.V., Tsutsui, N.D., Holway, D.A. & Case, T.J. (1999). Behavioral and genetic differentiation between native and introduced populations of the Argentine ant. Biological Invasions, 1: 1-11.

Vanzweden, J.S., Brask, J.B., Christensen, J.H., Boomsma, J.J., Linksvayer, T. & D'ettorre, P. (2010). Blending of heritable recognition cues among ant nestmates creates distinct colony gestalt odours but prevents within-colony nepotism. Journal of Evolutionary Biology, 23, 1498-1508.

Velásquez, N., Gómez, M., González, J. & Vásquez, R.A. (2006). Nest-mate recognition and the effect of distance from the nest on the aggressive behaviour of *Camponotus chilensis* (Hymenoptera: Formicidae). Behaviour, 143: 811-824.

Viana, A.M.M., Frézard, A., Malosse, C., Della Lucia, T.M.C., Errard, C. & Lenoir, A. (2001). Colonial recognition of fungus in the fungus-growing ant *Acromyrmex subterraneus subterraneus* (Hymenoptera: Formicidae). Chemoecology, 11, 29-36.

Wiley, R.H. (2013). Specificity and multiplicity in the recognition of individuals: implications for the evolution of social behavior. Biological Reviews of the Cambridge Philosophical Society 88: 179-195.

Permissions

All chapters in this book were first published in Sociobiology, by Universidade Estadual de Feira de Santana; hereby published with permission under the Creative Commons Attribution License or equivalent. Every chapter published in this book has been scrutinized by our experts. Their significance has been extensively debated. The topics covered herein carry significant findings which will fuel the growth of the discipline. They may even be implemented as practical applications or may be referred to as a beginning point for another development.

The contributors of this book come from diverse backgrounds, making this book a truly international effort. This book will bring forth new frontiers with its revolutionizing research information and detailed analysis of the nascent developments around the world.

We would like to thank all the contributing authors for lending their expertise to make the book truly unique. They have played a crucial role in the development of this book. Without their invaluable contributions this book wouldn't have been possible. They have made vital efforts to compile up to date information on the varied aspects of this subject to make this book a valuable addition to the collection of many professionals and students.

This book was conceptualized with the vision of imparting up-to-date information and advanced data in this field. To ensure the same, a matchless editorial board was set up. Every individual on the board went through rigorous rounds of assessment to prove their worth. After which they invested a large part of their time researching and compiling the most relevant data for our readers.

The editorial board has been involved in producing this book since its inception. They have spent rigorous hours researching and exploring the diverse topics which have resulted in the successful publishing of this book. They have passed on their knowledge of decades through this book. To expedite this challenging task, the publisher supported the team at every step. A small team of assistant editors was also appointed to further simplify the editing procedure and attain best results for the readers.

Apart from the editorial board, the designing team has also invested a significant amount of their time in understanding the subject and creating the most relevant covers. They scrutinized every image to scout for the most suitable representation of the subject and create an appropriate cover for the book.

The publishing team has been an ardent support to the editorial, designing and production team. Their endless efforts to recruit the best for this project, has resulted in the accomplishment of this book. They are a veteran in the field of academics and their pool of knowledge is as vast as their experience in printing. Their expertise and guidance has proved useful at every step. Their uncompromising quality standards have made this book an exceptional effort. Their encouragement from time to time has been an inspiration for everyone.

The publisher and the editorial board hope that this book will prove to be a valuable piece of knowledge for researchers, students, practitioners and scholars across the globe.

List of Contributors

I Okita
Gifu University, Gifu, Gifu, Japan

K Murase
The University of Tokyo, Bunkyo-ku, Tokyo, Japan
Tokyo University of Agriculture and Technology, Fuchu, Tokyo, Japan

T Sato
Tokyo University of Agriculture and Technology, Fuchu, Tokyo, Japan

K Kato
Shizuoka Prefectural Research and Coordination Office, Shizuoka, Shizuoka, Japan

A Hosoda
Hamamatsu Gakuin University Junior College, Hamamatsu, Shizuoka, Japan

M Terayama
The University of Tokyo, Bunkyo-ku, Tokyo, Japan

K Masuko
Senshu University, Kawasaki, Kanagawa, Japan

FJ Zorzenon
Instituto Biológico, Unidade Laboratorial de Referência em Pragas Urbanas, São Paulo, SP, Brazil

AE de C Campos
Instituto Biológico, Unidade Laboratorial de Referência em Pragas Urbanas, São Paulo, SP, Brazil

SM Almeida
Universidade do Estado de Mato Grosso, Nova Xavantina, Mato Grosso, Brazil
Universidade do Estado de Mato Grosso, Cáceres, Mato Grosso, Brazil

SR Andena
Universidade Estadual de Feira de Santana, Feira de Santana, Bahia, Brazil

EJ Anjos-Silva
Universidade do Estado de Mato Grosso, Cáceres, Mato Grosso, Brazil

SRS Cardoso
Universidade Estadual Paulista (UNESP), Botucatu, São Paulo, Brazil
Instituto Federal do Tocantins, Araguatins, Tocantins, Brazil

LC Forti
Universidade Estadual Paulista (UNESP), Botucatu, São Paulo, Brazil

NS Nagamoto
Universidade Estadual Paulista (UNESP), Botucatu, São Paulo, Brazil

RS Camargo
Universidade Estadual Paulista (UNESP), Botucatu, São Paulo, Brazil

HF Cunha
Universidade Estadual de Goiás, UnUCET, Anápolis, GO, Brazil

JS Lima
Universidade Estadual de Goiás, UnUCET, Anápolis, GO, Brazil

LF Souza
Universidade Estadual de Goiás, UnUCET, Anápolis, GO, Brazil

LGA Santos
Universidade Estadual de Goiás, UnUCET, Anápolis, GO, Brazil

JC Nabout
Universidade Estadual de Goiás, UnUCET, Anápolis, GO, Brazil

R Jaffé
Instituto de Biociências, Universidade de São Paulo (USP), São Paulo-SP, Brazil

A Szczuka
Nencki Institute of Experimental Biology, Warsaw, Poland

B Symonowicz
Nencki Institute of Experimental Biology, Warsaw, Poland

J Korczyńska
Nencki Institute of Experimental Biology, Warsaw, Poland

A Wnuk
Nencki Institute of Experimental Biology, Warsaw, Poland

EJ Godzińska
Nencki Institute of Experimental Biology, Warsaw, Poland

NS Little
USDA-ARS, Southern Insect Management Research Unit, Mississippi, USA

NA Blount
Mississippi State University, Mississippi, USA

MA Caprio
Mississippi State University, Mississippi, USA

JJ Riggins
Mississippi State University, Mississippi, USA

AA Wachkoo
Department of Zoology and Environmental Sciences, Punjabi University, Patiala, India

H Bharti
Department of Zoology and Environmental Sciences, Punjabi University, Patiala, India

AL Szalanski
University of Arkansas, Fayetteville, Arkansas, USA

AD Tripodi
University of Arkansas, Fayetteville, Arkansas, USA

RF torres
Universidade Federal da Grande Dourados, Dourados, Mato Grosso do Sul, Brazil

VO torres
Universidade Federal da Grande Dourados, Dourados, Mato Grosso do Sul, Brazil

YR súarez
Universidade Estadual de Mato Grosso do Sul, Dourados, Mato Grosso do Sul, Brazil

WF antonialli-junior
Universidade Estadual de Mato Grosso do Sul, Dourados, Mato Grosso do Sul, Brazil

R. Henry L. Disney
University of Cambridge, Department of Zoology, Cambridge, United Kingdom

Marcos A. L. Bragança
Universidade Federal do Tocantins, Palmas, Tocantins, Brazil

AP Mello
Universidade Estadual da Paraíba, Campina Grande, Paraiba, Brazil

BG Costa
Universidade Estadual da Paraíba, Campina Grande, Paraiba, Brazil

AC Silva
Universidade Estadual da Paraíba, Campina Grande, Paraiba, Brazil

AMB Silva
Universidade Estadual da Paraíba, Campina Grande, Paraiba, Brazil

MA Bezerra-Gusmão
Universidade Estadual da Paraíba, Campina Grande, Paraiba, Brazil

S Hozumi
Chigakukan secondary school, Mito 310-0914, Japan

K Kudô
Niigata University, Niigata 950–2181, Japan

H Katakura
Hokkaido University, Sapporo 060–0810, Japan

S Yamane
Ibaraki University, Mito 310-8511, Japan

A Somavilla
Instituto Nacional de Pesquisas da Amazônia, Petrópolis, Manaus, AM, Brazil

K Schoeninger
Instituto Nacional de Pesquisas da Amazônia, Petrópolis, Manaus, AM, Brazil

AF Carvalho
Universidade Federal de São Carlos, São Carlos, SP, Brazil

RST Menezes
Universidade Estadual de Santa Cruz, Ilhéus, BA, Brazil

MA Del Lama
Universidade Federal de São Carlos, São Carlos, SP, Brazil

MA Costa
Universidade Estadual de Santa Cruz, Ilhéus, BA, Brazil

ML Oliveira
Instituto Nacional de Pesquisas da Amazônia, Petrópolis, Manaus, AM, Brazil

VM Ramos
Universidade do Oeste Paulista, Presidente Prudente, SP, Brazil

F Cunha
Universidade do Oeste Paulista, Presidente Prudente, SP, Brazil

KC Khun
Universidade do Oeste Paulista, Presidente Prudente, SP, Brazil

RGF Leite
Universidade do Oeste Paulista, Presidente Prudente, SP, Brazil

WF Roma
Universidade do Oeste Paulista, Presidente Prudente, SP,
Brazil

C Ruiz
Universidad Técnica Particular de Loja, Loja, Ecuador

W de J May-Itzá
Universidad Autónoma de Yucatán, Yucatán, México

JJG Quezada-Euán
Universidad Autónoma de Yucatán, Yucatán, México

P de La Rúa
Universidad de Murcia, Murcia, España

MS Modanesi
Universidade Estadual Paulista, Botucatu, SP, Brazil

SM Kadri
Universidade Estadual Paulista, Botucatu, SP, Brazil

PEM Ribolla
Universidade Estadual Paulista, Botucatu, SP, Brazil

DP Alonso
Universidade Estadual Paulista, Botucatu, SP, Brazil

RO Orsi
Universidade Estadual Paulista, Botucatu, SP, Brazil

JO Macías-Macías
Universidad de Guadalajara. Centro Universitario del
Sur, México

JM Tapia-Gonzalez
Universidad de Guadalajara. Centro Universitario del
Sur, México

F Contreras-Escareño
Universidad de Guadalajara. Centro Universitario de la
Costa Sur, México

F Báthori
Department of Evolutionary Zoology and Human Biology,
University of Debrecen, Debrecen, Hungary

WP Pfliegler
Department of Genetics and Applied Microbiology,
University of Debrecen, Debrecen, Hungary

A Tartally
Department of Evolutionary Zoology and Human Biology,
University of Debrecen, Debrecen, Hungary

HC Resende
Universidade Federal de Viçosa, Viçosa, MG, Brazil
Universidade Federal de Viçosa, Campus Florestal, MG,
Brazil

TM Fernandes-Salomão
Universidade Federal de Viçosa, Viçosa, MG, Brazil

MG Tavares
Universidade Federal de Viçosa, Viçosa, MG, Brazil

LAO Campos
Universidade Federal de Viçosa, Viçosa, MG, Brazil

HR Allen
Virginia Tech University, Blacksburg, VA, United States

DM Miller
Virginia Tech University, Blacksburg, VA, United States

FG Vossler
Laboratorio de Actuopalinología, CICyTTP-CONICET/
FCyT-UADER, Diamante, Entre Ríos, Argentina

GA Fagúndez
Laboratorio de Actuopalinología, CICyTTP-CONICET/
FCyT-UADER, Diamante, Entre Ríos, Argentina

DC Blettler
Laboratorio de Actuopalinología, CICyTTP-CONICET/
FCyT-UADER, Diamante, Entre Ríos, Argentina

FMJ Sommerlandt
University of Würzburg, Würzburg, Germany

W Huber
University Vienna, Vienna, Austria

J Spaethe
University of Würzburg, Würzburg, Germany

F Rodrigues
Universidade Federal do Vale do São Francisco, Petrolina,
PE, Brazil
Universidade Federal Rural do Semiárido (UFERSA),
Mossoró, RN, Brazil

MF Ribeiro
Empresa Brasileira de Pesquisa Agropecuária (Embrapa),
Petrolina, PE, Brazil

L Costa
ARCADIS Logos, Divisão de Meio Ambiente, São Paulo,
Brazil

RM Franco
ARCADIS Logos, Divisão de Meio Ambiente, São Paulo,
Brazil

LF Guimarães
ARCADIS Logos, Divisão de Meio Ambiente, São Paulo,
Brazil

A Vollet-Neto
ARCADIS Logos, Divisão de Meio Ambiente, São Paulo, Brazil
Universidade de São Paulo, São Paulo, Brazil

FR Silva
ARCADIS Logos, Divisão de Meio Ambiente, São Paulo, Brazil
Universidade de São Paulo, São Paulo, Brazil

GD Cordeiro
ARCADIS Logos, Divisão de Meio Ambiente, São Paulo, Brazil
Universidade de São Paulo, São Paulo, Brazil

CBS Lima
Universidade Federal do Recôncavo da Bahia, Cruz das Almas, BA, Brazil

LA Nunes
Universidade Estadual do Sudoeste da Bahia, Jequié, BA, Brazil

MF Ribeiro
Empresa Brasileira de Pesquisa Agropecuária, Petrolina, PE, Brazil

CAL de Carvalho
Universidade Federal do Recôncavo da Bahia, Cruz das Almas, BA, Brazil

LAM Medina
Universidad Autónoma de Yucatán, Mérida, Yucatán, México

AG Hart
University of Gloucestershire, Gloucestershire, U.K

FLW Ratnieks
University of Sussex, Brighton, U.K

RS Pinto
Universidade Federal do Maranhão, São Luís, MA, Brazil

PMC Albuquerque
Universidade Federal do Maranhão, São Luís, MA, Brazil

MMC Rêgo
Universidade Federal do Maranhão, São Luís, MA, Brazil

C Oliveira-Abreu
Universidade de São Paulo, Ribeirão Preto, Brazil

SD Hilário
Universidade de São Paulo/IBUSP, São Paulo, Brazil

CFP Luz
Instituto de Botânica de São Paulo, São Paulo, Brazil

I Alves-dos-Santos
Universidade de São Paulo/IBUSP, São Paulo, Brazil

AG Silva
Universidade Federal do Maranhão (UFMA), São Luís, MA, Brazil

RS Pinto
Universidade Federal do Maranhão (UFMA), São Luís, MA, Brazil

FAL Contrera
Universidade Federal do Pará (UFPA), Belém, PA, Brazil

PMC Albuquerque
Universidade Federal do Maranhão (UFMA), São Luís, MA, Brazil

MMC Rêgo
Universidade Federal do Maranhão (UFMA), São Luís, MA, Brazil

TG Pikart
Universidade Federal de Viçosa, Viçosa, MG, Brazil
Universidade do Estado de Santa Catarina, Lages, SC, Brazil

PG Lemes
Universidade Federal de Viçosa, Viçosa, MG, Brazil

WC de C Morais
Universidade Federal de Viçosa, Viçosa, MG, Brazil

JC Zanuncio
Universidade Federal de Viçosa, Viçosa, MG, Brazil

TMC Della Lucia
Universidade Federal de Viçosa, Viçosa, MG, Brazil